INSTRUMENTAL IN WAR

HISTORY OF WARFARE

VOLUME 28

INSTRUMENTAL IN WAR

Science, Research, and Instruments Between Knowledge and the World

EDITED BY

STEVEN A. WALTON

BRILL
LEIDEN · BOSTON
2005

On the cover: Two civilian scientists/engineers inspecting a WWII-era aerial bomb with what appears to be an X-ray machine at the Watertown Arsenal, Watertown, New York. Illustration courtesy of the U.S. National Archives and Records Administration (NARA), Photographic Division, College Park, Maryland, Record Group 156, Records of the Office of the Chief of Ordnance, 1816-1967, photo 156-WAA-2-21.

This book is printed on acid-free paper.

Library of Congress Cataloging-in-Publication Data

Instrumental in war : science, research, and instruments between knowledge and the world / edited by Steven A. Walton.
p. cm. — (History of warfare, ISSN 1385-7827 ; v. 28)
Includes index.
ISBN 90-04-14281-9 (alk. paper)
1. Military research—History. 2. Military art and science—Technological innovations. 3. Scientific apparatus and instruments—History. 4. Military art and science—Instruments. I. Walton, Steven A. II. Title. III. Series.

U390.I57 2005
355'.07—dc22

2004062920

ISSN 1385-7827
ISBN 90 04 14281 9

PRINTED IN THE NETHERLANDS

CONTENTS

LIST OF ILLUSTRATIONS

Walton

Lynch

Mauskopf

Fleming

Scharfenberger

McBride

Grier

Kroker

ACKNOWLEDGEMENTS

Projects such as this seem to come from nowhere and acquire a life of their own, and I have to extend my deepest thanks to many of our contributors here for keeping the fire lit under me once it had been kindled. The genesis of the volume came at the 2001 History of Science Society conference in Denver, where after hearing a talk that just barely touched on the use of instruments in war—or rather, feeling that it *should* have touched more on them—I mused aloud to Deborah Warner and Kenton Kroker that there ought to be a study of the intersection of the two ideas. They agreed, I offered, and so here it is—or at least a first salvo. There is, however, much work to be done, so although we ventured out first on this endeavor, there is plenty of room on the road for other drivers.

Thanks also to Kelly DeVries who encouraged the project from the beginning for the *History of Warfare* series; I am confident in his assessment that this is a worthy project. Julian Deahl, Boris van Gool, and Marcella Mulder at Brill shepherded it through the publication process, often demystifying the world of publishing and also encouraged this far-ranging study for the series. Doug Womelsdorf compiled the index (mostly) without protest, making that process go very smoothly once the authors indicated key terms for me. And the numerous anonymous readers who I tapped to read drafts of these chapters deserve many thanks for their positive, and sometimes constructively incisive comments on the individual contributions.

Some of these papers were presented in embryonic format at the 2002 Society for the History of Technology (SHOT) meeting in Toronto, some at the 2003 Society for Military History (SMH) conference in Knoxville, one at a Folger Shakespeare Library seminar. Some were initially intended for the 2002 History of Science Society (HSS) meeting in Milwaukee; the 2003 HSS conference in Cambridge, Mass. provided some final contact with contributors to this project. Some of the introductory chapter was tried out on my colleagues at the Science, Technology, and Society (STS) Program at Penn State when I first arrived there—thanks for the clarifications there, as well as for the venue. I was also pleased to have made Peter Galison's acquaintance at a workshop on agnatology in the history department at Penn State organized by Londa Shiebinger and Robert Proctor. Finally, a hearty thanks go out to attendees of the summer 2002 meeting on "Scientific Instruments and the History of Collections" at the Museum for the History of Science at Oxford University with whom many pleasant hours were spent contemplating, life, the universe, Kepler's drinks dispenser, and above all, scientific instrument collections. The contributors to the volume I have met everywhere from graduate school to conferences as far-flung as Paris, Boston, and Granada, but, as is becoming more and more common these days, I owe a debt of gratitude to my virtual colleagues—academics, collectors, and antiquarians (in the best sense of the word)—who agreed to be part of this collection and with whom I have corresponded by electronic mail but have yet to meet face-to-face. It was said that the world wide web would expand the power of the individual. Perhaps. But it has most certainly expanded the range of research possibilities for the scholar. Maybe it, too, is an instrument . . .

VOLUME CONTRIBUTIONS

> *Neither the bare hand nor the understanding left to itself are of much use. It is by instruments and other aids that the work gets done, and these are needed as much by the understanding as by the hand. And just as instruments improve or regulate the movement of our hands, so instruments of the mind provide suggestions or cautions to the understanding.*
>
> – Francis Bacon, *Novum Organum*, aphorism 2*

This volume spans nearly the last five hundred years, and looks at scientific instrumentality in American, British and French military activity. Initially, this project had been envisioned as a sort of compendia that gave relatively even coverage to the last half millennia, chronicling the development of scientific instruments *in* warfare, or possibly to the idea of warfare itself *becoming* scientific through the use of instrumentation. This goal rapidly became unworkable—partially because the resultant volume would have run to thousands of pages; partially because we could not find people working on the topic over the entire span of the period. The later eighteenth and early nineteenth century, for example, when science began to manifest itself as a properly institutionalized profession and in which modern military institutions were born, seems to be understudied in terms of the use of scientific instruments (material or theoretical models) by the armies of the American and French Revolutions or in the nineteenth century European military (r)evolutions. More can obviously be done on these topics, and the scope could be extended to other European, colonial, or non-Western contexts. We will all welcome such work.

Arranged chronologically, these essays hope to show the progression of instrumental thought in war and action from about 1600 to the present. The volume opens with three contributions examining early modern armies' acceptance of the newer scientific models and ideas of what we generally term the "Scientific Revolution." Steven Walton's contribution looks at how individual military men—primarily gunners—saw physical mathematical instruments as the key

* Francis Bacon, *The Novum Organum*, trans. and ed. by Peter Urbach and John Gibson (Peru, Ill., 1994), p. 43.

to their military advancement. Investing a great deal of importance and energy into these rules, quadrants, and other instruments, gunners from the mid-sixteenth century onward ultimately made an only marginally successful power play for status within armies. They did, however, rather successfully instate the broad use of mathematical rules, instruments, and behaviors into the then high-tech field of artillery. Although gunners were unsuccessful at a personal level, societal expectations of instrumental use was changing at the time, and Bill Lynch's essay provides some evidence of how their power play developed in his look at the mid-seventeenth century military surveying endeavors by the British in their subjugation of Ireland. Here we have an example of warfare that *triggered* an instrumental use of rapid cadastral surveying by circumferentor and chain (hence both conceptual and physical instruments) that then became the norm for colonial surveying in the eighteenth and early nineteenth centuries. In the eighteenth century, as Jamel Ostwald shows, this instrumental outlook had become *de rigeur* for the armies of Enlightenment France, so much so that despite Vauban's insistence that his 'clockwork siege' was merely a metaphor and that local conditions would alter and affect his progress, people nonetheless came to believe this instrumental metaphor; by the time Clausewitz entered the scene, he needed *yet another* scientific metaphor—friction—to more correctly explain war. These three papers open up the three ways in which instruments can be instrumental in war: materially as symbols, practically as tools, or ideologically as models (a parallel with Baird's tripartite division discussed in the Introduction).

Turning from the theoretical apparatus one needs to plan warfare, Seymour Mauskopf offers us a glimpse at the laboratory benchtop where specific military substances are tested. He also provides an example of "classic" instrumentalism in the military context: he explains how the French military munitions engineer, Paul Vielle, developed the bomb calorimeter and recording manometer for testing gunpowder in the 1880s. Here we have an example of military sciences turning to what we think of as an indispensable tool of the chemical laboratory for information and standardization of one of its main and precious munitions. But Mauskopf happily shows us that this standard chemical apparatus and in many ways the theory around it was *developed for* the need to test that very gunpowder. In the other direction, another instrument arising from the scientific tradition, the telegraph which arose from the work of Faraday, Maxwell, and Œrsted (S.F.B. Morse just put it all together with a language in a nice neat package), soon became a military instrument for command and control. That story, especially in the U.S. Civil War, is well known. James Fleming instead examines the role of the military telegraph in the creation of the U.S. National Weather Service after the Civil War. From the very earliest days of the telegraph, the Army Signal Office pressed it into service to monitor national security, but also included reports on the weather across the country. Not surprisingly, weather and meteorological information have always played an important part in military strategy; more surprisingly, however, Fleming also shows that meteorological rationale provided the basis for building the post-Civil War U.S. military telegraph system in the first place.

If the Army used telegraphs at all its posts around the country for concentrating information, the nineteenth century saw the development of numerous proving grounds around the country for disseminating information on new technologies and methods of pursuing war. Everything from cannon to torpedoes to uniforms were put through their paces at camps and arsenals across the country. Today places like Aberdeen Proving Grounds for ordnance, Edwards Air Force Base (or Groom Lake) for experimental aircraft, and China Lake for chemical weapons come to mind, but Gerard Scharfenberger offers an archaeological look at one of the more important late nineteenth-century testing sites, the Sandy Hook Proving Grounds at the southwestern entrance to New York harbor in New Jersey. Scharfenberger was one of the team that did unexploded ordnance sweeps on the beaches of the proving grounds—which is now a National Recreational Area!—discovering numerous remnants of shells and ordnance systems as well as one case of a maritime rescue system that was tested there. His contribution reminds us that documents and museum instruments are not the only entrée into questions we might ask of these topics. William McBride continues the testing *topos* by examining the Engineering Experiment Station, set up in 1903 adjacent to the U.S. Naval Academy in Annapolis, Maryland. Largely unrecognized as one of the first purpose-built naval engineering stations in the country (preceded only by the 1869 Naval Torpedo Station in Newport, Rhode Island, although research and testing had always happened at various naval yards such as Brooklyn, Washington, and Philadelphia), the EES provided the navy with testing information on engines, lubricants, metals, and all manner of equipment for the steam navy. Although largely unremarked, this component of scientific research and instrumentation is truly the interface between invention and use for many real-world technologies, and the monumental shift between a sail- and steam-powered navy tested this transition severely. Concluding this section focussed on specific testing stations, David Grier offers another way of looking at them, this time telling the story of testing under pressure—through the eyes of mathematicians set to the task of calculating ballistic firing tables for the U.S. Army artillery during WWI. Oswald Veblen (Thorsten's less well-known nephew) served the army in this capacity before going on during the Depression to become one of the leading American mathematicians. In fact, Grier reminds us, mathematicians from Galileo to Newton to Laplace and Gauss (Napier as well) have frequently been called into national service (or put themselves in the service), putting their mathematical skills to military ends. His story focuses on the need to very quickly develop instrumentation and computational resources at a new proving ground which was largely still a muddy pit, and move beyond the theoretical world of mathematical ballistic theory to the concrete world of "mass production" of data for practical use.

From here we return to Europe to compare the French and British attitude towards scientific development of gyrocompasses during the interwar period in Sébastien Soubiran's essay. Comparing the national cultures at work in the two countries, both interested in developing systems that allowed battleships to fire accurately at their targets even while changing course, Soubiran shows that the interplay between scientific instrument firms, naval officers, civilian scientists, and national military establishments can produce remarkably different results depending on the priorities and attitudes of each nation. In the end, the outbreak of WWII made decisions for their research programmes for both countries, but their attitudes towards those programmes during the luxury of peacetime reminds us that instrumental rationality in one context may not be so rational in another. Kenton Kroker provides an excellent comparison of an American case where civilian scientists worked themselves into military research during WWII to test potential aviator's abilities to withstand the rigors of combat flying. His look at Hallowell Davis' work with electroencephalography (EEGs to you and I) for the Navy's training school at Pensacola shows that the scientific outlook, if pushed too far, may break down. Obsessed with being able to be scientific and bring testing instruments to bear on the problem of who would make successful aviators led Davis and his team to either over- or underestimate the abilities of their subjects—in fact some ace flyers and instructors could not pass the test while other unfit pilots passed with "flying colors." In the end, Kroker's story is one of the failure of an instrumental approach, as EEGs dropped from the *atelier* of aviation selection immediately after WWII.

Rounding out our volume are two contributions by noted scholars on the post-WWII American climate of scientific research and instrumental development. Deborah Warner looks at Cold War gravimetry, the measurement of minute variations in local gravity across the globe. This seemingly esoteric discipline was of crucial importance in the era of ICBMs, which used inertial guidance systems that would be affected by these local variations as they flew half way around the globe. Thankfully it did not come to that in the Cold War, but Warner shows how military instrumentation was a double-edged sword after WWII, with concerns about funding, access, secrecy, and publication swirling about the military-industrial complex and where civilian scientists and the military entered into a continuing pact. Incidentally, hers is also a wonderful example of a civilian instrument (this time from the petroleum exploration industry) more successfully adopted into service and refined by the military. Finally, Peter Galison has happily consented to include a reprint of an important yet neglected essay he wrote on physics research during the Cold War at Stanford, Princeton, and Harvard, showing that the Cold War in fact *made* physics what it is today (or what it was in 1988, when the essay originally appeared, and before the superconducting supercollider project was terminated, ending the postwar "Big Physics" bubble). He has also added a brief postscript to suggest areas of further research for 20th century science, and it is hoped that this volume will stimulate similar research on contemporary and historic military activity in war and peace from the point of view of scientific research, instrumentation, and instrumentalism in general.

LIST OF CONTRIBUTORS

James R. Fleming, a graduate of Penn State, Colorado State, and Princeton universities, is Professor of Science, Technology and Society at Colby College, Fellow of the AAAS, Founder and First President of the International Commission on History of Meteorology, and author of *Meteorology in America, 1800–1870* (Baltimore, 1990) and *Historical Perspectives on Climate Change* (Oxford and New York, 1998). His full curriculum vita is available online at <www.colby.edu/profile/jfleming>

Peter Galison is the Mallinckrodt Professor of the History of Science and Physics at Harvard University. A former MacArthur Fellow and Max Planck Award winner, Galison works on the interrelations between physics and the physicists and the cultures they inhabit, asking how we know that experiments are correct and how experimentation, instrumentation, and theory interact. Author of numerous books and articles, *Einstein's Clocks, Poincaré's Maps: Empires of Time* has just appeared (Norton, 2003), completing a trilogy of books begun with *How Experiments End* (Chicago UP, 1987) and *Image and Logic* (Chicago UP, 1997).

David Alan Grier is currently an Associate Professor of Computer Science and International Affairs at George Washington University in Washington, DC. He has published extensively on the development of computation and the institutions that support computation in publications ranging from the *American Mathematical Monthly* to the *Washington Post*. Currently a member of the IEEE Computer Society Publication Board and the Editor-in-Chief of the IEEE *Annals of the History of Computing*, his most recent book is *When Computers Were Human* (Princeton UP, 2004).

Kenton Kroker is Assistant Professor of Science & Technology Studies, York University, Toronto, Canada. He has published articles on the history of allergy, electrophysiology and epidemic encephalitis (the topic of his current research). His first book, *The Sleep of Others* (forthcoming, University of Toronto Press), analyzes the evolution of sleep as an object of biomedical investigation.

William T. Lynch is associate professor in the department of Interdisciplinary Studies at Wayne State University, Detroit, Michigan. He earned a Ph.D. in Science and Technology Studies at Cornell University in 1996. He is the author of *Solomon's Child: Method in the Early Royal Society of London* (Stanford University Press, 2001); "Beyond Cold War Paradigms for Science and Democracy" *Minerva*, 41.4 (2003): 365–79; "The Ghost of Wittgenstein: Forms of Life, Scientific Method, and Cultural Critique," *Philosophy of the Social Sciences*, forthcoming; and "The Utility of the Present in Reconstructing Science's Past: Historical Counterfactuals and Contemporary Possibilities," *Scientia Poetica*, forthcoming.

William M. McBride is an associate professor of history at the United States Naval Academy and former Shaeffer Distinguished Humanist while on the faculty at James Madison University. He has a Ph.D. in the history of science and technology from Johns Hopkins and was an Olin Fellow in military and strategic history at Yale University. A former practicing engineer, he holds a M.Sc. in aerospace and ocean engineering from Virginia Tech and a B.Sc. in naval architecture from the Naval Academy. His most recent book is *Technological Change and the United States Navy, 1865–1945* (Johns Hopkins University Press, 2000).

Seymour Mauskopf did his B.A. at Cornell University, and his Ph.D. at Princeton University in the history of science. His fields of research include the history of chemistry [*Crystals and Compounds* (1976), *Chemical Sciences in the Modern World* (1993)] and the history of marginal science (parapsychology) [*The Elusive Science*, with Michael R. McVaugh (1980)]. Currently, he is investigating the role of scientists in the development of munitions. In 1998, he received the Dexter Award for Outstanding Contributions to the History of Chemistry from the American Chemical Society. He has taught history of science at Duke University since 1964.

Jamel Ostwald received his Ph.D. in History from The Ohio State University in 2002. His dissertation, "Vauban's Siege Legacy in the War of the Spanish Succession, 1702–1712," analyzes the European-wide legacy of siegecraft left by the famous French military engineer Sébastien Le Prestre de Vauban, part of a planned broader analysis of early modern military culture and the conventions of warfare.

The dissertation highlights the triumph of the decisive mindset of military commanders over the Enlightened form of warfare championed by military engineers such as Vauban, and is currently being revised for publication with Brill. A post-doctoral fellow at George Mason University, Dr. Ostwald has also published "The 'Decisive' Battle of Ramillies, 1706: Prerequisites of Decisiveness in Early Modern Warfare" in the July 2000 *Journal of Military History*. His website <www.ostwald.hispeed.com> includes a collaborative website (EMWWeb) dedicated to research on early modern military history.

Gerry Scharfenberger is an archaeologist with The Louis Berger Group, Inc. in East Orange, New Jersey and an adjunct professor of history and archaeology at Monmouth University. With degrees from Rutgers (B.A.), Hunter (M.A.) and CUNY Graduate Centre (Ph.D. [ABD]). Scharfenberger serves on the executive board of the Committee on Northeast Historical Archaeology (CNEHA) and is a member of the Society for American Archaeology (SAA), Society for Historical Archaeology (SHA), Archaeological Society of New Jersey (ASNJ), Iowa Archeology Society (IAS), and the Battlefield Restoration and Volunteer Organization (BRAVO).

Sébastien Soubiran recently finished his Ph.D. at the University of Paris VII in history of science and technology. He is now temporary lecturer at the University Louis Pasteur in Strasbourg, France, and in charge of a program at the *Mission culture scientifique et technique* to collect and preserve the university's physics heritage. He is also an affiliated researcher of the *Institut de recherche interuniversitaire sur les sciences et la technologie* (IRIST), working on physics and physicists since WWII at the University of Strasbourg. His own work focuses on the intersections of scientific and military cultures and technological development in France and Great Britain.

Steven A. Walton is an assistant professor of Science, Technology, and Society at The Pennsylvania State University in State College, Pennsylvania. With a Cornell undergraduate and Caltech graduate degree in mechanical engineering, and taking his Ph.D. in history of science and technology from the University of Toronto, his work concentrates on the intersection of the history of technology (especially the use of commercially produced objects) and the history of scientific ideas (and the users' understandings of the objects) within military contexts. He is currently working on projects on renaissance gunnery, early nineteenth-century American ordnance manufacture at the West Point Foundry, and post-Civil War scientific engineering at the Naval Torpedo Station, Newport, Rhode Island.

Deborah Jean Warner is Curator of the Physical Sciences Collection in the National Museum of American History, Smithsonian Institution. She has written widely on the history of scientific instruments; established and edited *Rittenhouse*, a journal of the American instrument enterprise; and, with Robert Bud, edited *Instruments of Science. An Historical Encyclopedia* (Garland, 1998).

INTRODUCTION

Not only are . . . instrumental methods more elegant; they are potentially more efficient, and are more and more used . . . where efficiency counts.[1]

Although known to most as the designer of the geodesic dome, R. Buckminster Fuller remains relatively underappreciated for his visionary writings on our technological future as well as for his poetic inspirations. Meditating once in his epic poem, "No Secondhand God", in 1940—before the full extent of WWII had been realized, it should be remembered—Fuller imagined a pilot in a naval battle off Norway who, while ultimately victorious, does not succeed through individual skill or heroic action, but rather through his instruments:

I think of such of the aviators and sailormen as
are in command of their faculties
on both sides at this moment.
Though you have been out in
a froth-spitting squall
on Long Island sound or
in an ocean liner on a burgeoning sea
you have but a childlike hint of
what a nineteen-year-old's reaction is
to the pitch black shrieking dark out there
in the very cold northern elements
of unloosening spring
off Norway's coast
tonight
15,000 feet up, or
fifty under or
worse,
in the smashing face of it and
here I see God

Although one might expect this deity to be the cruel, dangerous, or redemptive one that both sides invoke in conflict,[2] instead, Fuller proclaims,

---

[1] B.L. Clarke, "What is Analysis?" *Industrial and Analytical Chemistry (Analytical Edition)* 19.11 (1947): 822, quoted in Davis Baird, *Thing Knowledge: a philosophy of scientific instruments* (Berkeley, Cal., 2004), p. 100.

[2] Compare Mark Twain's *War Prayer* from 1904/5. "O Lord our God, help us

> I see God in
> the instruments and the mechanisms that
> work
> reliably,
> more reliably than the limited sensory departments of
> the human mechanism.

Thus he sees "God's fundamental orderliness/ticking along on those dials" and chides those that may be "befuddled by self or/ by habit," or those that listen to others, or those that simply through "sheer chaos of unbelief in/God" fail to pay attention to their instruments, for they shall die. On the other hand, "he who unerringly/interprets these dials/will come through."[3]

Such a clear faith in the importance and indeed the necessity of instrumentation in modern combat could not be made more clear. The contributions in the pages that follow may not be quite as

---

to tear their soldiers to bloody shreds with our shells; help us to cover their smiling fields with the pale forms of their patriot dead; help us to drown the thunder of the guns with the shrieks of their wounded, writhing in pain; help us to lay waste their humble homes with a hurricane of fire; help us to wring the hearts of their unoffending widows with unavailing grief; help us to turn them out roofless with little children to wander unfriended the wastes of their desolated land in rags and hunger and thirst, sports of the sun flames of summer and the icy winds of winter, broken in spirit, worn with travail, imploring Thee for the refuge of the grave and denied it—for our sakes who adore Thee, Lord, blast their hopes, blight their lives, protract their bitter pilgrimage, make heavy their steps, water their way with their tears, stain the white snow with the blood of their wounded feet! We ask it, in the spirit of love, of Him Who is the Source of Love, and Who is the ever-faithful refuge and friend of all that are sore beset and seek His aid with humble and contrite hearts. Amen." Mark Twain, *The War Prayer* (New York, 1951).

[3] R. Buckminster Fuller, *No More Secondhand God and Other Writings* (Carbondale, Illinois, 1963), pp. 3–4. His message has political messages akin to Twain's embedded within it as well: "Observe some paradoxes. On the one hand the Germans and the Russian preaching the rule of the masses . . . are in fact the sole demonstrators of the importance of individualism . . . And as individuals they believe precisely in the survival superiority of mechanized man. But the organized church uncomprehending the mechanical extension of man says that such belief is pagan and the totalitarians accept the designation as true and therefore proceed with wanton conviction to behave in the appropriate manner of a social outcast. Thus 'licensed' they invent vast and ofttimes ludicrous prevarications as ballistic instruments. They lie with as eager enthusiasm as that applied to gunpowder and plane making. [While for the Allies,] the machine is to them and their church but an inanimate, garagey sort of money pump." (Fuller pp. 7–9) For the story of the instrumentation that made some of Fuller's pilots ability possible, see Frederick Suppe, "The Changing Nature of Flight and Ground Test Instrumentation and Data: 1940–1969," in Peter Galison and Alex Roland (eds.), *Atmospheric Flight in the Twentieth Century*, Archimedes 3 (Dordrecht, 2000), pp. 67–106.

unabashedly optimistic about the presence of God in the instruments, but they all share the belief in the power of instrumentation to extend the range of human cognition and to delve into areas normally closed to our senses. The authors all share another common belief: that scientific instruments—whether conceptual or material—in warfare or peacetime, are crucial in shaping our understanding of the world around us. As pattern-seeking animals, humans have the wonderful and sometimes terrible facility to see what they want to see and to perceive the world in the way that they often already know it to be. Much of the time, this is hardly a problem, and one could make the strong argument that in warfare, when one's life and the life of one's cause actually depends upon it, it is in fact a good thing at the personal level. In society at large, the faith in instrumentation is perhaps no better signified by our faith in the lie detector to ferret out criminals, despite the fact that no polygraph has ever been successful in finding a spy in our midst, and copious evidence that polygraph results are subjective and quite open to active manipulation by the 'subject'.[4] We have come to believe that scientific instruments tell the truth. While philosophers of science will debate the finer points of the 'reality' of inscription devices, these debates need not concern this volume as in most cases of military use of scientific instruments, the goals are clear and the actors rarely consider that their actions or the instrumental inscriptions may not in fact have bearing on the practical task at hand. Instruments are, as it were, instrumental in war.

Instrumental in War is a collection of essays which seek to understand the apparently natural, yet understudied connection between scientific instruments, scientific behavior, and scientific methodologies in military action or preparation.[5] These "instruments" may be as concrete as a graduated brass gauge or calibrated optical scope, as mechanical or computerized as a fire control system or metallographic test bench, or they may exist in the more abstract world as conceptual frameworks accepted by all to accomplish goals. For as Francis Bacon noted in his second aphorism, "just as instruments improve or regulate the movement of our hands, so instruments of the mind provide suggestions or cautions to the understanding."[6] That

[4] William Safire, "Lying 'Lie Detectors'," *New York Times*, 10 October 2002, A37.

[5] For ease of style, I am falling back on the old use of 'scientific' to subsume science proper, engineering activity, and (high-)technological development.

[6] See epigraph to the Volume Contributions, this volume, p. xv.

is, an "instrument" in our sense may also be a combination of action and conception in such a way that the aggregate becomes a real thing in the minds of the actors—it becomes instrumental to their military undertakings. This natural alliance between instrumentalism in the most concrete of its philosophical connotations and military planning, action, or reaction forms the core of all our investigations.

As Steve Shapin recently put it, "It is now difficult to imagine what the social institution of science would look like divorced from its military ties."[7] Yet considering first the concrete use of the term, the literature of scientific instruments proper *and* warfare is remarkably slim. Literature on scientific instruments *in* warfare is slightly more robust, but is rarely conceptualized as such—that is, studies or radar or the development of the Norden bombsight can be found in great number, but told as the stories of mere technologies (in the same categories of the myriad books on individual planes, ships, guns, or other military hardware), not as instrumental objects that melded ideas, scientific measurement, theories, research, development, and construction to the aim of winning a war. Physical instruments are the realm of collectors who have largely eschewed warfare for the more refined sciences, where great strides have been made in our understanding of bench-top physics, acoustics, optics, and other sciences; instrumentalism as a concept largely escapes concrete situational application in the philosophical literature, where military action is relatively scarce (morality in and of war, of course, occupies a prominent philosophical niche); and military historians are less likely to delve into the minutia of battlefield activities where the instruments might make their contribution, and even if they do, to analyze the objects or ideas as instruments would be very rare indeed. Yet considering the two concepts together is important, and this volume offers the first salvo at understanding the relationship between them.

War

Warfare is of crucial importance to understand humanity: endemic throughout history and apparently prehistory; presumably not some-

[7] Steven Shapin, "Science and the Public," in Robert C. Olby, *et al.*, *Companion to the History of Modern Science* (London, 1990), p. 1004.

thing that will ever be fully abolished; and one of the most important driving forces for our understanding and our manipulation of society and the world. To speak of the use of instruments of war takes us back to the earliest examples of technical equipment being used to great effect in battles. Biblical stories of machines of great power and allude to some specialized knowledge,[8] but more concretely, even schoolchildren learn the story of the famous Greek scientist and inventor Archimedes being called into service to defend his native Syracuse on Sicily against the invading Romans in 214 BC. In its defense, Archimedes constructed hoists, catapults, and other engines, but is particularly renown for his a great burning mirror that set the Roman ships on fire in the harbor, saving the city.[9] The Archimedian idea kept its grip on military engineers down though the ages, and we find men like the Italian engineer Taccola sketching just such a concave burning mirror to set a bark alight in his mid-fifteenth century *De rebus militariis*, even if a bit more fancifully than realistically.[10] And while historians of technology caution us not to easily conflate the intellectual work of science and the concrete work of technology—and especially to not assume the former necessarily generates the latter—it should be remembered that one of the more important and developed physical sciences in antiquity was that of catoptrics to which Euclid and Diocles turned their attention (the optics of reflection; as opposed to *dioptrics*, the optics of transmission/refraction, which largely fermented in the Arabic middle ages).[11] Archimedes burning

[8] Although sieges fill the old testament, only a few passages speak obliquely of the technology. For example, II Samuel 20:15 says that "All the troops with Joab came and besieged Sheba in Abel Beth Maacah. They built a siege ramp up to the city, and it stood against the outer fortifications [and] they . . . batter[ed] the wall to bring it down."

[9] On this feat, see D.L. Simms, "Archimedes and the Burning Mirrors of Syracuse," *Technology & Culture* 18.1 (1977): 1–24 and his more recent augmentation, "Archimedes the Engineer," *History of Technology* 17 (1995): 45–111. A recent modern analysis suggests that any land-based concave mirror, even a literally multifaceted one made up of numerous polished *scudi* [shields], would have been unable to focus enough solar energy to ignite ships sails or hull at such a distance, although such sun-focusing multi-mirrored systems do work well to collect energy as is the case with the 1800-mirror, 10 megawatt solar power tower Solar One (steam; 1982–88) near Barstow, California, and its more efficient successor, Solar Two (molten salt; 1996–), near Albuquerque, New Mexico.

[10] Mariano Taccola, *Liber de Ingeneis ac edifitiis non usitatis*, edited by J.H. Beck (Milan, 1969), p. 134. See also D.L. Simms, "Archimedes' Weapons of War and Leonardo," *British Journal for the History of Science* 21 (1988): 195–210.

[11] Hero of Alexandria, the other great ancient engineer, also wrote a book on

mirror, then, may well be the very first scientific instrument used in war.[12] That one of the greats in the pantheon of ancient science is remembered through the application of a scientific instrument to a military moment suggests the importance of this *topos*. Thus, even though we have no accounts after Archimedes of armies actually employing such an instrument, the dream of applying scientific knowledge of the natural world to military purposes seems to have been with us for a very long time.

In the modern world, the use of scientific instruments in warfare seems like a truism. Partially this is due to the idea that warfare itself has become scientific, particularly in the twentieth century. After all, so notable a astronomer as George Ellery Hale said quite explicitly in his advocacy for the establishment of the National Research Council, "War should mean research," and the First World War became known as "The Chemists' War" and the Second, "The Physicists War."[13] It is only natural, then, that such metonymy would make us assume the critical role of scientific instruments. But that then raises the question of what one means when speaking of objects or ideas being "scientific" *and* also as "instrumental in warfare". In what sense are scientific instruments part of, or do they contribute to warfare? We might see such iconic projects as the Manhattan Project, so clearly scientific (albeit in reality a massive engineering challenge more than a scientific one)[14] as an obvious example, but in many ways, the scientific instruments used in it were not *directly* used in war. That is, although the *Enola Gay* was outfitted with many instruments to analyze the drop over Hiroshima, and Los Alamos, Chicago, Hanford, and Oak Ridge all bristled with instruments, as

catoptrics (see W. Schmidt [ed.], *Heronis Aleandrini Opera Quae Supersunt Omnia* 2 [Leipzig, 1900]). On optics in general see David C. Lindberg, *Theories of Vision from Al-Kindi to Kepler* (Chicago, 1970) and A. Mark Smith, "External Principles in Ancient and Medieval Optics," *Physis* 31 (1994): 113–40 and other works by both these authors.

[12] Seraphina Cuomo has recently suggested that the scientific behavior of catapults was well understood and exploited by the ancient engineers as early as the 4th century BC. See her "The Sinews of War: ancient catapults," *Science* 303.5659 (6 February 2004): 771–2.

[13] Daniel Kevles, *The Physicists* (Cambridge, Mass., 1987), ch. 8, 9, 20, and *passim*.

[14] Fermi and others had known that fissile materials could *theoretically* make a bomb from the moment Hahn, Strassmann and Meitner split the atom in 1935 if not before; the challenge was to figure out how much material was needed, how to configure it, and how to get it to the target. See among the multitude of Manhattan Project literature, Robert Serber and Richard Rhodes, *The Los Alamos Primer* (Berkeley, Calif., 1992).

far as the military aviators were concerned, the actual bombing run on August 6, 1945 was little different than any other bombing run apart from the physical size of the payload and the curious flight parameters they were instructed to use. But we would not want to deny that all these instruments were crucial to the prosecution of the military technology.

Alternately, consider digital computers. While initially developed specifically for military purposes[15] and for the last 50 years indispensable for planning and execution of both scientific and military research, development, and operations, they are hard to fit into the same mould as the development of acoustic rangefinding or submarine detection in World War I, where both the advancement of science and the location of enemy artillery or submarines proceeded hand in hand.[16] And only recently has anyone to my knowledge clearly and unambiguously claimed connections between instrument companies and government-sponsored military activity: "the history of the precision instruments industry (since the end of the nineteenth century at least) is best understood within the context of their military applications, not only in the United Kingdom and France but also throughout the world,"[17] and in "the development of instrumentation . . . the importance of the role played by the military, particularly in the context of World War II, cannot be overestimated."[18]

Nevertheless, neither philosophers, scientists, instrument makers, nor military practitioners always recognized this congruence. In the mid-nineteenth century, arguably a high point of exquisite precision instrument manufacture, scientific instruments seemed to be applicable to every field *except* military ones. Take, for example, J.F. Heather's *Treatise on Mathematical Instruments*.[19] It covers all manner of instruments from

[15] John Mauchley and John Eckert built ENIAC at the University of Pennsylvania on a contract with the Army's Ballistic Research Laboratory in order to develop a numeric calculator to compute ballistic firing tables, specifically those for understanding air-to-air sidewise firing trajectories—U-Penn's Moore School needed these to complete work on their electromechanical fire directors. See Stan Augarten, *Bit by Bit: an illustrated history of computers* (New York, 1984), pp. 110, 120–24.

[16] Kevles, *The Physicists*, ch. 9.

[17] Mari E. Williams, *The Precision Makers: a history of the instruments industry in Britain and France, 1870–1939* (London, 1994), p. 9. See also her "Training for Specialists: the precision instruments industry in Britain and France, 1890–1925," in Robert Fox and Anna Guagnini (eds.), *Education, Technology, and Industrial Performance in Europe, 1850–1939* (Cambridge and Paris, 1993), pp. 227–50.

[18] Baird, *Thing Knowledge* (note 1), p. 79.

[19] J.F. Heather, *A Treatise on Mathematical Instruments*, 2nd ed. (London, 1851).

simple straight edges and drawing compasses to complex optical instruments ranging from a prism to compound microscopes and columnators, all manner of surveying instruments (again from simple surveyor's chains to theodolites and sextants), and even a section on astronomical instruments. But no mention of military uses. In a book published in *Mr. Weale's Rudimentary Scientific Works for Beginners* series, the inclusion of simple instruments makes perfect sense, and one can even understand the description of more complex instruments in fields in which the average purchaser of the one-shilling book might find employment or recreation (there is even a short entry on the decidedly uncommon goniometer, an instrument to measure diffraction angles in crystals). Still, although Heather was a professor at the Royal *Military* Academy at Woolwich, where instrumentation and scientific knowledge held a high place in the curriculum from the late eighteenth century onwards and we can see a tacit assumption that scientific instrument *must obviously* have a place in military matters, these connections are not explicitly advertised to the rest of society.

The definitional imprecision and slipperiness of whether a conceptual model or a material object is or is not a scientific instrument, much less what its role *in* warfare is begs the question of how these instruments may in fact be instrumental in war. Van Helden and Hankins set a fourfold division of the effects of instruments which may begin to help: instruments confer authority (for both practitioner and by legitimizing the science); they are created for specific audiences (and their audiences in turn shape them); they bridge pure science (theory) and popular culture (belief); and their role changes when used on organisms (suggesting a cyborgian approach).[20] We may take these distinctions as stable and relevant for our purposes as well. Certainly our authors find examples of authority, audience-specificity, and elements of human factors design, although the differential meanings to different audiences does not extend well into a military context as the public rarely gets to use Apache helicopter laser targeting to find a parking space or torpedo testing water tunnels as water slides.[21] Still, society generally does look up to science and

[20] Albert van Helden and Thomas L. Hankins, "Introduction: Instruments in the History of Science," in van Helden and Hankins (eds.), *Instruments, Osiris*, 2nd series, 4 (1994), pp. 1–6 at p. 5. Parenthetical insertions added.

[21] Although recent pop culture develoments like the U.S. History Channel's cable television program, "Tactical to Practical", that showcases the civilian uses of military technology, may suggest a blurring of these lines.

often defers to military might (or at least respects it despite ideological disagreements), so the fusion of the two can be seen as a powerful force.

There are at least four distinct modes for the use of objects or idea sets encoding scientific ideas in military contexts. In some cases, an instrument is adopted for military use, having found development in a non-military context; SONAR or the concept of a clockwork siege provide clear examples. Ideas or objects may be instrumental in that they may directly facilitate the construction of military technology; the instrumentation and theory in the Manhattan Project offers an case in point. The rise of scientific testing in the nineteenth century provides another option, where theories of the strength of materials or even goiniometers might allow an arsenal (the Watertown Arsenal in New York comes to mind in the case of the U.S. Army, but both Grier and McBride offer examples here) to produce or verify the performance of war materiel. Finally, theories or embodies knowledge may find parallel research developments in purely scientific fields and the persecution of military goals; here acoustic rangefinding and detection for submarines, either in WWI or in the Cold War ocean floor listening posts offer an example. Thus, while it is hard to develop a one-size-fits-all definition or framework, we can come to know a military scientific instrument when we see one.

Beyond this, Davis Baird's recent division of instruments into models, working knowledge, and "encapsulated knowledge" offers a convenient way to consider the combined theme of this volume.[22] In his *Thing Knowledge*, he points out that physical models of scientific concepts—orreries, model waterwheels, or Watson and Crick's stick-and-ball DNA double helix—may function exactly like theories in that they can provide explanations, predictions, refutations, or confirmations. By contract, his "working knowledge", embodies the idea that instruments, once built, bear "knowledge of a kind of material agency." That is, once these items exist (again, whether solidly physical or purely intellectual), operators become skilled in their use and are able to generate knowledge and effects that could not have existed without them. This regular, repeatable, tacit knowledge may be largely inarticulable—Baird uses the analogy of riding a bicycle where we can't really say *how* we do it, we just *do* it—but is nonetheless generative as well as regulating on the behavior of the users. Finally,

[22] For this paragraph, see Baird, *Thing Knowledge* (note 1), pp. 12–16 and ch. 2–4.

encapsulated knowledge appears when "effective action and accurate representation work together in a material instrument to provide measurement." I would argue that his concept would also work well for immaterial instruments (that is, conceptions) that provide non-metric measurement (in the colloquial, "a yardstick for action"), and his analyses are broad enough that this should not rattle his framework much if at all.

Consequently we must speak of scientific instruments as a class of idea-objects which embody some principle or theory that is considered mathematical or scientific. Further, we must then make the distinction between whether these come from science used instrumentally specifically for warfare, civilian scientific developments used by or for the military, military productions which later evolved into instruments for science (that is, for the discovery of underlying properties and behaviors of the natural work that have no *necessary* relation to warfare), or an instrument which can be used in either science or in warfare but whose functional use in the two realms are unrelated (e.g., such simple things as calorimeters to investigate latent heat of gunpowder or for heating oil or ideas such as friction which can be measured to improved driveshaft performance or contemplated to diagnose "rough spots" in logistics).[23] For our purposes, those items that can reasonably be seen as mathematical or scientific can qualify as instruments if they are used in pursuit of a military goal, whether that be in active conflict or in preparation during peacetime.

Qualifications

Previous studies of scientific instruments have curiously either omitted or minimized the military components of the subject. Collectors guides rarely include any military scientific instruments at all,[24] many surveys either omit them by categorical exclusion or selective attention,[25]

[23] For an important contribution to this debate, see Michael Smithurst, "Do the Successes of Technology Evidence the Truth of Theories?," in Roger Fellows (ed.), *Philosophy and Technology* (Cambridge: Cambridge University Press, 1995), pp. 19–28.

[24] This is the case with the more general introductory guides: Gerard L'E. Turner, *Antique Scientific Instruments* (Poole, UK, 1980) explicitly uses *philosophical* categorization for the objects in his book, which by necessity omits military topics.

[25] See Turner's more recent *Scientific Instruments 1500–1900, an introduction* (London, 1998) and his much more complete *Nineteenth-Century Scientific Instruments* (London, 1983) for examples of the former; and for an example of the latter situation, Nigel Hawkes, *Early Scientific Instruments* (New York, 1981).

and some, while including many an instrument used for military purposes fail to conceive of this category as a meaningful one.[26] Other authors provide taxonomies which make no obvious room for military instruments, although such instruments could appear in many other areas.[27] Still, we should not fault the field too much, for Albert van Helden opened the important 1994 issue of *Osiris* with Koyrés argument with Crewe and diSavio's translation of Galileo's Two New Sciences which revolved around whether or not the history of scientific instruments was of any relevance to the history of science itself.[28] If even historians of science could not agree that they wanted to look at instruments as interesting entities (rather than shiny museum pieces at best) as recently as 50 years ago, then should military historians really be faulted for not paying too much attention to either the physical instruments, or even more understandably, conceptual ones? Well, we believe the answer would be yes. Not that we fault past historians; we seek to remedy this shortcoming.

If we consider this neglect in terms of the objects themselves, one needs to ask what qualifies as a military scientific instrument. Does a sextant purchased by an Admiral make it a military sextant? Or if the military lets a contract to a general testing firm that uses a dynamometer to test Jeeps, are they military dynamometers? Classes like testing and navigation are accepted categories in their own right, so it seems reasonable to let the instruments be identified primarily through them. But at the conceptual level, if a scientific idea is pressed into service to explain a military action, does it too, become

[26] For example, G. L'E. Turner and D.J. Bryden, *A Classified Bibliography of the History of Scientific Instruments* (Oxford, 1997), has a category for 'Military Instruments", but it only includes 6 entries (and 4 by one author) in 120 pages. A.V. Simcock, *A Supplement to a Classified Bibliography of the History of Scientific Instruments* (Oxford, 1998) adds but 5 more. Robert Bud and Deborah Jean Warner (eds.), *Instruments of Science: an historical encyclopedia*, Garland Encyclopedias in the History of Science 2/Garland Reference Library of the Social Sciences 936 (New York, 1998) do not have index entries for "military" or "warfare" although to be fair, their index is based on concrete rather than abstract terms.

[27] Henri Michel, *Instruments des sciences dans l'art et l'histoire* [trans. as *Scientific Instruments in Art and History* by R.E.W. Maddison and Francis R. Maddison (New York, 1967)] offers the following: Basic Elements (measuring, drawing, and calculating), Measuring the Earth (topography, geography, and navigation), Measuring the Heavens, Measuring Time (gnomics vs. chronometry), and Physical Measurements. He does include a number of fine military instruments under the first (gunners' calipers, no. 12) and second categories (gunners' levels, nos. 22 and 23).

[28] Van Helden and Hankins, "Introduction" (note 20), p. 1.

instrumental in war? Does this distinction shift if it is used metaphorically ("The General boiled over") versus technically ("The radiators boiled over")? Other inconclusive examples include the fusion of a scientific idea with a military idea to produce a merely entertaining effect, as in cannon noon-dials: a fusion of a miniature cannon, a sundial, and an optical magnifying glass set in such a way such that the magnifying glass focuses the sun's rays on the primed touch-hole of the cannon at exactly high noon.[29] Or the case where a military reality became a standard demonstration in a chemistry classroom, as with the electrically fused, small military-style mortar to demonstrate the explosive power of expanding air.[30] Van Helden quite rightly recognized that "instruments must be seen in the context of the science done with them, and this context usually grows around the instrument,"[31] which in some ways begs the question in saying that they are in some ways inseparable from the context, especially if we allow instrumental ideas as a topic for investigation as well.

Deborah Warner has asked the question of what a scientific instrument even is,[32] but for the sake of argument, we shall use the philosophical 'duck' test first: if it walks like a duck, looks like a duck, and quacks, for now, we'll call it a duck. So as a first cut, if we find an operational object or coherent set of beliefs or practices that have some relation to something we consider "scientific" (or, historically speaking, "mathematical" or "philosophical", as Warner points out), operating in a military context, then its fair game for this analysis. Further, in many cases in this military context, we will see that the instrumental idea-objects are often used to make life easier for the scientists, engineers, and soldiers, rather than necessarily advancing the theoretical underpinnings of the science. This meaning of instrumentalism, far from being denigrated,[33] is crucial for action in the

---

[29] See, for example, Turner, *Nineteenth-Century Scientific Instruments* (note 25), p. 345.

[30] *Ibid.*, pp. 193–4.

[31] Albert van Helden, "The Birth of the Modern Scientific Instrument, 1550–1700," in John G. Burke (ed.), *The Uses of Science in the Age of Newton* (Berkeley, Cal., 1983), pp. 49–84 at 51–52.

[32] Deborah Jean Warner, "What is a Scientific Instrument, When Did it Become One, and Why?" *British Journal for the History of Science* 23 (1990): 83–93. Van Helden, just cited, seemed unpuzzled by the question, for he quite contentedly answered that very question a few years previously.

[33] Van Helden even went so far as to say that "medieval European instruments all aimed at convenience (a sure sign of the derivative nature of medieval astronomy)" and claimed that accuracy and convenience are necessarily mutually exclusive (note 31, p. 54).

real world, again, an area crucial to military endeavors. But that takes care of our intellectual boundaries—we still have not adequately solved the question of why previous studies have not been done in this nascent field, even where there has been a great deal of work on instrumentalism in the last half century.

Some of the blame lies in the concentration of instrument(al) studies that speak of the theoretical side over the practical (that of "representation" over that of "intervening" in Hacking's conception).[34] The post-Enlightenment belief that philosophical knowledge is somehow unsullied (or unsulliable) by the "real" world likely also has something to do with it. Warfare, however, is about as real as it gets. Consequently, the natural philosophers who in some sense owned the ideas and at least partially claimed ownership over instruments—again physical or mental—were frequently disinclined to speak of their use in the baser pursuits. At times this prejudice carries over into modern writing on those instruments: the description that "Galileo developed a sector between 1597 and 1599 for use as a general purpose calculator" completely ignores that the work he published on its development was the *Operations of the Geometric and Military Compass.*[35] And throughout the nineteenth and twentieth century, many of the great names in science lent a hand to military pursuits: Robert A. Millikan, Albert Michaelson, and George Ellery Hale, not to mention important physicists in WWII including, J. Robert Oppenheimer, Hans Bethe, Richard Feynman, and Enrico Fermi (contrary to popular belief, Albert Einstein had nothing whatsoever to do with the Manhattan Project; the FBI considered him a security risk, and he likely found weapons research distasteful if not repugnant). But in the marshalling of science for war, instrumental pursuits were either obvious (the theodolite for military surveying), assumed (measuring chromatographic dispersion for optical technology like aerial cameras), or implied (inclinometers could be used for navigation or rangefinding). We should remember in this regard that throughout the nineteenth century the core curriculum at the U.S. Military Academy at West Point was engineering.

---

[34] Ian Hacking, *Representing and Intervening: introductory topics in the philosophy of natural science* (Cambridge, 1983).

[35] Turner, *Antique Scientific Instruments* (note 24), p. 58. Galileo Galilei, *Operazioni del compasso geometrico, et militare* [*Operations of the Geometric and Military Compass, 1606*], trans. Stillman Drake, Publications of the Dibner Library 1 (Washington, D.C., 1978).

One look at the scientific instrument industry makes the connection obvious. Trade cards of early firms show that they were quite proud to be considered as makers for military pursuits. Great Britain provides the most replete set of examples: John Browne, a mid eighteenth-century "compass Maker and Ship Chandler" from Wapping, happily advertised "all Sorts of Instruments for Sea and Land" alongside "Guns, Gunpowder, Shot, and all sorts of Ship Chandlery and Turnery Wares";[36] George Lee and Sons were manufacturers of "Mathematical, Optical, and Nautical Instruments to the Honble Corporation of the Trinity House & the Admirality," with offices on Ordnance Row near the Tower of London and in Portsea and Southsea, the two main ports of the British Navy; and Henry Porter styled himself an optician and mathematical instrument maker to the Admiralty and War Office first, and only then to the Royal Geographic Society, Christ's Hospital, and the Swedish and Norwegian Governments (who doubtless bought his instruments for their navies as well).[37] Authors of treatises on instruments, too, would advertise their connections to the military establishment if possible. George Adams, a prominent mid-18th century instrument maker, published a description of the nautical quadrant he had invented and listed himself as "Mathematical Instrument Maker to His MAJESTY's Office of ORDNANCE, at *Tycho Brahe's Head*, the Corner of *Racquet Court*, in *Fleetstreet, London.*"[38] Between 1748–1772, Adams supplied many hundreds of instruments to the Ordnance Office, as well as repairing their existing stock of theodolites, quadrants, plane tables, drawing instruments, compasses, and shot gauges (sheet-brass instruments with various sized holes to determine the standard caliber of cannonballs).

From the mid-nineteenth century onward it becomes unsurprising to discover a tighter bond between scientific instrumentation and war. Many of the technologies in war were increasingly made, tested,

---

[36] Turner, *Antique Scientific Instruments*, p. 37.

[37] Turner, *Nineteenth-Century Scientific Instruments* (note 25), p. 20, 22. Not surprisingly, many others—including J.D. Potter of London—could also claim to make instruments for the Admiralty (p. 95), suggesting just how important instrumentation was to the Navy.

[38] George Adams, *The Description and Use of a New Sea Quadrant, for Taking the Altitude of the Sun from the Visible Horizon* (London, 1848), title page reproduced as fig. 2.12 in John R. Millburn, *Adams of Fleet Street, Instrument Makers to King George III* (Aldershot, UK, 2000). For Adams work for the Ordnance Office, see pp. 51–54. Brahe, remember, was known for his exquisite and enlarged astronomical instruments.

and refined by the aid of science, which by then was truly becoming applicable. Whether in the development of stronger alloys for ordnance, new nitrocellulose propellants, optical or acoustic devices, or in aeronautical developments only made possibly through instrumentation and testing (remember that the Wright brothers had built a wind tunnel at Kitty Hawk to test wing and propeller shapes), the objects, methodology, and metaphors of science came to drive military technology. There were numerous cases where scientific ideas were thought immediately applicable to military purposes. Historically, we have already mentioned Galileo's military sector, but to this we might add various experiments on pneumatics from the mid-eighteenth through the end of the nineteenth centuries that promised to develop air guns for the military.[39]

In this century, with the advent of electronics, computers, and the Military-Industrial Complex (for all its effects), military hardware has become more and more scientific. Examples can be rattled off quite quickly but consider, for example, the great developments in submarine warfare since WWI, where instrumentation is crucially necessary when you can't see your adversary. SONAR is the most obvious instrument that allows submarines to accomplish their tasks, but there are numerous other twentieth century innovations even more obviously identified as scientific instruments. Acoustic arrays initially deployed by seismologists on the ocean floor to track undersea earthquakes in the Pacific became a crucial element of anti-submarine warfare (ASW). Bathythermographs, for example, allow submariners to measure the thermal profile of the ocean, and then with an understanding of thermoacoustic sound propagation in the water, to find enemy submarines or even to evade them by tailoring their own acoustic signature.[40] And if one were to list most major manufacturers of scientific equipment—Texas Instruments, Raytheon, IBM, Hughes, and Data General—the military connections are obvious.

Finally, the rhetoric of conflict at the dawn of the twenty-first century repeatedly emphasizes the scientific, precision nature of warfare. Laser-guided bombs, GPS enabled troops, and night-vision IR goggles all take combat to the enemy with precision, power and progress,[41]

---

[39] Turner, *Nineteenth-Century Scientific Instruments* (note 25), pp. 105–6.

[40] Gary Weir, *An Ocean in Common: American naval officers, scientists, and the ocean environment* (College Station, Tex., 2001).

[41] These are three of van Helden's criteria for a "modern" scientific instrument

regardless of the overall tactical success in "major combat operations." There is no suggestion—nor desire—on the part of the volume contributors that such scientific instrumentation should cease, and although it may be a bit like Alfred Nobel believing that his dynamite would make war incomprehensible, it may not be incomprehensible to think that these developments might possibly lead to less conflict in the future. If nothing else, they will certainly lead to more analysis.

(note 31, p. 49)—the other is that scientific instruments are seen as the "natural" or "proper" to study nature, an assumption we have already noted for their use in military matters. The connection to 21st century warfare is mine.

CHAPTER ONE

MATHEMATICAL INSTRUMENTS AND THE CREATION OF THE SCIENTIFIC MILITARY GENTLEMAN

Steven A. Walton

Introduction

Every high school graduate "knows" that Galileo showed that the flight of a cannonball is a parabola. And every graduate also intuitively understands that this was the beginning of scientific warfare, even if their teachers never made that link explicit. But, although others have been at some pains to show that Galileo's trajectories actually *would* work (closely enough) for high-angle mortars because in that case the effect of lateral air resistance is relatively negligible,[1] it is clear that his work was of more intellectual than military utility. More perspicacious students might know further that Niccòlo Tartaglia claimed to have invented mathematical ballistics and Daniel Santbech offered an alternate solution two generations before Galileo's groundbreaking (or perhaps air-breaking) work, but these, too were more idealized geometry than practical gunnery.[2] So why, then, do we continue

[1] Brett Steele, "Siege Mortars and Galileo's Ballistics," paper presented at Society for the History of Technology (SHOT) meeting, Lowell, MA, October 1994.

[2] See, among others, A. Rupert Hall, "Gunnery, Science, and the Royal Society," in John G. Burke (ed.), *The Uses of Science in the Age of Newton* (Berkeley, Cal., 1983), pp. 111–41 and his important *Ballistics in the Seventeenth Century* (New York, 1968); Evelyne Barbin and Michèle Cholière, "La trajectoire des projectiles de Tartaglia á Galilée," in *Mathématiques, arts et techniques au XVII*[ème] *siècle* (Mans, 1987), pp. 40–147; Stillman Drake and James MacLachlan, "Galileo's Discovery of the Parabolic Trajectory," *Scientific American* 232.3 (March 1975): 102–10; and numerous important works by Ronald H. Naylor, including "Galileo's Early Experiments on Projectile Trajectories," *Annals of Science* 40 (1983): 391–94; "Galileo: the Search for the Parabolic Trajectory," *Annals of Science* 33 (1976): 153–72; and "Galileo's Theory of Projectile Motion," *Isis* 71 (1980): 550–70. For Santbech, see Andreas Kleinert, "Zur Ballistik des Daniel Santbech," *Janus* 63.1–2–3 (1976): 47–59. For more general outlines, see István Szabó, "Die Anfänge der äußeren Ballistik," *Humanismus und Technik* 14.3 (1970): 1–28 and Pierre Thuillier, "De l'art á la science: La découverte de la trajectoire parabolique," *Recherche* 18 (1987): 1082–89.

to believe that by the end of the sixteenth century, gunnery as an art took on the trappings of science?

One avenue to understand how a technology becomes scientific is to consider the component parts of the technology under scrutiny that are themselves scientific. That requires that we have a effective definition of *scientific*, which is in itself problematic.[3] For present purposes, when dealing with early modern military technologies, a fairly simple distinction may be made between those simply 'used' technologies and those 'understood' those technology intellectually, and in particular, mathematically. These latter may be seen to have been "scientific" for their time. Pikes and muskets, for example, were understood practically and tactically, but no attempt was made to understand the plunging forces or explosion chemistry necessary for them to work;[4] macro-level theories for deployment may have been posited, but little theoretical understanding of the objects themselves was forthcoming; ultimately, no attempt was made to understand intellectually *how* they as objects worked. By contrast, I will argue that early modern artillery[5] was not only put to use practically, but it was also an object of intellectual inquiry, although not along the lines usually assumed. Specifically, early modern artillerists came to a numerical understanding of their weapons and installed this knowledge onto wooden, brass, and steel instruments and then turned around and used these instruments to justify and advance their very professional existence. Consequently, it was through the gunners' attempt to know

For a recent interesting adjunct to Galileo's ballistic parabolas, see David Topper and Cynthia Gillis, "Trajectories of Blood: Artemesia Gentileschi and Galileo's parabolic path," *Women's Art Journal* 17.1 (1996): 10–13.

[3] Although the literature on this matter is vast, for a recent appraisal, see Sungook Hong, "Historiographical Layers in the Relationship between Science and Technology," *History and Technology* 15 (1999): 289–311; Harvey Brooks, "The Relationship between Science and Technology," *Research Policy* 23 (1994): 477–86; and the classic paper by Otto Mayr, "The Science-Technology Relationship as a Historiographic Problem," *Technology & Culture* 17.4 (1976): 663–73.

[4] One short-lived exception to this distinction was in the attempt by Henry Percy, the 9th Earl of Northumberland, to understand these ideas through a Ramist dichotomy where he did recognize that different kinds of force were at play with different weapons. He did not, however, pursue this line of reasoning to any practical end. See Steven A. Walton, "The Classification of Arms: Henry Percy's Ramist Ideas of Weapons," *Journal of the Arms and Armour Society* 18.1 (2003): 25–39.

[5] Although 'artillery' in the early modern period meant all forms of missile weapon—from sling to longbow to musket to cannon—here it is used in the modern, more restricted, sense of heavy artillery: gunpowder weapons of greater than 2-inch bore mounted on wheeled carriages for land or sea service.

their weapon in a quantitative manner and their sometimes success in reducing their performance to calibrated rules that makes early modern artillery one of the first "scientific" military technologies.

In the early modern era, where scholars still debate whether there was a "scientific revolution" (or revolutions, or whether it should be capitalized), it is perhaps dangerous to even speak of "scientific" instruments. For in the absence of theories we would admit as *science*, can the instruments used in pursuit of that theory themselves be scientific? From that point of view, a Paracelcian balance graduated in the influence of Mercury and Saturn would qualify. To avoid splitting hairs let us agree that one quality common to most instruments that appear to be scientific to us is that they were and are quantitative, even if the theory that codified those quantities in the past might be discredited of altered beyond recognition today. In effect, we ought rightly to speak of *mathematical* instruments in the early modern world. Yet this terminology has its pitfalls as well, for just as many instruments fail to be examples of science in that the theory behind them is hidden—and in fact that is the very definition of an instrument: that which compresses and black-boxes a theoretical relationship for easier practical effecting of a task—similarly do many mathematical instruments either hide theory or rely on little theory beyond a scale, whether linear, exponential, or later, logarithmic. More will be said on the issue of theory vs. scales below, but here let us be clear: mathematical instruments are scientific in that they consolidate considerable amounts of *scientia* or knowledge into one convenient package so that the user need not use or even know, for that matter, the theories embodied in them. More importantly, regardless of knowledge of the embedded theory, to use them was for the practitioners to behave scientifically. Looking at the concept of instrumentation and the seventeenth-century Scientific Revolution from the other end[6] one self-evident feature appears to be greater and greater reliance of quantitative measurement.[7] Consequently, mathematical instruments of the

[6] A. Mark Smith, "Knowing Things Inside Out: The Scientific Revolution from a Medieval Perspective," *American Historical Review* 95.3 (1990): 726–44.

[7] On this vast subject, see W.P.D. Wightman, *Science in a Renaissance Society* (London 1972), ch. 10, "Accent on Quantity", as well as Robert K. Merton's classic argument in *Science, Technology, and Society in Seventeenth-Century England* (New York, 1970). For the modern interpretation, see M. Norton Wise (ed.), *The Values of Precision* (Princeton, NJ: Princeton UP, 1997).

Renaissance may safely be claimed as the precursors of modern scientific instruments, often with a high degree of continuity, even if the theoretical element might seem seemingly slim or absent early in their development.

The question, though, is where the fusion of mathematical instruments and military activities came from. The earliest fusion of military activity and mathematico-scientific instruments may be the development of geometrical reasoning applied to fortification design—the so-called *trace italienne* system appearing in the first years of the sixteenth century although with fifteenth and even fourteenth century antecedents.[8] However, the flowering of geometrical fortification design did not occur until well into the seventeenth century, and interestingly, both effective mathematical ballistics and rigorously geometrical fortification design are really products of the eighteenth-century development of mathematical technique, despite the products of this "science" having worked well for 2 centuries.

But as artillery became an important part of warfare in the later sixteenth century in a cat-and-mouse interaction with this fortification, the offensive side of the equation began to crystallize into a more widely and thoroughly mathematical approach.[9] In the fifteenth century pictorial works appeared explaining the new gunpowder technologies to practitioners and interested observers alike (although the lavishness of existing copies suggests the elite status of the observers, in distinction to the non-elite status of the practitioners).[10] By the middle of the next century most kingdoms, cities, and sometimes even rather small towns employed either full- or part-time gunners to defend

---

[8] Albrecht Dürer set out geometrical rules as early as the later 15th century, and Kelly DeVries has recently summarized the pre-*trace* reactions to gunpowder in "The Impact of Gunpowder Weaponry on Siege Warfare in the Hundred Years War," in *The Medieval City Under Siege*, edited by Ivy A. Corfis and Michael Wolfe (Woodbridge, UK, 1995), pp. 227–44. In general, see John Bury, "Early Writings on Fortifications and Siegecraft: 1502–1554," *Fort* 13 (1985): 5–48 and for the earliest English case, see Robert Corneweyle, *The Maner of Fortification of Cities, Townes, Castelles and Other Places, 1559* (Richmond, UK, 1972).

[9] That is, the peak of geometrical rules for fortification came quite late in the 16th century and really was more characteristic of the 17th. By comparison, mathematization of gunnery began as early as the 1450s, and was more widely and uniformly practiced by the mid-16th century. More importantly when thinking of the shift writ large, however, fortifications numbered in the scores at most across Europe, while every state, town, ship, and port had gunners. For the Whiggish, internalist view, see Rupert Hall, "Gunnery, Science, and the Royal Society" (note 2).

[10] Ranier Leng, *Ars Belli: taktische und kriegtechnische Bilderhandschriften und Traktate im 15. und 16. Jahrhundert*, Imagines Medii Aevi 12 (Wiesbaden, 2002).

their castle or town walls. These gunners may have initially been simply told to touch a burning match to the hole on top of a 'hook-gun' on the rampart and fire into the masses of any attackers foolish enough to approach too closely,[11] but whether or not the actual practice demanded a theory of heavy artillery, one started developing.

In what might reasonably be called a power play for status, heavy ordnance gunners (hereafter, simply 'gunners') sought by a number of means to ally themselves with "the mathematicks."[12] This field consisted at its root an understanding of the four basic arithmetical operations along with simple plane geometry: angular measurement, right and similar triangles, and how to scale between them. In advanced stages, understandings of proportional relationships, (proto)trigonometric calculations, and possibly even some rudimentary polar geometry might be added, but gunners rarely if ever needed to move beyond the first stage. It was that continuity between the stages, however, that seems to have attracted the gunners. As was the case well into the nineteenth century, the most accomplished and therefore highest status mathematicians of the day were the natural philosophers well versed in astronomy who had a command of a full range of arithmetic, geometry (plane and spherical), and trigonometric functions and interpolations, as well as positions of stars and celestial bodies. Not forgetting that astronomers were usually astrologers looked up to by court and commoners alike (recall that Kepler cast horoscopes for his royal patrons), it is not hard, therefore, to see gunners understanding mathematics as a key to social and professional advancement, as well as advancement of the understanding of the world.

In many ways mathematics at the time was dually defined by mathematical education in the theory and practice of numbers, and also a facility with instruments purpose-made to reduce difficult (for

---

[11] Bert Hall, "The Changing Face of Siege Warfare: Technology and Tactics" in Corfis and Wolf, *Medieval City Under Siege* (note 8) and his *Weapons and Warfare in Renaissance Europe: gunpowder, technology, and tactics* (Baltimore, 1997).

[12] Interestingly, light artillery—those who became identified with Musketeers in the following century—did not tread this path. Like heavy gunners, these military men tended to specialize at an early date, they were hired directly by the head polity, tended to operate rather independently, and yet they did not 'theorize' their profession much. The clockwork-like Maurician reforms do not evidence of such a move as they were at the system level (performance of the group), not the object level (performance of the individual weapon), as with heavy artillery. One could argue that Maurician reforms *could* have worked in the age of bows or slings as well.

the day) mathematical relationships to simple practice.[13] The fact that these instruments embodied mathematical theory, however simple, set them but one step removed from natural philosophy which was itself becoming more mathematically oriented in some branches. In allying with that field, the gunners felt they could buoy their status within society and the state. Military engineers' experience differed from country to country, with Italy professionalizing first in the early sixteenth century, France, Spain, and Germany shortly thereafter, and England catching up at the very end of the century, but the rise of the mathematical practitioners led to a concerted effort by the engineers themselves to redefine the profession of military engineer so that by the later seventeenth century, the concept of a "scientific military gentleman" would personify the military technician.[14]

In addition, scientific gunnery is often assumed to necessarily mean ballistics *á la* Galileo, but distinct from his work on *external* ballistics (the flight of the cannonball), developments on *internal* ballistics had been ongoing for over a century. Gunners were concerned with geometrical relationships between the diameter and length of the bore and the charge of powder and the ultimate range of the gun, as well as arithmetical relations for the constituents of the powder as early as the 1450s. German artillerists in particular, began to try to think of their *art* (as the profession had indeed been in the pre-gunpowder period) in a more numerical and graphical manner, but it was not until the middle of the next century that we have a clear agreement that artillery should be a mathematically-based science. In the early phase, there were occasional gunners, such as Martin Mercz in southern Germany who attempted to define how much powder should be used for a shot using a geometrical compass-proof based on the bore diameter.[15] But the penchant for mathematization can be seen in the proliferation of gunpowder recipes where the cookbook approach

---

[13] A.J. Turner, "Mathematical Instruments and the Education of Gentlemen," *Annals of Science* 30 (1973): 51–88.

[14] Mario Biagioli, "The Social Status of Italian Mathematicians, 1450–1600," *History of Science* 27 (1989): 41–95; Mary J. Henniger-Voss, "Between the Cannon and the Book: Mathematicians and Military Culture in 16th-Century Italy," Ph.D. dissertation, Johns Hopkins University, 1995. The term "scientific-military gentleman", however is mine.

[15] I must thank Pamela Long for bringing this manuscript to my attention. For Mer(c)z, see her *Openness, Secrecy, Authorship: technical arts and the culture of knowledge from antiquity to the Renaissance* (Baltimore, 2002), pp. 120–21, and Leng, *Ars Belli* (note 10), pp. 199–202, 206, 455.

rapidly developed into published books with lists of all sorts of powder, distinguished by their relative proportion of ingredients.[16] Master gunners who might be called upon to mix or prove these recipes also wanted to ensure their proper use in various cannon, so once a recipe had been determined, each cannon was marked as receiving a certain amount of powder (here there seems to have been virtually no theoretical guidance; charges were nominally the same as the weight of the ball, although slightly less for larger balls). To insure that the proper load was given, the master gunners were taught how to set out ladle sizes, again using compass-proofs, so that each cannon had a matched ladle, and a given number of scoops would load the appropriate charge. Again, although we do not often think of ladles as mathematical instruments, in this case that occupied many a gunner from the end of the sixteenth century on, the instrumental behavior is quite clear.

Although I explicitly do not claim that gunnery was part of developments in natural philosophy in the sixteenth century, it is worth noting that as natural philosophy increasingly applied mathematical methods to previously qualitative phenomena in the later sixteenth century, at the interface of these two hitherto mutually distinct disciplines of mathematical instruments and military technology instrument makers also affected in wood, brass, or gold numerical relationships for various investigations and quantification of war. Despite recent assertions that the "Scientific Revolution" was nothing special,[17] from the viewpoint of mathematical instrumentation, we can say with absolute certainty that the second half of the sixteenth century did see a profound acceleration in the number, types, and precision of mathematical instruments available to mathematical practitioners from Italy to England. As the discipline grew, instruments of all sorts appeared for use in the aiming, loading, and firing of cannon and in an interlocked feedback loop, so too did claims of the usefulness of "the mathematicks" for using that artillery.

[16] The very first printed English work on gunpowder, Peter Whitehorne's, *Certain waies for the orderyng of Souldiers in battelray* (London, 1562), which was an companion volume to Whitehorne's translation of Machiavelli's *Art of War*, offers over two dozen recipes for gunpowder and many more for various firework powders.

[17] Steven Shapin, *The Scientific Revolution* (Chicago, 1998). Against this, for artillery, see Mary Henniger-Voss, "How the 'New Science' of Cannons Shook up the Aristotelian Cosmos," *Journal of the History of Ideas* 63.3 (2002): 371–97.

Instrumental Instruments

The concept of "instrument" has a convenient double meaning: it is both the material object which measure the world ("devices [which] *respond to* a physical quantity or phenomenon . . . rather than accomplish an effect"),[18] but then it is also more generally any idea, concept, procedure—or another person—that can provide an easier way to accomplish a task.[19] Thus did Roger Williams, one of Queen Elizabeth's most experienced soldiers, say of guns that "it is an errour to thinke that experimented Souldires are sodeinlie made like [fragile] glasses, in blowing them with a puffe out of an iron instrument."[20] And it is worth remembering in the context of the history of technology, that exactly this time (1578), Jacques Besson published his *Theatrum Instrumentorum et Machinarum* (in French, titled *Théâtre des instrumens mathématiques*), full of numerous machines for all sorts of tasks, frequently military ones.[21] In fact, when we browse the use of the term 'instrument' in the early modern period—at least in English—other than musical instruments, it not infrequently refers to military hardware.[22]

The first level of analysis for military hardware in this period must be to determine what actually became instrumental in the military

[18] *Oxford English Dictionary*, sb. "instrument" 2a, emphasis added.

[19] At the time, the 9th Earl of Northumberland would speak of all subordinates in an army as the "Marshall's instruments in the field"; see Alnwick Castle, Alnwick, Northumberland, MS 512, p. 537. J.J. Simmons, used the term this way in his "Early Modern Wrought-Iron Artillery: Macroanalyses of Instruments of Enforcement," *Materials Characterization* 29 (1992): 129–38.

[20] Roger Williams, *Briefe Discourse of Warre*, sig. B3v [in John X. Evans (ed.), *The Works of Sir Roger Williams* (Oxford, 1972), p. 11].

[21] As our modern term "instrument" has taken on the silent implication of either measurement (spectro*meter*), inscription (spectro*graph*), or precision observation (micro*scope*) and hence *action*, this genre has become known as the "Theatre of Machines", omitting the original sense of instrumentality, preferring instead to concentrate on the mechanical nature of the inventions by of Besson (1576), Ramelli (1588), Böckler (1603) and others. That many of these machines were more fantastic than practical finds a certain parallel in the mathematization of gunnery where 95% of all active gunning was done at point blank range. That is the range where no math is needed.

[22] *Oxford English Dictionary*, sb. "Instrument" 2a: as early as the 14th century it refers to the equipment for a siege such as ladders, scaffolds, pikes and slings; by the 17th it is a synonym for 'gun'; Shakespeare speaks more generally of "bend[ing] the fatal instruments of war / Against his brother and his lawful king" (*Henry VI*, pt. iii, V.1.88–89). The connection between musical instruments and mathematical instruments, as both involved numerical ratios and were based on the *quadrivium*, would make an interesting linguistic study.

arts in both senses of the word outlined above. First, in the intellectual sense, engineers, natural philosophers, and experimentalists (recalling that these areas were rarely mutually *in*clusive) sought to codify numerical relations between such things as the weight of shot, the weight of a gunpowder charge, the size of the cannon, the distance to target, and elevation angle of the gun into a formula (*any* formula) to make the gunners' job exact. Consequently, from the early sixteenth century onward, in gunnery treatises—manuscript or print—one finds lists of data citing that for a cannon of a certain diameter bore, a gunner should use a ball of so many pounds and a weight of powder of so many pounds, which will throw the ball so many paces.[23] By the end of the century, these relationships had been codified into tabular form[24] and then eventually inscribed on rules as described below. The gunners never did find a closed analytical solution (something which, by the way, still eludes modern science), but the process defined gunnery as an art which *ought* to be mathematical and instruments to aid this belief proliferated.[25]

Incidentally, on the other side of the gun sight, fortification engineers from the early part of the century onward began to codify the geometrical relationships between wall section and plan—angles, heights, depths and widths—into orderly relationships to one another, trying to merge fortification design with the orderly, Euclidean practice of surveying. Despite nearly continual revision, refinement, and

---

[23] Initially the ball weights were only given for stone, but quickly lead and iron weights were added as casting technologies made them (and iron particularly) the common ammunition. Even late into the 17th century, stone weights were specified in table and on rules, even though their use had largely been abandoned, and lead was often specified for balls of impressively (that is to say uncommon in the extreme) diameters.

[24] Elizabeth Tebeaux, *The Emergence of a Tradition: technical writing in the English Renaissance, 1475–1640* (Amityville, N.Y., 1997), pp. 70–73, argues that early printing actually made the text more instrumental in terms of offering the presentation of tabular data. In this, she rather misses the point that it was simply much harder to align text on a compositors board to produce a table with rank and file aligned—which after all, is crucial for the effective function of the table—than it was to do with ruler and pen in the manuscript tradition. Further, she makes the unwarranted suggestion that modern tables are derivative of the Ramist tradition (hierarchical dichotomous division); Ramus introduced the typographical decision tree, not so much the table.

[25] *Instrument* also carries connotations of being finer and more precise than a tool, often because it uses quantitative methods. There are some tools which are quantitative that are not generally considered instruments (*e.g.*, a torque wrench), as well as many instruments which are not quantitative (*e.g.*, a scalpel), but in the context of Renaissance gunnery, instruments were always quantitative—that was their advantage after all.

alteration over three centuries to find the "perfect" fort outline—remember that the great Vauban himself went through three systems alone—here, too, mathematics sought to make the fortress engineer's job exact and that process defined fortification as an art which *was* mathematical. For fortification as well, numerous instruments were developed to aid in these calculations. Ruler, compass, and protractor were initially used to design the outlines but then when it came time to break ground plane tables and early theodolites marked out in 360° helped set out fortification plans on the ground. Simple graduated quadrants with plumb bobs might be used to set inclines of walls. But in the end, the art developed a number of highly specialized instruments with which the fortification engineer could work his art.[26] One of the most impressive was the *radio latino* (figure 4—described in more detail below in its capacity as a gunnery instrument as well) which could be used to take or set the inside, outside, acute, obtuse, oblique, and all other angles in a fortification, as well as determine scale ratios in the manner of a sector.

Gunnery Instruments

When firing a cannon, the gunner needed to know a number of external variables—range to the target and relative elevation differences—and then needed to find a combination of the internal variables—type of cannon (which determined the bore diameter and barrel length), type and amount of gunpowder, and the material (weight) of the cannonball—to place an accurate shot on his target. With these and either a rule of thumb, manuscript or printed table, or increasingly a gunnery instrument, he could then set the barrel elevation, measure the powder charge, and possibly even estimate the flight time, which could be useful if it were an incendiary with a timed fuse designed to rain fire down up on the target. This scenario, however, makes the whole process sound much more analytical and mathematically

[26] For example, a Holland Circle (similar to theodolite) by Jacob de Steur now in the Smithsonian Institution's National Museum of American History (cat. no. PH*317346) uses both gradient scales used in military engineering (400 gradiens in a circle rather than 360 degrees, offering 100 grad per quadrant for easier division) and irregular scales marked "anguli centri" and "mechanica" to sent out polygonal fortification features for regular polygonal forts with from 3 to 12 sides. I thank Deborah Warner for calling this instrument to my attention.

soluble than it was in reality in the sixteenth and seventeenth century. Often one to three ranging shots were made and the gunner simply 'zeroed in' on the intended target. But increasingly, at least in the rhetoric of gunnery texts, instruments came to be the arbiter of presumptive precision for the gunners.[27]

The first and simplest gunnery instrument is a quadrant, a 90° arc of wood with one long leg and a plumb line affixed at the center of the circle. With the long leg inserted in a gun barrel, a gunner could easily read off the elevation angle, whether in degrees or in 'points'.[28] Niccoló Tartaglia, Italian mathematician and courtier in the early sixteenth century, claimed to have invented the quadrant in his *Novo Scientia*, and although pure *braggadocio*,[29] Tartaglia's description does establish the common use of numeric elevation knowledge for gunnery as early as the 1530s. One of the earliest English gunners manuals (1564) makes it very clear: its seventh rule of gunnery was "to knowe hou to sheut in Reull or quaderent at anne to[w]nes or anne banketh or at anne hell or doune the hell or doun the to[w]nes or stepell," making instrumental knowledge one of the standards for an educated gunner.[30] Other common instruments for gunnery in the sixteenth century included calibrated rules for converting bore diameters into shot weights in different materials, calipers for measuring external diameters of shot and internal diameters of bores, graduated sights with fine screw adjustments for adjusting cannon elevation for given ranges (and compensating for crosswinds, although this skill was very rare) and for lateral leveling of the carriage, and the aforementioned geometrically-constructed powder ladles matched

[27] Charles J. Purdon, "The Study of Ballistics. Part 1: Gunpowder & Early Instruments," *Arms Collecting* 35.3 (1997): 75–83 and "The Study of Ballistics. Part 2: Pendulum Alternatives," *Arms Collecting* 35.4 (1997): 114–20.

[28] There are 8 points per quadrant or 32 points per circle (1 point=11¼°)—a measure related to compasses, as in 'sailing 3 points to the wind'. Quadrants are also used to find the elevation of stars, which provides sailors with latitude, so this connection should come as no surprise.

[29] For Tartaglia's claim, see Cyprian Lucar's translation, *Three Bookes of Colloquies Concerning the Arte of Shooting in Great and Small Peeces of Artillery* (London, 1588), p. 1, as well as Stillman Drake and I.E. Drabkin, *Mechanics in Sixteenth Century Italy: selections from Tartaglia, Benedetti, Guido Ubaldo, & Galileo* (Madison, Wisc., 1969), pp. 63–65. Quadrants are commonplace elevation-taking instruments as early as the 8th century West and probably in antiquity as well; see *e.g.*, Hugh of St. Victor, *Practical Geometry*, trans. Frederick A. Homann, S.J. (Milwaukee, Wisc., 1991).

[30] Richard Wright, "Book of Great Artillery," London, Society of Antiquaries of London MS 94, fol. 4r.

to individual cannon. By the end of the century it is clear that although gunners prided themselves on learning the mathematical relationships needed for operating ordnance, with the instruments available they had reduced the 'art' to a mundane practice not unlike baking today ("take 3 parts . . .").[31]

One great expositor of the *art militaire*, for England at least, was Cyprian Lucar. His 1588 translation of (or, better, his homage to) Tartaglia's *Questi et Inventioni* (1542) was accompanied by his own *Lucar Appendix* which distilled various Continental authors' works into a compilation of nuggets of gunnery information. In this melange, he added not only marginal annotations as finding-aids, but also clear, naturalistic woodblock illustrations displaying instruments rather than processes. Illustrations were a considerable investment in printing a book, and Lucar's willingness to include so many speaks to his belief in the importance of proper depictions of those instruments. Not only ought a gunner "have a ruler and a payre of compasses to measure the heigth & length of every peece his concavity, and the length, depth, and wideness of every ladle,"[32] in the book we find images of calipers, quadrants, levels, gunners' staffs (sometimes called the "Gunners' Javelin"),[33] and even a complex "ruler" that measures elevations and allows the gunner to mount his cannon appropriately. (figure 1) A long section on surveying shows how the gunners also became the primarily topographical survey body for the English crown (hence the *Ordnance Survey* maps for the British Isles to this day). It is clear, then, that by the time of the Armada, gunnery was a fully developed instrumental practice.

Serving this industry, instrument-makers turned out a staggering variety of devices to measure angles, elevations, and proportions, and which contained scales of all sorts of relationships. There were sectors to measure shot and with tables of numerical relations engraved on them. There were more complex devices that took multiple angles at the same time—in aeronautical terms, the roll, pitch and yaw of the cannon—or included range-finding tables or scales upon them.

[31] Although not the place to develop here, the concept of recipe-concoction was also important to gunners in making gunpowder, fireworks, and other incendiary mixtures. In fact, renaissance gunnery was not so much about making artillery a *theoretical* science as it was about making it a *cookbook* science.

[32] Cyprian Lucar, *A Tretise named Lvcar Appendix* (London, 1588), p. 2.

[33] Thomas Smith, *The Arte of Gunnerie* (London, 1600), pp. 67, 70–71, which says it should be 8 feet long as a javelin or a 6 feet as a staff.

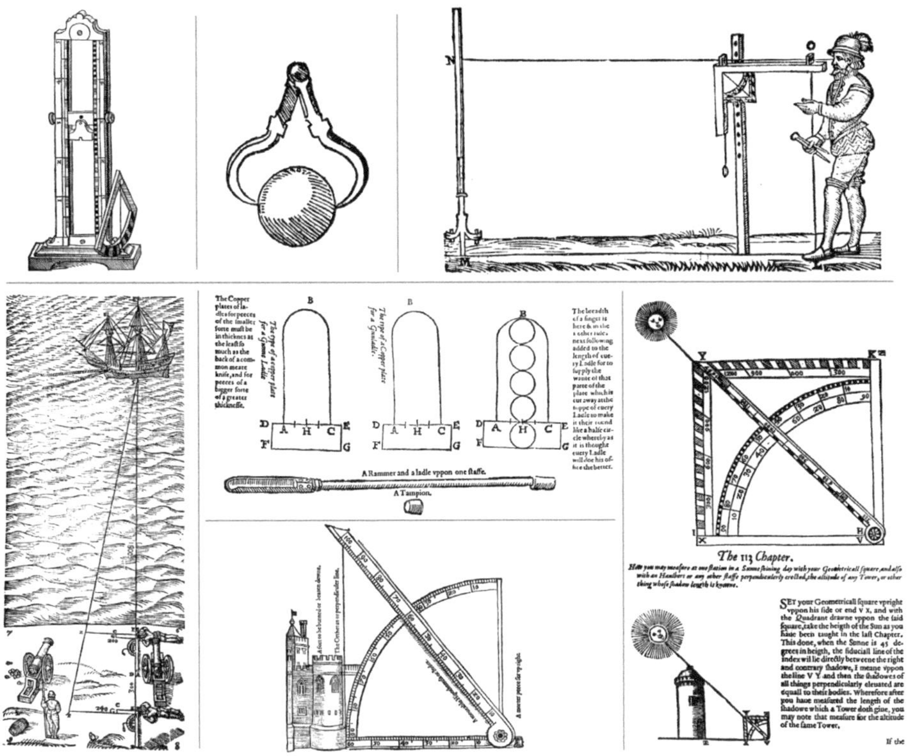

Figure 1. A series of gunnery instruments from Cyprian Lucar, *A Tretise named Lvcar Appendix* (1588). *Top*: a combination sighting ruler and elevation/depression quadrant (left), shot calipers (center), and a sighting quadrant and staff sighting to a gunner's javelin planted in the ground (right). *Bottom*: triangulation to find the range to a ship (left), ladle patterns and a ladle (center top), a mortar quadrant (center bottom), and taking tower elevation by means of a solar quadrant (right).

How common these ever were is hard to tell. The manuals are silent on them, and yet they survive in wonderfully elaborate and expensive versions (mostly made in Germany and signed by well-known instrument makers such as Christopher Treschler or Viktor Stark), or in instruments like Swiss army knives that combine levels, scales, quadrants, and even powder testers into single instruments.[34] Their existence is evidence that the instrument makers were responding to a demand from wealthy patrons, and that demand implies that for these social elites, instrumentation exemplified was the way to become well-versed in artillery.

The most common and likely useful of these instruments were gunners' rules made in wood, iron, and for the upscale market, inlaid ivory wood or gilded brass. Simple rules with exponential scales for the artilleryman, gunners' rules allowed him to know what weight cannonball, whether made of iron, lead, or stone, would fit a given muzzle. (figure 2) Some could be used to measure the muzzle directly, others in conjunction with dividers. Numerous fine examples survive in museum collections in gilt brass, and in some cases we still have multiple examples from the same workshops (often in Germany) indicating that a considerable market must have existed for even these upscale examples.[35] Gilt brass was for princes and courtiers or perhaps a wealthy or favored master gunner; common gunners, though, clearly also made use of gunners' rules as wooden and brass examples found in shipwrecks clearly attest.[36]

Making such a rule, was, however, not a skill that was explicitly relayed to the gunner in the various manuscript or printed gunners' manuals or in notebooks that survive. To a modern eye, it would not be that hard—simply transfer the weight values from the ubiquitous tables to a straight stick of wood or metal (*e.g.*, for a scale of iron shot, at a length of 5 inches, mark 18 lbs. for a culverin with a bore

[34] For these instruments, see Jim Bennett and Stephen Johnston, *The Geometry of War, 1500–1750* (Oxford, 1996) and Henri Michel, *Instruments des sciences dans l'art et l'histoire* [translated as *Scientific Instruments in Art and History* by R.E.W. Maddison and Francis R. Maddison (New York, 1967)].

[35] For example, the Mensing Collection of astronomical instruments at the Adler Planetarium (Chicago, Ill.) includes a large number of military instruments, including gunners rules. These rules, mostly from upper Germany have matching scales and two seem even to be of the same 'model' although clearly not meant to be paired. It is assumed, therefore, that some level of quantity production was in effect.

[36] Colin Martin, "De particularizing the particular: Approaches to the investigation of well documented post medieval shipwrecks," *World Archaeology* 32.3 (2001): 383–99 and Alex Hildred, "A Gunner's Rule from the *Mary Rose*," *Journal of the Ordnance Society* 15 (2003): 29–40.

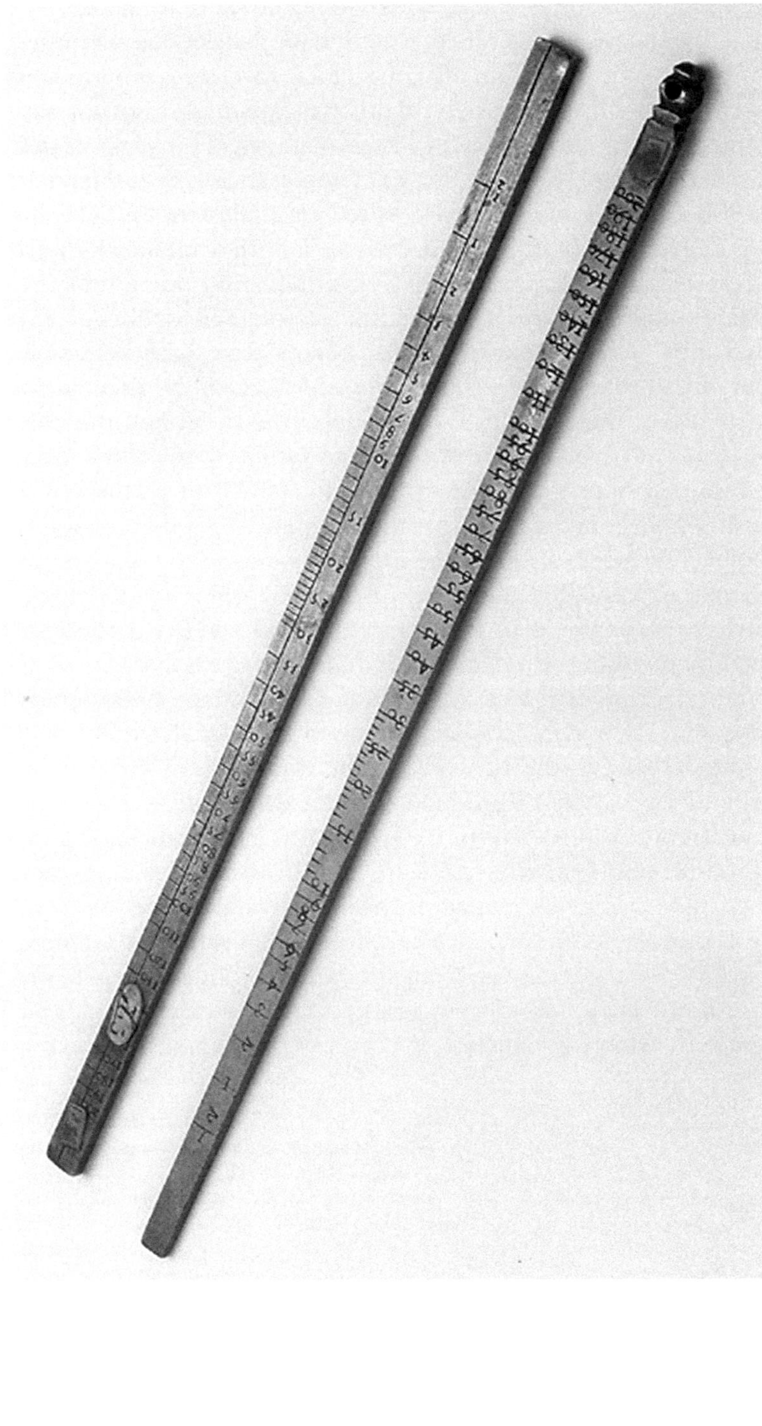

Figure 2. Two sixteenth-century gunner's rules from the Istituto e Museo di Storia della Scienza, Florence. From the EPACT database, Oxford.

diameter of 5 in.)—but this raises the question of whether the gunners would have had the simple mathematical skills to make that conceptual leap. Presumably they would have, since many printed and manuscript gunnery manuals do include directions for making quadrants,[37] although interestingly, workaday quadrants do not survive, despite their constant mention in the manuals. This suggests that they were rarely made or used, for as Thomas Bedwell commented in 1589, "it is very rare in any service to mount a peece of Ordenaunce above ten degrees of the Quadrant."[38] The fact that quadrants were discussed at length in the manuals shows the importance attached to their knowledge content as indicative of the skill of the profession. That they may or may not have been made further suggests that while of theoretical interest but little practical utility, their inclusion is evidence of the theoretical and intellectual striving of the gunnery profession at this time. With regard to gunners' rules, however, we may assume, then, that the fact they do survive in quantity and their construction is not detailed in the manuals suggests that linear measures (even if marking out a cubic relationship on the scale itself) were assumed to be within the grasp of potential gunnery students—they clearly were assumed to be literate after all—whereas the construction of a quadrant (angular measurement) was not.

That the quadrant had legs in both the world of the philosopher-astronomer and the navigator-practitioner suggests intellectual bonds may be found between the two fields. The multitude of instruments produced and the variety of production levels evidences that gunnery instruments penetrated society in both breadth and depth, indicating a pan-European mathematical practice.[39] This then raises the question of whether adopting mathematical instruments would necessarily aid in the professionalization, or perceived professional status, of gunners. Success in this strategy was far from guaranteed, but in the attempt to increase both status and efficacy (and perhaps efficiency), the area became one in which a number of players could claim legitimacy.

[37] For example, see "The Secrets of Gunmen," Oxford, Bodleian Library, Ashmole MS 343, fol. 133^{v}–134^{r}, and Smith, *Art of Gunnerie*, pp. 56–58.

[38] Thomas Bedwell, *Avrea Regula Coss, Nova Geometrica, or Rule of Proportion Geometrical*, 1633 (Oxford, Bodleian Library, MS Laud misc. 618), fol. 3^{r}.

[39] In this case, the impetus is not the transfer of knowledge that trains up individuals in various countries, but very clearly the transfer of individuals—the trope of princes employing foreign gunners is especially well established—who helped standardize the knowledge and practices across Europe, as well as the sale of instruments across all of Europe.

Through the instrumentation of fortification and gunnery, both common practitioners with no claim on elite status as well as elite gentlemen with hereditary status yet no necessary skill in the field converged on an ideal of the "scientific military gentleman". The personal instruments served as both foci upon which to fix the technical expertise as well as the badge of office and proficiency. In the world of military affairs, both then and today, insignia mark individual soldiers. Today we might look for the general's stars or captains' bars—in the early modern period fewer men were thus identified: the Marshall had his baton and the captain his banner (held by, not surprisingly, the *ensign*). Gunners wanted their insignia as well, and they seem to have chosen instruments as that insignia.

The evidence for this is circumstantial, yet compelling, for no treatise survives that says the gunner is formally identified by his ruler or calipers. They are simply too low in the military hierarchy to need an official badge of office (although others have suggested that in a Germanic fifteenth century context, a feathered cap set the master gunner apart from his mates or common gunners)[40] but this would mesh nicely with the suggestion of a programme undertaken by gunners to increase their status through the use of mathematics. For example, the 1564 self-portrait of Richard Wright, English novice gunner, suggests that the identifying mark of a gunner was indeed a rule, as he depicts himself nattily dressed, proudly holding aloft a foot-long gunners' rule in his right hand. (figure 3) Further, relatively common in the Continental context are gunner's stilettos, daggers with shot scales (*i.e.*, those that would be found normally on the gunners' rule) engraved on one, two, or three faces of the blade. And these personal weapons, often finely produced, do appear to have been the identifiable sidearm of at least Italian and Spanish gunners at the turn of the seventeenth century.[41] Further, some instruments designed for fortification and gunnery were specifically designed not only as instruments, but a display sidearm as well. The *radio latino*, invented about 1583 by Latino Orsini, and of which at least three examples survive, had a handle much like Wright's rule, but included a sharp tip like the stilettos and folded up neatly to fit in its own scabbard. (figure 4) It could be unfolded to find internal and

[40] Hall, *Weapons and Warfare* (note 11).

[41] See Marcello Terenzi, *Considerazioni su di un tipo di pugnale detto stiletto da bombardiere* (Rome, 1962)—copy in the National Museum of American History library, Smithsonian Institution, Washington, DC.

Figure 3. A self-portrait of the novice gunner holding aloft a gunner's rule as a badge of office. Courtesy of the Society of Antiquaries of London.

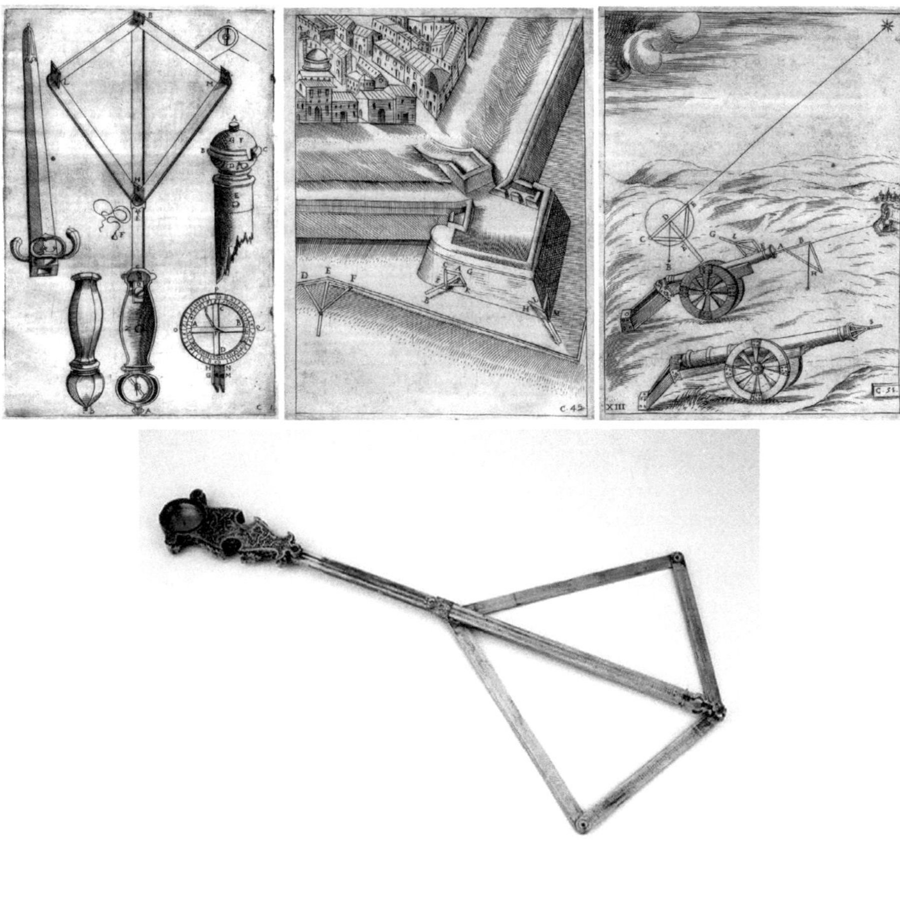

Figure 4. The *radio latino* as designed by Latino Orsini and described by Egnatio Danti (top left) and its use in fortification (top center) and gunnery (top right). Bottom: an existing example by Giovanni Maria Mancini in the Museum of the History of Science, Oxford (Italy c. 1600). From the EPACT database, Oxford.

external angles on fortifications, used to calculate gunnery ranges, or set upon a gunners' staff as a rangefinder, among other uses. Bartolomeao Crescenzi designed his "military proteus" to fold up into a dagger from its various assembly modes where it could be used to calculate ratios and angles in navigation, fortification, warfare, and even architecture (including painting, sculpture, and perspective, if we are to believe the title page).[42] One can imagine the self-importance a master gunner and fortification engineer might attach to such an instrument, and how he would expect others to see such an instrument as indicative of his knowledge and position.

By the seventeenth century, master gunners and ordnance officials would see themselves as identified by their instruments, as for example was Johann Carl, the "Master of the Ordnance and *Engineer*" in Nuremberg. In an engraving commissioned for his seventy-fifth birthday (figure 5), Carl is shown holding dividers in one hand and a rule in the other, with a model cannon on the table before him and a complex quadrant on the wall behind. Included is a panegyric poem by the noted poet Sigmund von Birken:

> When, for home and for altar, the cannon fire [lit. Earth-Thunder] plays upon the foe, this German *Archimedes* invents instruments.
> When the fiery arrow of devotion flies to God in heaven, there stands a prayer-temple [created] by the hand of this German *Hiram*.
> Look and honor this picture; the work shows you his spirit.
> The former intellect he gathered from the [30 Years'] War and out of Holland,
> The latter [piety] he inherited from his father, that it would live on in him.
> Fame shall praise this son and father, for as long as art shall be art.[43]

---

[42] Many thanks to Jim Bennett of the Museum of History of Science at Oxford for drawing my attention to the *radio latino*. For this instrument, see J.A. Bennett and S.A. Johnston, *Geometry of War* (note 34), nos. 60 and 84, and the EPACT database, items nos. 299, 454, and 453 (online at http://www.mhs.ox.ac.uk/epact). For the description of the work, see Latino Orsini, *Trattato del Radio Latini* (Rome, 1583, 1586). The military *proteus* is described in Silvia De Renzi, *Instruments in Print: books in the Whiple Collection* (Cambridge, 2000), pp. 27–28. For the original work, see Bartolomeo Crescenzi, detto Romano, *Proteo militare di Bartolomeo Romano divisio in tre libri* (Naples, 1595).

[43] Herbert Langer, *The Thirty Years' War* (New York, 1990), fig. 153:

"Johann Carl Zeüchmeister und *Ingenieur* in Nürnberg, Ward Geboren Xo 1587. 13. Januari.

Wann, vor Heerd und vor Altar soll der Erden-Donner spielen
auf die Fiend': er fündet Stücke diesen Teütsche *Archimed*.

Figure 5. Portrait of Johann Carl, *Zeugmaster und Ingenieur* of Nuremberg, aged 70 (1657) by Jacob Sandrart. From Langer, *The Thirty Years War* [1990], fig. 153.

Linking Carl to Archimedes speaks to his skill in inventing mathematical gunnery instruments, including a proportional compass and sector. His piety and equation with King Hiram of Tyre (who built a temple to the Lord in Jerusalem; see 2 Samuel 5:10–11) refers to his role as architect of St. Trinitatis in Regensberg (now Ratisborn) in 1630 seen in the picture on the wall behind him. This dual role as *Zeugmeister* and engineer in Nüremburg links instrumental knowledge and individual piety. While the latter symbolized the devotion to God and country, the former seems to be the specific intellectual attribute of his profession.

Thus, across Europe, gunners turned increasingly to the mathematics to bolster their profession. In doing so, they created a market for mathematical instruments adapted specifically to the operation of heavy ordnance. This field, then, seems to have been the first military area that can have reasonably said to have used instrumentation to become scientific in a concerted fashion. But quite beyond the physical instruments gunners may have made for themselves, or that princes may have bought as princely *Kunstcabinet* decoration, the drive towards instrumentation went much deeper—to the very core of what was considered important in the field of gunnery itself.

Education schemes

Under Humanist revivals of the sixteenth century and the expansion of courtly culture, education became an indispensable element of gentlemanly bearing and stature in the Renaissance. Military education had always been the bailiwick of the noble classes, but under the influence of Humanism, the rise and expansion of the bureaucratic state, and the ascendancy of the middle classes, such pursuits no

Wann der Andacht Feüerpfeil will zu Gott gen Himmel hielen:
 Durch des Teütschen *Hirams* hände dorten Bete-Tempel steht.
Schaü und ehre dessen Bild: seinen Geist das Werck dir weiset.
 Jene Witz Er, aüs dem Kreiege ünd aus Holland hat geholt:
 Diese Er vom Natter erbte, der in Ihm noch lehen wollt.
Fama, weil, künst, Kunst wird senn, diesen Sohn ünd Vatter preiset."

Carl's importance is attested by a poet of von Birken's statute penning the poem as well as the famous sculptor Jacob Sandrart having done the engraving. Many thanks to Barton Browning, Penn State University, for his assistance on translation of this idiomatic German poem.

longer exclusively defined nobility. And yet neither did they disappear. While Oxford, Cambridge, and Paris did not develop officer training programs, other educational endeavors along more "practical" lines were set up. Some nations developed specific military schools in the sixteenth century, while others developed informally. More intriguing for the question of instrumentation and war, however, is how such developments reflect attitudes in society quite outside of direct military necessity and how much the quantitative objects under discussion here fit into these movements.

As early as 1570 in a plan for "Queen's Elizabeth's Achademy", Humphrey Gilbert recommended mathematicians be on staff to teach their "Awditory" [*i.e.*, listeners—students] two days a week:[44]

> Also there shall be placed two Mathematicians; and the one of them shall one daye reade Arithmetick, and the other day Geometry, which shall be onely employed to Imbattelinges, fortificac*i*ons, and matters of warre, with the practiz of Artillery, and use of *all manner of Instruments belonging to the same.* And shall once every moneth practize Canonrie (shewing the manner of undermininges), and trayne his Awditorie to draw in paper, make in modell, and stake owt all kindes of fortificac*i*ons, as well to prevent the mine and sappe, as the Canon, with all sortes of encampinges and Imbattelinges and shall be yearely allowed for the same 100*li.*
>
> Also this Engineer shall be yearely allowed for the powder and shotte which shall be employed for the practize of Canonry and the use of mines 100*li.*
>
> Also there shall be under him one Usher, who shall teach his schollers the principles of Arithmetick and shall be yearely allowed for the same 40*li.*
>
> Also there shall be one other Usher, who shall teach his schollers the principles of Geometrie and shall be yearely allowed for the same 40*li.*

Not only fortification with its clearly geometrical base, but artillery too was mathematical; in fact it *was* mathematics.[45] To Gilbert at least,

[44] H. Ellis, "Copy of a Plan proposed to Queen Elizabeth by Sir Humphrey Gilbert, for Instituting a London Academy," *Archaeologia* 21 (1827): 506–20 at pp. 511–12, emphasis added. In fact, for all this instruction and the practicalities in the field, the students would have actually dug practice entrenchments, built battlements, and practiced their cannonry. £100 a year was quite a large sum in 1570, roughly equivalent to the annual income of a low-level courtier. Overall, the proposal is largely geared for martial training: even the reader of moral philosophy was to divide his readings into civil and martial policy, the Academy being designed for the young men to "study matters of action meet for present practize both of peace and warre" (pp. 510, 518).

[45] This same conflation/confusion continued into the 18th century. See, for

the proper use of the quadrivium was clearly military. It is unclear whether both mathematicians were paid £100 per annum, or each one £100, but clearly, even conservatively at £50 the practical, instrumental elements of military science were highly valued.

Although nothing ever came of this petition (William Cecil, the Lord Treasurer, did see and endorse it), some 20 years later Thomas Hood accepted a position as the publicly-financed "Mathematicall Lecturer to the Cittie of London" in 1588. In his acceptance speech he speci-fically mentioned "the Gunner witnesse in planting his shot" among the beneficiaries of the lectureship. Further, he said that it had "pleased . . . divers grave, wise and pollitick men, giving encouragement therunto . . . to erecte a lecture for the mathematicall science, a knowledge most convenient for militarie men."[46] Certainly in numerous gunnery manuals one of the principles that all gunners must "consider of, . . . practice, and learne," included "to know the use of all Geometricall instruments belonging to the profession of a gunner."[47] These lectureships, along with John Dee's taxonomy of the mathematical arts from the same year as Humphrey Gilbert's proposed Academy which included artillery, fortification, and troop mustering among the mathematical arts, all point to a common conception of these activities as both mathematical and "meet for a Gentleman."[48] Such practice would by necessity require instruments and by the 1590s it is clear that a gunnery school was operating in the Artillery Garden outside Bishopsgate, undoubtedly including instruction in the "instruments belonging to the same."[49]

---

example the confusion over whether artillerists were engineers (*ingénieurs*) during the French Revolution in C.C. Gillispie and K. Adler, "Exchange: Engineering the Revolution," *Technology & Culture* 39.4 (1998): 733–54, at pp. 739–40 (Gillispie) and 749 (Adler).

[46] F.R. Johnson, "Thomas Hood's Inaugural Address," pp. 105, 100. For Hood's career, see also Stephen Johnston, "Mathematical Practitioners and Instruments in Elizabethan London," *Annals of Science* 48 (1991): 319–44. pp. 330–341.

[47] Smith, *Arte of Gunnerie* (note 33), p. 97.

[48] Dee's taxonomy includes numerous mathematical arts, including gunnery, in a broad schematic of the field; see John Dee, *The Mathematicall Praeface to the Elements of Geometrie of Euclid of Megara (1570)*, edited by Allen G. Debus (New York, 1975).

[49] On the Artillery Garden, see Steven A. Walton, "The Bishopsgate Artillery Garden and the First English Gunnery School," *The Journal of the Ordnance Society*, 16 (2004): 41–51.

Instrument evaluation

Some active military men also saw their position defined by an attempt to understand how the various instruments proposed by contemporary writers stacked up against one another. By engaging in active evaluation of alternatives, they also exemplify this "scientific military gentleman" approach to knowledge and practice in the sixteenth and seventeenth century. Edmund Parker, a gunner for Queen Elizabeth in Ireland (†1602), read many of the contemporary printed manuals on military science and tried out their ideas. He had read William Bourne's *Arte of Shooting in Great Ordnaunce* and Cyprian Lucar's *Three Books of Colloquies* both of which described instruments to find out whether the bore of a cannon was concentric with its outside surface of the barrel (important to compensate for when sighting along the barrel). Tartaglia had proposed an assembly of two parallel pieces of wood, connected by two shorter pieces, "somewhat more than half the thicknes of the Peece at the tail."[50] With one leg in the bore the gap between the outside of the gun and the other piece of wood lets the gunner see if the bore is coaxial or not. Bourne provided an alternate, double-ended scissor-like caliper to accomplish the gauging of the barrel thickness using two pieces of wood, "double the length of the hollow or concavity of the piece," fixed on a hinge in their center.[51]

Parker invented a third option using a pole centered in the bore by tapered plugs and a string to estimate whether it was truly on the axis of the barrel. He jotted in his notebook that,

> This I hould far better than tartaglies or bornes deuice for that ther instrument beinge so longe and so weacke will swaige to and frowe wher by no gret sertentye may be geuen to it ther resen is good if the instrument wer as good but this other waye is withe out all dout if the guner worke Cuningelye or eles blame the workeman and not the deuice 1596.[52]

[50] Tartaglia, *Three Bookes of Colloqvies* (note 29), bk. I, colloquy 23, pp. 43–5.

[51] William Bourne, *The Arte of Shooting in Great Ordnaunce* (London, 1588), pp. 8–9. Parker rightly notes that these instruments would be prone to "swaige" since those to test the shortest cannon would need to be at least six or seven feet long, and one of Bourne's to measure a whole cannon would be at least 18 feet long. Further, with no mention of cross-bracing, Lucar's would be prone to racking and distortion.

[52] Edmund Parker, "How to know if the Concauitie of anye pece of ordinane be trulye bored or placed in the mides of her mettell", London, Lambeth Palace Library, MS 280, fol. 11. It is doubly interesting that he dated this entry, suggesting a certain desire to mark his priority of invention.

One element then, of the "scientific military gentleman," was a willingness to engage with the technical equipment of the day with an eye towards its improvement. As such, these military men were active participants in the ongoing development of their art whether publishing or not. Parker's invention further emphasizes that the mathematical instruments of the day need not have been strictly quantitative; any device which allowed measurement or gauging of the relevant physical characteristics of the system were seen as useful and important developments for the field as a whole.

This process of the creation of the gunner's instrumental identity reached its apogee about the turn of the seventeenth century. By that time, gunners were employed by the crown as a matter of course in England, but at the same time, they had not achieved the corresponding status gains they appeared to have been seeking in the 1580s and 90s. Thomas Bedwell is in some ways the exception that proves the rule. A Cambridge graduate as well as an engineer in Elizabeth's actions in the Low Countries in the 1580s, Bedwell sought and gained a permanent position in the Elizabethan government explicitly through his self-advertisement of his instrument-making abilities. He offered a water-clock, a timberers' rule, and a gunners' rule to the Lord Treasurer as proof of his ability. It worked. This self-described "Master of Art and sometime fellow of Trinity College, Cambridge", was appointed Keeper of the Ordnance Stores at the Tower of London in 1589.[53] His gunners' rule, which he described in an exceedingly academic manner, tried to fuse the two realms of practice and scholarship, and by doing so from one of the chief national appointed military posts a gunner could hold, Bedwell believed that instrumental gunnery could be made once and for all better than respectable—it could be made a "science" in something approaching our modern sense of the term. Interestingly, he took as

[53] See Stephen A. Johnston, "Making Mathematical Practice: Gentleman, Practitioners and Artisans in Elizabethan England," Ph.D. dissertation, Cambridge University, 1994, pp. 190–202. Bedwell's letter to the Lord Treasurer is much like Leonardo da Vinci's letter to Ludivico Sforza in offering *practical* benefits for the Renaissance Ruler; Leonardo only mentions painting and sculpture as an afterthought, having listed military and civil engineering abilities galore. Note, however, that most military men who are not just 'grunts' also seek to make their lives easier through evaluation and innovation, so this cannot be enough; see the article on innovative WWII Pacific-theatre wind-powered washing machines: Arthur G. Sharp, "Windwashers," *Invention & Technology* 18.2 (2002): 62–63.

his starting point neither the position of the natural philosopher nor of the active gunner; his approach was a hybrid one, but one which centered upon the use of the gunners' instrument as the signifier of knowledge and skill.

This treatise, which was apparently never printed for reasons lost to posterity, Bedwell titled "AUREA REGULA COSS, NOVA GEOMETRICA", or in English, "The Golden Algebra, a New Geometry."[54] In it he promised that it would be a "Rule of Proportion Geometrical Like to that of Algebra . . . For shooting in greate Ordinance." It would, he claimed—like so many other gunnery texts of the day—"Teach howe farr eny peece of Artillery shooteth at Every Degree of Random And howe to Mount any peece of Ordinance by degrees." The secret for Bedwell, however, was that the gunner was to do so "with a Ruler onely: and more conveniently then by the Ordinary Quadrant." This Ruler accomplished four tasks: it replaced the quadrant, allowing the user to work more safely at the breech end rather than the muzzle; like all other tables of the day, it calculated ranges from elevations; it allowed the back-calculation of the necessary elevation for a desired range; and it reset a piece to hit an already marked target, "either by the Lynes serving for the degrees, or by the inches ordinarily used of Gonners." (fol. 3^r)

His work opens without dedication or further ado, announcing "The exposition and use of a rare Invented Ruler for the perfect shooting in great Ordinaunce &c." The invention, Bedwell says came from his realization that carpenters were unable to easily measure timber by hand and how "tedious and difficult" it was to do the same by arithmetic and so he made a one foot ruler that simplified the task so that any man able to count to 20 could measure timber. Consequently, he then set himself to find a "Methode (longe desired and sought of manye Mathematicians and Engeniers) for mounting great ordinaunce by degrees." This confluence of timber gauging and ordnance mounting through mathematics is indicative of the increasing instrumentation of the time. As mathematical understanding increases, practitioners and instrument makers realized that similar mathematical formulations (as in this case a tangent scale) would serve similar purposes for disparate tasks. Interestingly, this appears to be the case where

[54] Bedwell (note 38). 'Coss' is from the Italian *cossi*, "thing", referred to the unknown quantity in an equation at the time, and what we now call algebra was then known as the "Rule of Coss" in English.

the instruments themselves embodied this knowledge more than any clearly set out theoretical exposition of the geometrical intersection of lines. Consequently, we have a particularly good example of what Davis Baird has recently called *thing knowledge*, where scientific instruments provide more than the sum of their inscribed numbers; instruments become part of the knowledge base themselves.[55]

Bedwell also added an appendix of "Certaine Questions, touching greate Artillary, etc. Philosophically examined." In this work he was clearly courting both sides of the scholar-craftsman divide: at one point he cried *Doctorum pace dixerim* ("Peace be with the doctors," that is, the Doctors of Philosophy at the universities) after claiming that "I dare undertake no man shall ever Invent a better Rule for those uses," suggesting his deference to the higher-status natural philosophers; at another he makes sure we know that his methods do not "vary . . . from the common practize that hathe been made by some English Gonners," assuring us of his groundedness in the real world (fol. 2^{r-v}). It should be noted that Bedwell may not have been entirely unique in this regard of trying to set gunnery instruments on an academic footing,[56] but as an employee of the Ordnance Office, he is the one who chose to invest importance in the gunners' instrument in order to legitimate his own position and gunnery itself. That action more than any other indicates the position the instrument held for the gunner of the day. That Bedwell's move ultimately did not succeed ('failed' is perhaps too strong an indictment) and the treatise was never issued[57] suggests that this sort of power claim using

[55] Davis Baird, *Thing Knowledge: a philosophy of scientific instruments* (Berkeley, Cal., 2004).

[56] Other people such as Thomas Harriot bridge the practitioner/scholar divide and tried to make military matters academically respectable—in his case it was a Latinate exposition of the *trace itallienne* fortification (see British Library, London, Add. MS 6788, fols. 55–65) and some concrete, although ultimately flawed, investigations of ballistics. on the latter, see Steven A. Walton, *Thomas Harriot's Ballistics and English Renaissance Warfare*, Durham Thomas Harriot Seminar Occasional Paper no. 30 (Durham, 1999); Matthias Schemmel, "England's Forgotten Galileo: a view on Thomas Harriot's ballistic parabolas," in José Montesinos and Carlos Solís (eds.), *Largo campo di filosofare: Eurosymposium Galileo 2001* (La Orotava, 2001), pp. 269–280; and Schemmel, "Thomas Harriot as an English Galileo: the force of shared knowledge in early modern mechanics," *Bulletin of the Society for Renaissance Studies* 21.1 (2003): 1–10.

[57] The title page of the manuscript has a footer that mimics a typographic imprint: "Liber Guilielmi Laud Archiepi Cantuar: et Cancellary Vniuersitatis Oxon 1633", but whether this means this was Archbishop Laud's personal copy or somehow authorized by him (which would be odd from the point of view of the material contained in the book, and all the more so since Laud was chancellor of Oxford which Bedwell was a Cambridge man) or just copied out for him is unclear. The lack of a dedication

instruments was not quite strong enough to convince society of their inherent worth; the fact that many elites apparently owned fine gunnery and military instruments suggests the rhetoric was at least partially successful, even if the career strategy was not.

Bedwell adds one other interesting element for our story of the creation of the scientific military gentleman and the rise of instrumentation as part and parcel of that identity. He explicitly engages in a form of craft protection, partially as a "state secret", but also as a gunners' secret. He says, "I have purposedlie not set downe vppon the Ruler any Tytles, markes or Ciphers to declare the vses of the lines or of the mooving plate or scale whearby any man that shall finde this Ruler by chaunce, shall never knowe to what end or use it is made . . . except he shall vnderstand it by this description or by tradition . . . chieflie for that I would not haue this Invention over redily to be knowen or spread into other nations." (fol. 3^r). This is interesting in light of recent work by Pam Long on craft secrecy and is one of the few explicit enunciations of national intellectual protection I know of.[58] But it also suggests that, like a trained craftsman today who "knows his trade", Bedwell assumed that a gunner of his day given one of his rules would intuit how to use it from the scale markings even without labels. This implies that rules and instrumentation must have been commonplace at the time.

Conclusion

By inventing, using, or promoting the use of gunnery instruments, scientific—or at least mathematical—knowledge was displaced into and onto inanimate technologies. And by applying these inanimate technologies to warfare, the status and apparent knowledge of gunners and engineers was increased. Warfare went through a profound

in a work of this period is curious. It is particularly more strange in that this work was clearly a play for preferment. But perhaps since it was written in the late 1580s and the sole surviving copy exists in a 1630s copy, with political climates having changed as they had, the original dedication was omitted in the later copy. Internal evidence in the surviving manuscript copy shows that it was originally issued with (now lost) plates: "If I have made some of the plates differing in some small poinctes from this fourme (as at the first I did): yet a man may by this declaracion easily knowe how to use and applie that small variety." (fol. 4^v)

[58] See Long, *Openness, Secrecy, Authorship* (note 15), pp. 11–12 and ch. 6.

change in the two centuries following the death of Henry VIII (1543), and while commentators have debated the meaning of the growth of army size and the changes in infantry tactics that characterized the period form the Eighty Years War through the later eighteenth-century wars across the Continent, it seems indisputable that the general staff and planners of Wellington were regarded as much more "scientific" than their predecessors in the reign of Queen Elizabeth. The soldiers, too, were more technical, or at least technical branches of the service were now formalized and expected. All advanced nations had military academies where the fundamentals of military practice, especially including a mathematical and instrumental basis, were taught. Thus, from humble beginnings of individual gunners seeking to ally themselves with the prestige and knowledge of mathematical instruments was one cornerstone of the modern military state laid.

Ironically, instruments which by their very nature allow *unskilled* and *unknowledgeable* users to perform complex tasks, became the badge of *skilled* and *knowledgeable* practitioners. Ultimately, this strategy was only temporarily successful for the practitioners themselves, perhaps because these instruments allowed no extension of the art, merely performance of it. But in using technologies as the arbiter of status, both elites and commoners defined the discipline itself as the bellwether for the development of technological subjects, or at least technological "behavior" in the early modern period. Perhaps it was this "conflict of interest" between skilled and unskilled or semiskilled practitioners which prevented the artillerists from gaining the status they sought (in England at least). That they sought it through instruments rather than public visibility (as fencing masters), political appointment (as surveyors), or direct court patronage (as astronomers and navigators) suggests that in this interface between science and the military, instruments held a central and important place in the early modern conception of technical proficiency. That this fusion took place in the military arts may also signal something fundamental about the place of technology and warfare in Western society.

CHAPTER TWO

SURVEYING AND THE CROMWELLIAN RECONQUEST OF IRELAND

William T. Lynch

Following the Irish rebellion of 1641, the English Parliament financed troops for reconquest by offering not-yet-confiscated Irish land to private investors. While these "adventurers" provided only a fraction of the funds necessary, the scheme helped ensure that the eventual reconquest involved a massive seizure of Catholic land rather than a negotiated settlement, since such confiscations were necessary to satisfy the adventurers and to pay off army arrears. These financial contingencies led to the need for a unprecedented, massive survey of land, and early debates over settlement among the adventurers and Army officers turned on the method needed to carry out the survey. The resulting Down Survey (with maps laid "down" from 1654 to 1659) was directed by William Petty, physician general to the army and later Fellow of the Royal Society fellow, and carried out with the help of a pool of Commonwealth Army soldiers.

The survey was significant for its extent (measuring 11 million acres), setting instrumental and organizational precedents for future large-scale surveys carried out by military and civilian agencies in colonial situations. To avoid reliance upon a limited pool of skilled surveyors, soldiers were trained in the use of circumferentor and chain with mathematical surveyors employed in an oversight capacity. Surveying played an important role in meeting the Commonwealth state's military and political objectives and modifications in surveying practices were introduced by the need to use untrained soldiers in many surveying practices. By focusing on the transmission and standardization of skilled, material practices, we can understand better the significance of military surveys like the Down Survey that would otherwise be neglected by historians of technology concerned narrowly with technical innovations. Early military and colonial surveying played an important role in advancing surveying and cartography, often neglected by historians who focus on cutting-edge, mathematical techniques, ill-adapted to field practice. By simplifying the procedures employed for field work and mapping, and training unskilled labor for these tasks, Petty was able to make the best use of the available skilled surveyors in an oversight capacity. His approach demonstrated that large surveys could be carried out quickly, enabling new settlers to occupy conquered land, thereby facilitating British (and American and European) colonial expansion over the next few centuries.

The Down Survey

The background to the Down Survey of Ireland lies in the English Parliament's need to finance an army to reconquer Ireland following rebellion at the height of the struggles with Charles I. While King and Parliament were locked in the struggle leading to civil war, the Irish rebellion broke out on 23 October 1641, displacing an earlier generation of English planters from the estates they had seized from Irish landlords.[1] The Parliament was not eager to give the King money for an army to reconquer Ireland, fearing it could be used against his enemies in England. Consequently, Parliament offered 2.5 million acres of Irish land it claimed would be forfeited by the rebellious Irish to act as security for the raising of a private army. A committee of members of the House of Commons and private investors called "adventurers" (since they had *adventured* the money, that is "come to" the business venture with funds) collected the money and appointed the general and officers, leaving only the King to sign commissions. When he delayed commissions as the conflict with Parliament heated up in 1642, the civil war broke out and Parliament ordered Lord Wharton to lead 5,000 infantry and 500 horsemen to march against the King. This misuse of the troops—intended only for Ireland—kept future subscriptions low until Charles was tried and executed in January 1649. Cromwell marched on Ireland in August 1649 and the war continued in earnest for two years, although victory was not declared until 26 September 1653.[2]

[1] Jane Ohlmeyer, "The Civil Wars in Ireland," in John Kenyon and Jane Ohlmeyer (eds.), *The Civil Wars: a military history of England, Scotland, and Ireland, 1638–1660* (Oxford, 1998), pp. 73–102.

[2] The following section draws from Patrick J. Carish, "The Cromwellian Regime, 1650–60," in T.W. Moody, F.X. Martin, and F.S. Byrne (eds.), *A New History of*

The adventurers were offered land at cheap rates and discounts were offered to elicit further money. Ultimately, 1,533 subscribers enlisted for £306,718 worth of land, a joint-stock operation exceeded at the time only by the East India Company.[3] The government could not pay the army's wages and these arrears presented a serious obstacle to disbanding the army and settling the land on the adventurers (little money had been sent to Ireland from 1642–49).[4] Without pay, the soldiers had settled themselves on devastated Irish land and farmed land near garrisons since crops and cattle had been destroyed as a method of subduing the Irish. Taxation was doubled to support the military presence and many soldiers and native Irish could not afford to pay. When the adventurers' rates were extended to the soldiers themselves in lieu of wages in order to disband them, a process of transplantation of native Irish and Anglo-Irish landowners to the inhospitable terrain of Connaught was begun.[5]

On 26 September 1653 Parliament passed an act for settling Ireland, the government reserving to itself the towns, all Church land, and the counties of Dublin, Kildare, Carlow, and Cork to settle debts and favors. The adventurers were assigned lots in Munster, Leinster, and Ulster, set out by lot. Ten counties were assigned: Waterford, Limerick, and Tipperary in Munster; Meath, Westmeath, King's county and Queen's county in Leinster; and Antrim, Down, and Armagh in Ulster. The baronies in each county were to be divided between adventurers and soldiers so that adventurers, mostly merchants and tradesmen, would feel safe settling. The rest of the soldiers were to be settled throughout the remainder of Ireland, Connaught excepted. Almost all landowners were to be displaced, as were all lower orders who had engaged in rebellion, which included those compelled by their landlords. A landed petitioner who showed that had had held "constant

Ireland (Oxford, 1991), pp. 353–86, Robert C. Simington, *The Transplantation to Connacht, 1654–58* (Shannon, Ireland, 1970), and John P. Prendergast, *The Cromwellian Settlement of Ireland* (London, 1865).

[3] Carish, "Cromwellian Regime," p. 360.

[4] Karl S. Bottigheimer, *English Money and Irish Land: the 'Adventurers' in the Cromwellian settlement of Ireland* (Oxford, 1971); Hugh Hazlett, "The Financing of the British Armies in Ireland, 1641–9," *Irish Historical Studies* 1.1 (1938): 21–41; J.R. MacCormack, "The Irish Adventurers and the English Civil War," *Irish Historical Studies* 10.37 (1956): 21–58; John Morrill, "Postlude: Between War and Peace, 1651–1662," in John Kenyon and Jane Ohlmeyer (eds.), *The Civil Wars* (note 1).

[5] Peter Berresford Ellis, *Hell or Connaught!: The Cromwellian colonisation of Ireland, 1652–1660* (London, 1975).

good affection" to the English Parliament was exempted from transplantation; those who resided in or collected rent from Irish areas automatically forfeited good affection.[6]

In deciding how much land they had with which to work, Parliament was working almost entirely on guesswork. The only reliable survey that existed was the Strafford survey of Connaught and some nearby lands. The army could not be satisfied until the adventurers were given their land. No one knew how much land would ultimately be available following confiscation, nor for that matter the actual size of Ireland. Meanwhile, the soldiers were supposed to carry out the transplantation in the middle of winter. The soon-to-be-displaced gentry and tenants had stopped harvesting in panic and confusion, and no planter could expect to occupy land in time to till the harvest. Nevertheless, transplantation proceeded, with fathers sent to commissioners in Loughrea to be assigned land in Connaught, where they were to proceed and build huts for their families, who were to follow in May. Applications for delay or dispensation were heard and officers tried to prevent having their tenants and servants exiled. Irish tax-collectors were escaping with collected money to Connaught.

In this chaotic situation, the authorities needed to obtain accurate information about the land that was to change hands and ordered surveys to be carried out by juries empowered to take testimony on the boundaries of land and the standing of proprietors. The 1653 Act for Satisfaction called for a "gross survey" to be carried out under the authority of four senior officers to serve as Commissioners of the Commonwealth of England, for the Affairs of Ireland.[7] A "civil survey" was begun in 1653, establishing courts of survey in every county to determine the properties to be forfeited, but no maps were produced nor were actual measurements of lands taken. Here the idea was to get rough descriptions of the size and value of the

[6] Prendergast, *Cromwellian Settlement* (note 2), p. 29. Carish, "Cromwellian Regime," (note 2), p. 361, observes that the terms would imply "an almost universal confiscation of land held by Catholics."

[7] "An Act for the speedy and effectual Satisfaction of the Adventurers for Lands in Ireland, and of the Arrears due to Soldiery there, and of other publique Debts, and for the Encouragement of Protestants to plant and inhabit Ireland," 26 September 1653, reprinted in C.H. Firth and R.S. Rait (eds.), *Acts and Ordinances of the Interregnum, 1642–1660* (London, 1911), II:722–53, and William Petty, *The History of the Survey of Ireland Commonly Called the Down Survey*, ed. by Thomas Aiskew Larcom (1851, New York, 1967), pp. 353–68, at pp. 359, 354. (Hereafter: Petty, *Down Survey*).

land to be taken with the idea that surveyors would be paid later to make an exact admeasurement. An initial survey of estate boundaries only (that Petty refers to as the gross survey) was begun under Benjamin Worsley's management in 1653, but complaints about its accuracy and utility made it practically useless.[8] A Committee of Survey including Petty, Worsley, and army officers was established by the Commissioners on 8 September 1654 to consider these problems and Petty's proposals to rectify them.

The "Down Survey" was the result of these proposals. William Petty, then physician general to the army, sought a speedier survey including measurement of territorial boundaries that would also simultaneously allocate land to the army. Proposing to act essentially as a subcontractor, Petty offered to survey all the land for a flat fee of £30,000 or for fees tied to the area of land surveyed. Petty offered £6 per thousand acres; in the end, he was granted £7 3s 4d per thousand acres for profitable land and £3 per thousand acres for unprofitable land, church land, and Crown land. Petty would be responsible for hiring and training surveyors and soldiers to perform the work, tapping state funds as needed. His fee would be payable, minus expenses, upon completion of the survey in thirteen months. To allay fears that Petty was not up to the task, Petty put forth a bond of £4,000, guaranteeing his work. The money was provided by Sir Hardress Waller, one of the army officers serving on the council overseeing the survey process, who was to act as an intermediary between Petty and the government in exchange for one-sixth the profits.[9]

In effect, Petty had taken control of the survey of confiscated land from Benjamin Worsley, the Surveyor General and a rival of Petty's dating from their participation in the circle of economic and technological reform centered on Samuel Hartlib.[10] Key to Petty's success in gaining control was his promise to complete the survey and allocate the land in thirteen months. Petty received the contract on 11 December 1654, with the survey due by one year and one month

[8] J.H. Andrews, *Plantation Acres: an historical study of the Irish land surveyor and his maps* (Omagh, Northern Ireland, 1985), pp. 297–305; Petty, *Down Survey*, p. 7.

[9] Petty, *Down Survey*, pp. 9, 27, 31–34. Petty tells us that Waller did not, in fact, play the role of intermediary nor did he claim any profits.

[10] Charles Webster, "Benjamin Worsley: Engineering for Universal Reform from the Invisible College to the Navigation Act," in Mark Greengrass, Michael Leslie, and Timothy Raylor (eds.), *Samuel Hartlib and Universal Reformation* (Cambridge, 1994), pp. 213–235.

later on 11 January 1656. Since there were delays in obtaining lists of forfeited estates from the Civil Survey and accounts of the debt due to the Army, as well as disputes about procedure, the starting date was reset to 1 February 1655 and Petty was granted an extension to 1 March 1646. Properties were allocated piecemeal to army units, much to the dismay of Petty, who had intended to mark each new claimants' property boundaries on the ground as the survey was carried out. The survey was completed by March 1656, albeit without the entire allocation of land carried out, a casualty of debates among army units on dividing up the land. Petty then served as a commissioner for setting forth lands to the army through 1659, with the land essentially distributed by February 1657. A later survey of the adventurers' land, co-directed by Petty and Worsley, was commissioned in September 1656 and completed by 1658, which led to conflicts with private surveys carried out by the adventurers, which Petty was able to resolve in favor of the state survey.[11]

Colonial Surveying

The Down Survey built up territorial maps from cadastral (property boundary) surveys of land seized from Irish landowners judged disloyal. As such, it continued a process of colonization in Ireland already established by previous plantations. Moreover, the techniques of surveying, particularly the use of circumferentor and chain to measure the perimeter of properties, were similar to those employed in North America where they were used to similar effect. Suited to conditions where skilled surveyors were short in supply or had limited time, where clear interior landmarks to serve as a basis of triangulation were not as prevalent as in England, and where a legal context requiring that surveys be performed before the possession and development of land could proceed, these surveying practices played a crucial role in the effective settlement and commodification of land in colonial settings where the native population was lacking or displaced.

The surveying practices were required for the successful exportation

[11] Petty, *Down Survey*, pp. 23–29, 41–46, 63, 103, 105–106, 121, 151, 240–41, 256, 390–92. "The Day Book of the Proceedings of the Trustees appointed for Satisfying of the Army's Arrears," British Library (BL), Additional MS 35102. A few loose maps date as late as 1659; see BL Add. MS 72873.

of class and legal systems that, in turn, made possible successful colonization. As such, they allowed pent-up demand for land and class mobility to be met on the periphery of the emerging British empire, thereby alleviating social conflict at home. At the same time, they were crucial practices behind the effective settlement of land in colonial situations, less because they rhetorically claimed ownership of native land than because they helped settle land relatively quickly and minimized (though by no means eliminated) *internal* conflict among the settlers.

In addition, much the same role for surveying and cartography in the practices of colonization can be seen in other colonial contexts, such as India, even where the particular ensemble of technical practices was different, ultimately involving the first cutting-edge trigonometrical survey of an entire country rather than the quick but less accurate approach favored in Ireland and early America.[12] The situation on the ground in India also differed from that in Ireland and North America. In India, for example, the colonial situation involved a co-option of local elites and not a complete displacement of native landowners or the native population as a whole. In Ireland, to be sure, this pattern of co-opting the local elite within the authority of the colonizing English and Scottish settlers was a common pattern. However, the Cromwellian settlement initially envisioned a nearly complete exile of the Catholic population to the province of Connaught. The transplantation did not occur to nearly that extent, not least since prospective landowners would be left without tenants to work the land.[13] Nevertheless, the massive seizure and transfer of land that did occur suggests comparisons with the processes driving westward expansion in America.[14] Hence, the Down Survey and the consequent Cromwellian and Restoration settlements were significant examples of a distinctively colonial process, one that would not have been conceivable without the technical and organizational practices developed by Petty and his undersurveyors during the Army's occupation of Ireland.

---

[12] Matthew H. Edney, *Mapping an Empire: the geographical construction of British India, 1765–1843* (Chicago, 1997).

[13] Prendergast, *Cromwellian Settlement* (note 2).

[14] Karl S. Bottigheimer, "Kingdom and Colony: Ireland in the Westward Enterprise, 1536–1660," in K.R. Andrews, N.P. Canny, and P.E.H. Hair (eds.), *The Westward Enterprise: English activities in Ireland, the Atlantic, and America, 1480–1650* (Detroit, 1979), pp. 45–64.

Hence, while debates continue about whether Irish history should be understood in reference to the concept of colonialism, the applicability of this concept is clearly appropriate for understanding the kinds of land settlements consequent to military conquest and control. Much of the debate about colonialism in Ireland has concerned the political relationship with the home country that was taken for granted by settlers. While one may legitimately question whether developments in Ireland in the sixteenth and seventeenth centuries should be described in terms of the establishment of a "colonial identity" among, first, the old English settlers and, then, the new Protestant settlers, this is a different issue from understanding a process of appropriating land and resources and displacing traditional patterns of land use and economic and political forms by an expanding core society. Thus, the argument that the landed elite, whether Catholic or Protestant, English, Scottish, or Gaelic, understood Ireland to be one of the Three Kingdoms ruled by the same monarch can explain much that is distinctive about the political history of Ireland as compared to that of America or India.[15] This does not negate the fact that this distinctive political history took place against the backdrop of processes that not only *look like* what is called colonization elsewhere, but are in important ways, *precedents* for the very same processes exported elsewhere, with appropriate modifications for the distinctive historical contexts of those societies. Surveying and mapping the land to be exploited is a key technical requirement for this colonization, and colonization is by its very nature a military activity.

The idea that Ireland should be seen as subject to a process of colonialism is controversial among historians. Barnard suggests that Ireland does not neatly fit the pattern of colonization elsewhere since many marginal provinces *within* Europe have undergone incorporation by a nearby, powerful state that display characteristics similar to the Irish case. To include Ireland within the category of colonized countries would seem to require us to water down the category of colonialism to such an extent that we would no longer be able to distinguish internal European history from colonizing activities abroad.[16] This

[15] T.C. Barnard, "Crises of Identity among Irish Protestants, 1641–1685," *Past and Present* 127 (1990): 39–83. Compare Nicholas P. Canny, "Identity Formation in Ireland: The Emergence of the Ango-Irish," in Nicolas P. Canny and Anthony Pagden (eds.), *Colonial Identity in the Atlantic World* (Princeton, 1987); Nicholas Canny, *Kingdom and Colony: Ireland in the Atlantic world, 1560–1800* (Baltimore, 1988).

[16] Barnard, "Crises of Identity".

objection to viewing Irish history under the category of colonialism can be turned on its head, however.[17] Far from watering down the concept of colonialism, the argument that much of Europe underwent an internal process akin to colonization before being exported abroad has helped explain the causes and forms that colonization eventually took. For example, the enclosure of the commons and the incorporation of agricultural producers as wage earners in Europe preceded a similar process in the European colonies, with similar social consequences for the society.[18] Similarly, the patterns of finance, governance, and plantation familiar in Atlantic colonization have their roots in Italian, Spanish, and Portuguese colonization in the Mediterranean, often accompanying the forays east during the Crusades. Land for colonization was granted under feudal tenure (a military system at its core), by company charters (protected by the military or granted the right to be military forces themselves), or incorporated under home rule following settlement (secured by the military). These institutional forms and economic patterns were transplanted from East to West, as Mediterranean colonies gave way to Atlantic colonies. Sugar plantations, often making use of slave labor, founded first by Italian cities in Palestine in the twelfth century, spread to Italian colonies in Cyprus and Crete by the fourteenth century. From there, production spread to Sicily and to near Atlantic islands (*e.g.*, Madeira, the Canaries, the Azores, São Tome) controlled by the Spanish and Portuguese with Italian technical and economic assistance by the fifteenth century. Finally, the practice spread in the sixteenth century to Brazil, Barbados, Jamaica, Puerto Rico, Mexico, and Peru.[19] Thus, a thorough understanding of the causes of colonialism requires an understanding of the dynamic of economic and political transformation within Europe that provided patterns to be emulated abroad, as well as producing the very social dynamics whereby direct agricultural producers increasingly denied rights to land would be willing to become settlers in peripheral areas.

The effect on the colonized society was paradoxical. A new landed elite was created, as legal control of land continued as the basis of

---

[17] On this issue, see the discussion in Bottigheimer, "Kingdom and Colony," especially p. 60.

[18] Karl Polanyi, *The Great Transformation* (New York, 1944), chs. 13–14.

[19] Charles Verlinden, *The Beginnings of Modern Colonization* (Ithaca, N.Y., 1970), ch. 1.

wealth and power. However, land was subject to greater commodification than in the home countries, where land had been made more fitfully subject to alienation. The key problem became finding the labor to improve the land one claimed, given the greater ease that potential laborers themselves had in becoming landowners. This problem of labor supply led to slavery in the American South, where landholdings were correspondingly large, and to greater collective control and more compact settlements in New England, where surveys were more likely to be required before land could be settled. In South Australia and New Zealand, colonial administrators sought more orderly and compact settlements by pursuit of "first, a 'sufficient price' for land as the regulator to ensure land was alienated in proportion to the labour it could attract, and second, the employment of systematic cadastral surveys to ensure spatial order in the whole process."[20] In Petty's Ireland, both labor and land remained scarce, since English and Scottish tenants could not be interested in settling where security was unsure and most of the land granted following the survey went to officers and speculators (including Petty himself) who bought up the debentures of soldiers and investors.[21] As is made clear by this speculative process, however, commodification of land was promoted by the convertibility proposed between wages due the soldiers and the land ultimately offered them in lieu of wages on the same scale as that offered to investors who had funded the reconquest. The use of conquered land to finance the military, begun as an accident of history as the English Parliament was faced with the need to raise an army in the midst of its struggle with Charles I, provided an important means for developing a military-fiscal state capable of significant expansion. A key technical prerequisite of this colonizing state is the ability to survey and map land efficiently. This ability was first demonstrated on a large scale by the Down Survey in Ireland.

---

[20] Roger J.P. Kain, "The Role of Cadastral Surveys and Maps in Land Settlement from England," *Landscape Research* 27.1 (2002): 11–24, quote on p. 17. The irony is that the slave-based system in the American South was the result of a greater commodification of land than was common in New England, with the subsequent need to find sources of labor to exploit large land holdings in a context where free labor could aspire to own land. This suggests that historiographical approaches that see slavery as a pre-capitalist economic system in comparison to a capitalist North are misleading. On the debate about slavery and capitalism, see *e.g.*, Sidney W. Mintz, "Was the Plantation Slave a Proletarian?" *Review* 2.1 (1978): 81–98.

[21] Kevin McKenny, "The Seventeenth-Century Land Settlement in Ireland: Towards a Statistical Interpretation," in Jane H. Ohlmeyer (ed.), *Ireland from Independence to Occupation, 1641–1660* (Cambridge, 1995), pp. 181–200 at p. 198.

Circumferentor and Chain

Surveying in seventeenth-century England developed as the outcome of tensions between mathematical reformers who tried to subject surveying practice to geometrical discipline and practitioners who constructed maps to scale without benefit of measurement of angles. The three principle instruments that emerged as standard in the seventeenth century were the altazimuth theodolite (adapted from the astrolabe), the circumferentor (a compass with sights), and the plane table (where angles could be directly plotted on paper).[22] Two basic strategies for conducting cadastral surveys existed. In the first, a traverse survey of each straight segment of the property's perimeter was carried out using a plane table or circumferentor. In the second, a base station or two are set up either inside the property or on the perimeter where all the corners of the property can be sighted (often atop hills) and angles measured with a theodolite or a plane table with a sight. In this scheme, the intersection of rays allows the figure to be directly drawn on a plane table (or more rarely, determined by trigonometrical calculation).[23]

The circumferentor was widely used in Ireland and America (where it was often called the surveyor's compass), in part since it was easy to use and in part since triangulation from interior points depended upon easily identifiable landmarks that were often lacking in these contexts. Among textbook writers, John Love in 1688 noted that circumferentors were "used by most surveyors in America, where they lay out very large Tracts of Land," while Waddington observed that they were "much used in America, and in some other foreign countries for surveying woods and forests."[24] Though it was cheaper than a theodolite, it was less accurate. Eighteenth century writers noted that

[22] J.A. Bennett, "Geometry and Surveying in Early-Seventeenth-Century England," *Annals of Science* 48 (1991): 345–54 at pp. 346–49.

[23] A.W. Richeson, *English Land Measuring to 1800* (Cambridge, Mass., 1966), pp. 101–103, 114–117; Norman J.W. Thrower, *Maps and Civilization* (Chicago, 1999), pp. 91–92. On the plane table as a compromise between the geometrical agenda of the mathematical writers and the use of ratios in most surveying practice, see Bennett, "Geometry and Surveying," pp. 346–347.

[24] John Love, *Geodaesia: or the art of surveying and measuring land made easie* (London, 1688), p. 105, cited in Sarah S. Hughes, *Surveyors and Statesmen: land measuring in colonial Virginia* (Richmond, Va., 1979), p. 34; J. Waddington, *A Description of Instruments Used by Surveyors* (London, 1773), p. 307, cited in Andrews, *Plantation Acres* (note 8), p. 307.

with a circumferentor, "we can't be certain of its giving any particular angle so near as two degrees." This accounted for its use in America, rather than England, since "land is not so dear, and . . . it is necessary to survey large tracts of ground, overstocked with wood, in a little time."[25]

In Irish surveying at the time, reliance upon the circumferentor was commonplace, offering the advantage of convenience and portability. A chain was used to measure distance. The major errors associated with the use of the circumferentor included that the needle might not play freely, that the line of sight might deviate more than a few degrees from the horizontal, and that magnetic interference from natural terrain or soldiers' weapons would affect measurements. As for the chain, it might be misused in various ways, such as carrying it over hilly terrain, miscounting the links, or through stretching or binding of the links.[26] The Irish circumferentor from 1667 in the Oxford Museum for the History of Science is compact and collapsible, with detachable sights, a lid to cover the compass, and a compartment for a spare needle, and would work well in difficult environments.[27] (see figures 1 and 2)

Charles Webster suggests that this circumferentor most closely approximates Petty's simplified tools, using the card Petty produced with the instrument-maker Henry Sutton. Webster also argues that Petty likely used a decimal chain for measurements, as they were then becoming standard.[28] Support for Petty's use of instruments like these can be found in some versions of the map of Ireland developed for Petty's atlas, *Hiberniae Delineatio*, including an early version of the atlas produced by Petty for Sir Allen Brodwick, Surveyor-General of Ireland from 1660–1667. (figure 3) The maps are colored and different illustrations adorn the text, including cherubs

---

[25] William Gardiner and George Adams, respectively, quoted in Deborah J. Warner, "True North—and Why It Mattered in Eighteenth Century America," *American Philosophical Society* (forthcoming).

[26] Andrews, *Plantation Acres*, pp. 297–305.

[27] Museum for the History of Science, Oxford, inventory no. 33340, "Circumferentor with Magnetic Azimuth Dial, by W.R., Dublin, 1667." I am grateful for Stephen Johnston's assistance in examining this item.

[28] Charles Webster, *The Great Instauration* (London, 1975), pp. 439–40; Petty, *Down Survey* (note 7), p. 324. A nineteenth-century chain is shown in J.A. Bennett, *The Divided Circle: a history of instruments for astronomy, navigation and surveying* (Oxford, 1987), p. 194, fig. 227, with 100 8-inch links, with brass tags every ten links (housed at the Whipple Museum, Cambridge University).

Figure 1. A surveying instrument similar to those used in the Down Survey of Ireland. Courtesy of the Museum of the History of Science, Oxford.

Figure 2. Circumferentor with Magnetic Azimuth Dial, by W.R., Dublin, 1667. Courtesy of the Museum of the History of Science, Oxford.

Figure 3. An early version of William Petty's map of Ireland (1667). Courtesy of the British Library.

playing with the chain and circumferentor.[29] Petty did not use plane tables, whereby angles are directly drawn rather than recorded, no doubt since they did not allow for independent checking short of replotting in the field.[30] Bennett has written on the impact of mathematicians on surveying, who introduced sometimes very complex instruments which were resisted as too cumbersome by practicing surveyors.[31] In the early to mid-seventeenth century a compromise was reached which included the plane table but also the circumferentor and chain. Petty pushed surveying practice in the direction of simplification, so that surveyors would not need familiarity with trigonometry in order that soldiers could be easily trained to make measurements. Moreover, he favored simplification to make it easier to check the quality of work performed.

Simplification and quality control were crucial to Petty's plan to carry out such a large survey in so short a time. Critics of Petty's plan thought his combination of a civil and natural survey would take twenty years, not the thirteen months he proposed and—allowing for a few startup delays—achieved. Petty is quite explicit (writing in the third person) that the biggest obstacle to his plan is the craft nature of surveying and sought to circumvent this through division of labor:

> Whereas surveyors of land are commonly persons of gentile and liberall education, and their practise esteemed a mistery and intricate matter, farr exceedinge the most mechanicall trades, and withall, the makeinge of theire instruments is a matter of much art and nicety, if performed with that truth and beauty as is usuall and requisite. The said Petty, consideringe the vastnesse of the worke, thought of dividinge both the art of makeinge instruments, as alsoe that of usinge them into many

---

[29] *An Abstract of the Geometrical Surveyes. Made by Dr. William Petty. Presented to Sr. Allen Brodrick Kt. & Baronet, his Majesties Surveyor-General*, 1667 [BL Maps. c. 21, fol. 2]. A nearly identical black and white version of the map of Ireland is "A General Mapp of Ireland," [BL Maps 10805. (1) (7.)]. J.H. Andrews, *Shapes of Ireland: maps and their makers, 1564–1839* (Dublin, 1997), p. 140, reproduces a map matching this one from Petty's atlas; the map of Ireland in the modern reproduction of the atlas, William Petty, *Hiberniae Delineatio, published with Geographical Description of Ye Kingdom of Ireland*, ed. J.H. Andrews (Dublin, 1685; Shannon, Ireland, 1969), does not include the circumferentor (mentioned by Andrews in the introduction, p. 11). See Geoffrey Keynes, *A Bibliography of Sir William Petty F.R.S. and of Observations on the Bills of Mortality by John Graunt F.R.S.* (Oxford, 1971), pp. 50–55.

[30] Andrews, *Plantation Acres* (note 8), pp. 308–309.

[31] Bennett, "Geometry and Surveying" (note 22); J.A. Bennett and Olivia Brown, *The Compleat Surveyor* (Cambridge, 1982).

> partes, viz., one man made onely measuringe chaines, viz., a wire maker; another magneticall needles, with theire pins, viz., a watchmaker; another turned the boxes out of wood, and the heads of the stands on which the instrumentt playes, viz., a turnor; another, the stands of leggs, a pipe maker; another all the brasse worke, viz., a founder; another workman, of a more versatile head and hand, touches the needles, adjusts the sights and cards, and adaptates every peece to each other.[32]

Note that skill does not disappear from this picture, it merely gets redistributed. Petty still relied upon a workman of a "more versatile head and hand" to put it all together; also time scales, protractors, compass-cards, "being matters of accurate division, are prepared by the ablest artists of London," rather than in Ireland itself (where Petty's orders may have launched a tradition of Irish instrument construction). The resistance some surveyors felt to Petty's initiatives can be seen in the complaint lodged by the surveyors displaced by Petty's plan and by the prohibition that the Provost at Trinity College, Dublin, tried to place on their mathematicians' participation.[33]

But Petty did not only rely upon the expertise of the professions, such as surveyors and instrument-makers, but also depended crucially upon "mearesmen" with "local knowledge" of property boundaries:

> At the same tyme care was taken to know who were the ablest in each barrony and parish to shew the true bounds and meares of every denomination, what convenient quarters and harbors there were in each, and what garrisons did everywhere lye most conveniently for theire defence, and to furnish them with guards, and with all who were men of creditt and trade in each quarter, fitt to correspond with for furnishinge mony by bills of exchange and otherwise; and lastly, who were men of sobriety and good affection, to have an eye privatly over the carriage and diligence of each surveyor in his respective undertakinge.[34]

Here we see a key function of maps in land resettlement: translating from primarily oral, local knowledge to fixed, written representation that was no longer subject to reliance upon the knowledge and memory of the native population.

---

[32] Petty, *Down Survey* (note 7), p. xiv. Instrument makers joined guilds of their choice as there was no guild specifically for their craft. See Joyce Brown, "Guild Organisation and the Instrument-Making Trade, 1550–1830: The Grocers' and Clockmakers' Companies," *Annals of Science* 36 (1979): 1–34; Joyce Brown, *Mathematical Instrument Makers in the Grocers' Company, 1688–1800* (London, 1979).

[33] Petty, *Down Survey*, pp. xiv, 18–19, 21. The Provost's prohibition did not stop Miles Symner, professor of mathematics, from becoming "probably Petty's most trusted associate" during the survey (Webster, *The Great Instauration* [note 28], p. 440).

[34] Petty, *Down Survey*, p. xv.

In contrast to established practice in England and elsewhere up to this time, use of the circumferentor in Ireland and in America measured angles of surrounds in reference to magnetic north, rather than with respect to the previous linear measure.[35] The typical Irish surveyor's field book included a single measurement of degrees, implying that magnetic north was used as one point of reference along with a single landmark.[36] A printed compass card produced by Henry Sutton in 1653 (figure 4) may be similar to those used in Petty's survey the following year. Like the card for the 1667 Irish circumferentor (figure 2), it includes a scale for 360 degrees, as well as the scale to 120 that was standard in surveying instruments before Sutton. The card inverts indicators for west and east, since a surveyor aligning the sights to point to a position some degrees east of the viewing position would find the needle pointing to the correct angle measurement (using the ninety degree scale for that quadrant) in the counterclockwise direction (now labeled 'E' rather than 'W').[37] Petty claimed that one of his innovations was "removing some entanglements in the card wherein the needle plays"; it may be that Petty found this layout to be better suited for training soldiers.[38]

Petty's maps were plotted with reference to magnetic north rather than true north, although he benefited from the fact that magnetic north and true north were very close in Ireland in the 1650s. While he was aware of the problem of relating measurements to true north, Irish maps for the next century and a half would orient to magnetic north.[39] A similar situation occurred in America; only in the late

[35] Love, *Geodaesia* (note 24), p. 105; Andrews, *Plantation Acres* (note 8), pp. 304–05; Hughes, *Surveyors and Statesmen* (note 24), p. 34.

[36] Andrews, *Plantation Acres*, pp. 297–305.

[37] J.A. B[ennett], "Henry Sutton Thinking: A Reading of a New Acquisition," *Sphaera* 10 (Autumn 1999): 2.

[38] Petty, *Down Survey*, p. 17. It is known that Henry Sutton provided the graph paper for Petty's survey (p. 324) and, most likely, the scales, protractors, and compass cards that Petty tells us were "prepared by the ablest artists of London" (p. xiv) rather than manufactured in Ireland like the remaining parts. Correspondence with Hartlib indicates that Petty worked with Major Anthony Morgan in 1653–54 on perfecting "an Easy practicable Invention" for surveying (quoted in Lindsay Gerard Sharp, *Sir William Petty and Some Aspects of Seventeenth Century Natural Philosophy*, D. Phil. thesis, Wadham College, Oxford, 1977, p. 114). Petty had advertised plans for simplified dialing instruments in 1649 (p. 391).

[39] William Petty, *The Petty Papers: some unpublished writings of Sir William Petty*, ed. the Marquis of Lansdowne (London, 1927), I:83; Andrews, *Plantation Acres* (note 8), pp. 304–305.

Figure 4. A compass card by Henry Sutton that may have been designed in collaboration with William Petty for the Down Survey of Ireland. Courtesy of the Museum of the History of Science, Oxford.

eighteenth century did the U.S. government require orientation to true north, a consequence of difficulties relating new and old surveys subject to magnetic variation.[40] In a text from 1654, Henry Osborne, active in planning for the survey, gives angles from magnetic north for linear measurements in his proposed scheme for checking undersurveyor measurements by decomposing north/south and east/west components to see that they add to zero on a complete surround.[41] Osborne's textbook indicates the interest in enabling independent checks of accuracy among those planning the survey, mentioning a conversation with Worsley, the Surveyor-General. Following "several Experiments made for [his] own private satisfaction in the business of Survey," Osborne reflected upon how better bookkeeping and protracting might improve the exactness of the survey. Recalling a navigation handbook treating the subject, he drew out some rules from it for surveying.[42] The handbook in question is authored by Richard Norwood, first Surveyor of Virginia.[43] In addition to distances (in chains and links) and angles (in degrees and minutes), Osborne records the vector components along the south/north and east/west directions, using tables for sines. After taking all measurements, the north and south numbers should add to zero as should the east/west. Columns labeled X and Z keep running tabs, adding or subtracting depending upon direction. (figure 5)

The virtue of this method is that one may know "at the first sight how much the Plot wants of closing even to a Link; and this before protraction is made, which cannot be done in the usual way, until the whole be first actually laid down."[44] Moreover, in the standard way, if there is disagreement, it is not clear whether the error would be the fault of the admeasurer or the plotter. Finally, errors in the field can be identified as falling on north/south or east/west lines. Petty had angle and length measurements plotted on graph paper, with areas calculated by counting squares; presumably, the protractor would use his judgment if the perclose seemed too large. By mistake,

[40] Warner, "True North" (note 25).

[41] Henry Osborne, *A More Exact Way to Delineate the Plot of Any Spacious Parcel of Land* (Dublin, 1654); Webster, *The Great Instauration* (note 28), p. 440.

[42] Osborne, *A More Exact Way*, "To the Surveyor of Lands."

[43] Richard Norwood, *The sea-mans practice: contayning a fundamentall probleme in navigation experimentally verified* (1637; London, 1644); Hughes, *Surveyors and Statesmen* (note 24), p. 9.

[44] Osborne, *A More Exact Way*, "To the Surveyor of Lands."

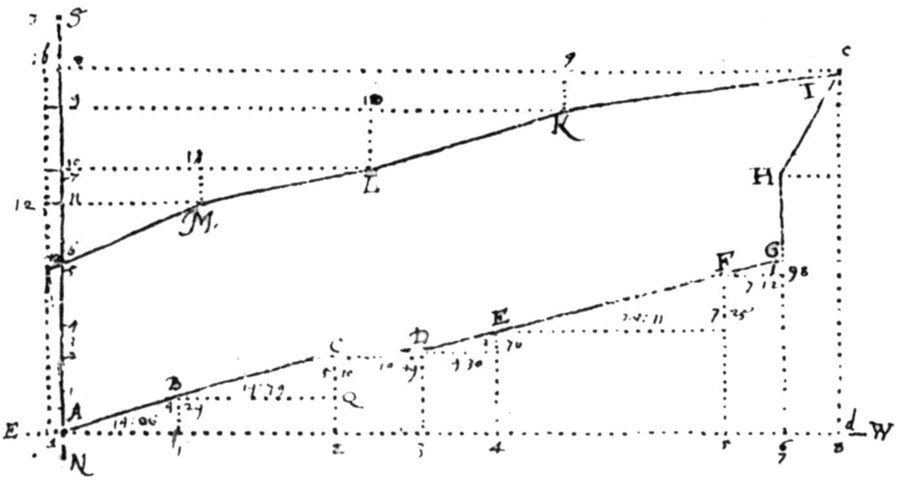

Diſtances.		Angles.	South.		North,		Eaſt.		Weſt.		X		Z	
14	:70	73 :00 S W	4	29					14	:06	4	:29	14	:06
20	:40	75 :30 S W	5	10					19	:79	9	:39	33	:85
10	:50	87 :00 S W	0	55					10	:49	9	:94	44	:34
9	:70	73 :30 S W	2	76					9	:30	12	:70	53	:64
29	:00	75 :30 S W	7	25					23	:11	19	:95	81	:75
7	:40	74 :30 S W	1	98					7	:12	21	:93	88	:87
10	:50	*South.*	10	60					00	. 00	32	:53	88	.87
14	:60	28 :30 S W	12	84					6	97	45	:37	95	:84
33	:95	82 :00 N E			4	:74	33	:61			40	:63	62	:23
25	:00	72 :00 N E			7	:74	25	:75			32	:89	38	:48
21	:40	78 :30 N E			4	:28	20	:97			28	:61	17	:51
21	:00	65 :30 N E			8	:71	19	:11			19	:90	1	:60
20	:00	4 :30 N W			19	90			1	60	00	:00	00	:00
238.	25		45.	37	45.	37	97.	44	97	:44	X		Z	

Figure 5. A scheme for checking for errors by field surveyors before plotting, developed during the Down Survey of Ireland. Osbourne, *A More Exact Way* . . . [1654].

one of these percloses remained on a final barony map. Petty opposed the reduction of the resulting polygon to multiple triangles, allowing the geometrical calculation of area, a standard technique that was seen as too complicated by Petty.[45] Osborne's use of a mathematical check prior to protracting probably reflects concern that plotting measurements with large errors would create problems in the field during the Down Survey, although it required more mathematical sophistication in using trigonometry tables to establish components for each measurement. The concern with quality control continued following Petty's completion of the survey as a committee charged with reviewing the work found it acceptable, while Worsley charged the survey with numerous deficiencies in comparison to the contract, to which Petty responded in detail to the satisfaction of yet another committee.[46] Concern with quality control and the emergence of the surveyor as an administrator are significant legacies of the Down Survey for future large surveys.

Petty employed about a thousand soldiers for two reasons: they constituted a ready pool of trainable labor and they were better able to deal with Irish terrain and people. Not surprisingly, the Down survey elicited hostility on the part of the displaced population; surveyors were killed and troops promised to Petty were often called away for other business.

> But the principall division of this whole worke was to enable certayne persons, such as were able to endure travaile, ill lodginge and dyett, as alsoe heates and colds, being alsoe men of activitie, that could leape hedge and ditch, and could alsoe ruffle with the severall rude persons in the country, from whome they might expect to be often crossed and opposed. (The which qualifications happend to be found among severall of the ordinary souldiers, many of whom, havinge bin bread to trades, could write and read sufficiently for the purposes intended.) Such, therefore (if they were but headfull and steddy minded, though not of the niblest witts), were taught . . . how to make use of their instruments, in order to take the bearinge of any line, and also how to handle the chaines, especially in the case of risinge or fallinge grounds; as alsoe how to make severall markes with a spade, whereby to distinguish the various breakings and abutments which they were to take notice of; and to choose the most convenient stations or place for observations, as well in order to dispatch as certaynty.

[45] William Leybourn, *The Compleat Surveyor* (London, 1653), pp. 27–28; Petty, *Down Survey* (note 7), pp. xvi, 17; Andrews, *Plantation Acres* (note 8), pp. 318–20.

[46] Petty, *Down Survey*, pp. 103–152.

Petty noted that field work, "being a matter of great drudgery (to wade through boggs and water, climb rocks, fare and lodge hard, &c.), he would instruct foot soldiers, to whom such hardships were familiar."[47] Since he had divided the art of surveying into components, the manual aspects of the task could be separated from those requiring special skills. Petty's biographer notes that "the crux of the matter was not simply to make the work of each artisan 'simple and easy,' but to use the very small supply of professional skill at his disposal exclusively for skilled work, while using unskilled labour (frequently in the form of foot soldiers) for the measuring."[48] Critics had objected that foot soldiers might stand to benefit from inaccurate surveys since titles to land were to be distributed to settle payment to the army. Moreover, papists in his employ might deliberately provide misleading information. Petty's defense relied on the claim that the rationalized division of labor he had established insured that misrepresentations would be discovered, since the task of field work was separated from that of protracting, casting, reducing, and mapmaking: "These men, being designed only for ffield worke, could abuse the State only in the length of their chaine, which would alsoe discover itselfe."[49]

Soldiers employed in the survey were also taught "how to judge of the vallues of lands, in reference to its beare qualities, and according to the rules and opinions then currant." Petty made use of the soldiers' prior skills whenever possible: those skilled at painting or drawing, "especially such as had beene of trades into which payntinge, drawinge, or any other kind of designinge is necessary, were instructed in the art of protractinge."[50] The key point to notice here is that the task of measuring distances and angles in the field is separated from the task of plotting these measurements on graph paper. The area of estates and larger boundaries was determined by visual inspection of plots on graph paper, which could be checked by those working on the project who were actually trained surveyors. This is less accurate than a calculation of areas by decomposing polygons into triangles. Petty prefers this approach, however, since it is quicker and less subject to error or fraud in that it allows for the possibility that

[47] *Ibid.*, pp. xv–xvi, 17–18.
[48] E. Strauss, *Sir William Petty: portrait of a genius* (London, 1954), p. 62.
[49] Petty, *Down Survey*, p. 20.
[50] *Ibid.*, pp. xv, xvi.

the maps could be replotted against the field measurements. Thus, a technically less elegant procedure may be more appropriate for large surveys lacking sufficient professional surveyors to carry out all the work.

Six professional surveyors "were contracted with to teach other men of meane and low fortune." Their task was to check the work of the soldiers: "noe ffield man should protract his worke, and he that protracted should be paid by salary [rather than by linear content], to the end he should have noe interest to admitt of any bad worke, and all this under the eye of one of these six artists, whome he stiled conductors." Petty was more concerned to counter the possible bias of surveyors by changing the way they were paid than he was to improve the accuracy of the plot through technical refinement. Petty's survey was controversial not least since he had usurped the existing plan by criticizing its method, dispensing with skilled labor wherever possible and changing the form of payment. Petty pointed out that the size of confiscated estates varied greatly (from 8,000 to 160,000 acres) and yet the surveyors had been paid by the area measured, which would increase faster than the labor of measuring the perimeter or "surround." Instead, he insisted that they be paid by "lineary content" (the length of their measurements) rather than by the "superficiall content" (the area surveyed), in order to remove the temptation to inflate the measure of area and to better correspond to the amount of work performed. Moreover, he paid the same rate for profitable as unprofitable land, so no one would be motivated to inflate the amount of profitable land.[51]

While Petty changed these incentives for workers in his employ, his own payment depended upon the area surveyed and he was paid more for profitable than unprofitable land. Petty argued that, as director of the survey, he was not in a position to inflate the area measured or return unprofitable land as profitable. Nevertheless, his critics argued that he had inflated the amount of profitable land. Petty's contract gave no good criterion for distinguishing profitable land and the binary nature of the choice forced difficult decisions with marginally profitable land. Nor could a surveyor be able to predict whether income from rents would exceed quitrents due to the crown. The irony of the situation is that a mistaken entry in the Down survey

[51] *Ibid.*, pp. 258–259, xiii, 312, xvii.

led to an overvaluation of quitrents due the crown for Petty's land in Kerry. He was harassed for the rest of his life by legal troubles over the title to his land and the taxes he owed to rent "farmers"—those who contracted to collect taxes for the Crown.[52] His reflections upon what accounts for the value of land—which he attributed to improvements made by labor—and his call for a land registry to fix the title for lands grew out of his experience with the Down Survey.[53]

Mapping Territory by Mapping Property

The Down Survey set a number of precedents for surveys in colonial settings outside Ireland, especially in America. First, the combination of territorial and property surveys was a significant legacy for future colonial surveys. Second, mapmaking was crucial to successful colonization, less through facilitating conquest or rhetorically claiming ownership, than through the role that surveys played in effectively enabling settlers to occupy the land quickly and legally, enabling effective control of land. Third, the Down Survey demonstrated the importance of speedy surveys of large areas in colonial settings, requiring the division of labor and careful management.

Petty combined cadastral surveys conducted by traversing the perimeter of individual estates with maps of political and natural boundaries, the first significant precedent for future colonial surveys.[54] The state cadaster became an important means for extending the power of the state in the sixteenth and seventeenth century. In England, the seizure of Church lands after the dissolution of the

---

[52] Marquis of Lansdowne (ed.), *The Petty-Southwell Correspondence, 1676–1687* (New York, 1967), pp. 4, 26–29, 32, 58, 75, 219–20; Strauss, *Sir William Petty*, ch. 8. See also the reference to "ye Excheqr [which] makes us pay ye Quitrents double & beforehnd" in Petty to Sir George Rowdan, October 1, 1667 [Huntington Library, San Marino, Calif., HA 15551].

[53] Petty, *Petty Papers* (note 39), I:77–90. For Petty's own efforts to make Kerry's mountainous land profitable through technological enterprises, see T.C. Barnard, "William Petty as Kerry Ironmaster," *Proceedings of the Royal Irish Academy* 82 c. 1 (1982): 1–32; T.C. Barnard, "Sir William Petty, Irish Landowner," in Hugh Lloyd-Jones, Valerie Pearl, and Blair Worden (eds.), *History and Imagination: essays in honour of H.R. Trevor-Roper* (London, 1981), pp. 64–69. On the idea of a land registry, see also Hartlib, *The Hartlib Papers* [CD-ROM] (Ann Arbor, Mich., 1995), "Ephemerides," 1655, 29/5/32A.

[54] Petty, *Down Survey* (note 7), p. xiv; Andrews, *Shapes of Ireland* (note 29), p. 122.

monasteries by Henry VIII, during the Civil War, and during enclosures of commons were contexts in which state property surveys were conducted.[55] A state interest in territorial maps evolved in parallel with the interest in state property records.[56] In Ireland, where colonization proceeded more rapidly than in America during the first half of the seventeenth century, "an effort to facilitate military conquest, administrative assimilation, and legal imperialism" led English administrators to commission "new maps of the 'moving frontier'" in the country.[57] At the same time, large landholders commissioned estate maps as part of an effort to secure the boundaries of their properties and lease land to encourage improvements, a move that encouraged surviving Gaelic landowners to pursue improvement, as well.[58] In effect, surveying land was a first step in introducing economic and technological improvements. Surveys encouraged the idea that land was a resource that required labor and technological innovation to fully exploit, an idea key to Petty's own latter work on "political arithmetic."[59]

Petty's innovation was a fusion of these two kinds of surveys, cadastral and territorial, in a mapmaking project that ultimately resulted in maps ranging from individual property survey maps to a map of the entire country.[60] In future colonial settings, the importance of property surveys by military or civilian agencies also shaped the production of territorial maps. Petty essentially mapped a large territorial area by building a map from surveys of estates confiscated from the Irish, relating the boundaries of the estates to each other and to town, parish, barony, and county lines. The Strafford Survey, an abortive project for plantation under the Earl of Strafford in the 1630s, set the stage for this fusion of property and political map-

---

[55] Roger J.P. Kain and Elizabeth Baigent, *The Cadastral Map in the Service of the State: a history of property mapping* (Chicago, 1992), pp. 8, 336–39.

[56] Robin A. Butlin, *Historical Geography* (London, 1993), pp. 94–96.

[57] Jane H. Ohlmeyer, "'Civilizinge of Those Rude Partes': Colonization within Britain and Ireland, 1580s–1640s," in Nicholas Canny (ed.), *The Origins of Empire: British overseas enterprise to the close of the seventeenth century* (Oxford, 1998), p. 140.

[58] *Ibid.*, pp. 140–41.

[59] Petty, *Political Arithmetick* and *The Political Anatomy of Ireland* in William Petty, *The Economic Writings of Sir William Petty Together with the Observations Upon the Bills of Mortality More Probably by Captain John Graunt*, ed. Charles Henry Hull (Cambridge, 1899), I:239–313, I:121–231; William Lynch, *Solomon's Child: method in the early Royal Society of London* (Stanford, 2001), ch. 6.

[60] Petty, *Hiberniae Delineatio* (note 29).

ping.[61] The survey had produced an index map of land intended for plantation in the province of Connaught and the counties of Clare and Tipperary and was likely carried out in a similar fashion. Since the entire area was considered for plantation, the surviving overview map does not have the gaps evident in Petty's barony maps, which leave unforfeited lands unsurveyed.[62] That also means that the Strafford Survey did not need to explicitly face the question of relating selected properties intended for survey and redistribution to political boundaries that lay in unforfeited areas. As Andrews puts it, "the Strafford Survey had shown every kind of territorial division from the parish upwards, but this coverage followed almost automatically from the unbroken continuity of the townland network set out for cadastral purposes in Connacht."[63]

In other words, the identification of an unbroken area for plantation in the Strafford survey led to the mapping of estates in maps of townlands, the smallest administrative units. There are about 60,000 townlands in Ireland today, and they range in origin from Gaelic clan lands to manors established by the Anglo-Norman colonization to more recent English plantations. They rarely correspond to urban towns in the modern sense and their administrative importance was marginal. Baronies constitute a number of townlands and there are 270 in Ireland today. From the Down Survey estate maps, Petty's surveyors produced barony maps, organized by county (larger units still), with terriers showing confiscated estates by size and divided into profitable and unprofitable portions.[64]

Since the Down Survey involved the seizure of land of the rebellious, (mostly) Catholic landowners among counties with many Protestants who would retain their estates, the issue of how to proceed with the survey was one that led to seemingly insurmountable logistical problems among those planning the land settlement in the army. In 1653, they had decided to survey on a barony basis, with all forfeited areas

[61] Hugh F. Kearney, *Strafford in Ireland, 1633–41* (Manchester, 1959), ch. 9.

[62] Trinity College, Dublin, MS 09, no. 69, reproduced in Andrews, *Shapes of Ireland* (note 29), p. 120, fig. 5.1.

[63] *Ibid.*, p. 122.

[64] These maps survive in bound volumes in the Petty Papers at the British Library and in a slightly different form in the Bibliothèque Nationale in Paris. I have examined the British Library copies, BL Add. MS 72868–72873, and the University of California-Los Angeles Map Library's copy of the reproduction of the Paris maps in *Hibernia Regnum* (Southampton, 1908).

to be subdivided later by cadastral surveys paid for by the recipients. This decision was reversed in 1654, when it was decided that estate surveys should be made, ignoring barony boundaries, which would make it difficult to relate these estates to administrative boundaries (in addition, many estates crossed barony boundaries).[65] The tension between the need to measure property boundaries to facilitate the division of spoils with the need to be able to relate these estates to administrative and topographical boundaries in a setting where these were not well known, indeed where the exact quantity of land with which they were dealing was unknown, seemed to paralyze the officers working on the settlement. This obstacle to effective colonization following reconquest was significant until Petty proposed measuring political and natural boundaries at the same time as cadastral surveys were made, even suggesting that they could be given to soldiers immediately after survey. This construction of territorial maps on a cadastral basis fit the colonial situation the English faced in Ireland (as also in America) and would not have been conceivable (or desirable) were it not for the land confiscations and reassignments that took place.

Petty's incorporation of the Down Survey's maps into his 1685 atlas of Ireland ensured that baronies would show up on maps of Ireland into the nineteenth century despite their marginal administrative importance; barony maps were a convenient basis for mapping and indexing the available land.[66] Though assigned a constable, baronies were less significant units than the smaller parishes contained within them or the larger counties. They were extremely significant, however, if your goal was to show land available for the taking in a way that would facilitate its redistribution and taxation. Consequently, the cluttered look of Petty's maps in his atlas (see an early version of the map of Ireland in figure 3), with its typology distinguishing, in ascending order, townlands, parishes, baronies and liberties, counties, and provinces, reflects its basis in the Down and Strafford surveys and the need to tie administrative boundaries to property in a colonial context following reconquest.[67] Still, as Goblet notes, Petty's atlas "was the first atlas of a whole country to be com-

[65] Andrews, *Shapes of Ireland* (note 29), pp. 122–23; Petty, *Down Survey* (note 7).
[66] Andrews, *Shapes of Ireland*, p. 125.
[67] For a modern reprint of Petty's atlas, see Petty, *Hiberniae Delineatio* (note 29).

piled 'in the field' and so perfect that it was only replaced more than a century later by the maps of the Ordnance Survey."[68]

Colonizing through Mapping

A second important precedent set for future colonial surveys by the Down Survey is the importance that mapmaking had for ensuring effective colonization following conquest. Without maps that coordinate property and government boundaries, it would not be possible to settle land upon individuals without significant *internal* conflict. It is not—as postmodernist cartography scholars would have it—that maps rhetorically establish English ownership of land in the face of native claims (*e.g.*, Irish or Native American). This makes the connection between mapmaking and conquest much too direct and inflates the power of maps, considered as mere physical artifacts, to assert ownership against competing systems of land use based upon traditional, sometimes orally maintained, knowledge not based in maps (or based on maps that do not record precise property boundaries).[69] While Callon and Latour's analysis of maps as "immutable mobiles," tools for acting at a distance, is certainly appropriate for considering the role that maps played in managing colonial endeavors from abroad, this does not really help us see its role in everyday colonial contexts, and certainly does not explain how initial, effective control and use of the land is established in the first place.[70] The implication of these approaches is that maps as material nodes in a knowledge network explain why colonial subjects were conquered, offered in opposition to the view that colonials lacked our rational modes of thought.

Rather than reifying maps as agents for displacing natives, however, I think we need to look elsewhere in explaining how conquest occurred.

[68] Y.M. Goblet, *Political Geography and the World Map* (New York, 1955), p. 5. See also Goblet, *A Topographical Index of the Parishes and Townlands of Ireland in Sir William Petty's Mss. Barony Maps (c. 1655–9) and Hiberniae Delineatio (c. 1672)* (Dublin, 1932).

[69] J.B. Harley, *The New Nature of Maps* (Baltimore, 2001); Denis Wood, *The Power of Maps* (New York, 1992); Richard Helgerson, "The Land Speaks: Cartography, Chorography, and Subversion in Renaissance England," *Representations* 16 (1986): 51–85; Richard Helgerson, *Forms of Nationhood* (Chicago, 1992); Bernhard Klein, *Maps and the Writing of Space in Early Modern England and Ireland* (London, 2001).

[70] Michel Callon, "On the Methods of Long-Distance Control: Vessels, Navigation and the Portuguese Route to India," in John Law (ed.), *Power, Action, and Belief* (London, 1986); Bruno Latour, *Science in Action* (Cambridge, 1987).

The implication of much postmodernist history of cartography is that the ideological role of maps was crucial for displacing native populations and legitimating Western rule. The practical aspects of mapping having to do with effective administration and control of territory—rather than the production of an artifact with cultural meaning—have been correspondingly neglected. The idea that the mapping eye plays a crucial role in imperialism and colonialism neglects the more direct role of technological and military superiority in conquest, which, of course, depends upon its own knowledge networks, allowing supplies and people to be mobilized. This ultimately would include military mapping operations and military reliance upon maps, though even well into the eighteenth century, military campaigns were unable to rely upon maps, employing local guides and sketch artists instead.[71] As a component of initial military control, maps did not play the role they play today. Maps did not facilitate conquest; conquest facilitated mapping, which in turn ensured effective settlement, use, and control of land. Settling large numbers of new settlers on conquered land, complete with legal protections ensuring that efforts to work the land would not be wasted[72] and that disputes would not turn inward, could ensure that colonial endeavors lasted and were less subject to military reversals. In short, maps helped export a particular legal and economic system to the new lands with former radicals and sectarians happy to adopt the role of landowners in new lands.

Thus, the need for speed in settling the army arrears in Ireland grew out of concerns that radicals in the army would thwart efforts by Cromwell and the landed elite to establish a political system protective of established interests. The land settled upon the army, even though bought up mostly by officers, effectively defused this concern. To encourage the adventurers behind the initial financing of the Army to settle, the ten counties were each split in half, encouraging the adventurers to feel secure by the presence of decommissioned soldiers in each county. Since the amount of land seized was so large, and the religious and national ramifications of the conflict so prominent in the popular imagination, the Cromwellian settlement was the key moment in transforming Ireland from a land where

[71] Martin van Creveld, *Technology and War* (New York, 1991), pp. 117–18.

[72] See Prendergast, *Cromwellian Settlement* (note 2), pp. 30–31, for a discussion of the effect that uncertainty over the exile of Catholics to Connaught had on planting crops.

Anglo and Gaelic Catholic Lords predominated to one where English and Scottish Protestants dominated. Between 1640 and 1688, the Protestant landowning share in Ireland rose from 41 to 78 percent, with most of that change occurring between 1652 and 1660.[73] As such, the case is closer to the case of colonization in America, as compared to earlier plantations in Ireland that did not lead to such a dramatic change in ownership or extend far beyond the Pale. The American pattern of a complete displacement and decimation of the native population (and the enslavement of some to work the land) differs from the Irish case, though the massive scale originally envisioned for the transplantation to Connaught of native Irish has some parallels to the American case. Petty's friend and fellow Commissioner for settling lands, Vincent Gookin, argued against transplantation of the native Irish, in part since they were needed to work the land and this need eventually tempered army prejudice urging a complete transplantation.[74] Vincent's brother Daniel urged a similar restraint with respect to Native Americans in New England at the time of King Philip's War—to no avail.[75]

Managing Large Surveys

A third precedent set by Petty's survey was the need for speedy surveys of large areas, leading to innovations in organizational patterns and management. While the surveys had to be reasonably accurate (Petty's measurements of area were ten to fifteen percent below actual values),[76] more important was the quick settlement of property titles on individuals. Petty's interest in rationalizing and managing tasks like this predated his involvement with the army in Ireland. In 1647–1648, Petty developed a plan to extract knowledge from skilled tradesmen through a kind of subcontracting, in order to make money and turn their various skills to a greater effect.[77] Petty's practical

---

[73] Bottigheimer, *English Money* (note 4), p. 3.

[74] Vincent Gookin, *The Great Case of Transplantation in Ireland Discussed* (London, 1655).

[75] Patricia Coughlan, "Counter-Currents in Colonial Discourse: The Political Thought of Vincent and Daniel Gookin," in Jane H. Ohlmeyer (ed.), *Political Thought in Seventeenth-Century Ireland* (Cambridge, 2000), pp. 56–82; Philip Ranlet, "Another Look at the Causes of King Philip's War," in Alden T. Vaughan (ed.), *New England Encounters* (Boston, 1999), pp. 139–165.

[76] Goblet, *Topographical Index* (note 68), p. vi.

[77] BL Add MS 72891, fol. 8.

emphasis on the organization and division of labor continued in the Army. In Petty's account of the Down Survey, he noted that he had intended to go to Ireland with Major-General Lambert, "a favourer of ingeniouse and usefull arts." Initially, the goal was to apply Petty's gift for inventions to a land that needed rebuilding. He tells us that he decided to accompany Lambert in 1652, "when the war there was near ended, and many endeavours used to regulate, replant, and reduce that countrey to its former flourishing condition, as a place most wanting such contrivances as tended to the above-mentioned ends" since he "had formerly gained some reputation in the world" for such inventions.[78] When Lambert did not come to Ireland after all, Petty accompanied Lieutenant-Generall Fleetwood as physician to the army. The idea of Ireland as a laboratory for trying out schemes for technological and economic improvements was a common one at the time among Baconian reformers like Petty.[79]

His first project in Ireland was to streamline the army's apothecary and this emphasis on efficiency and management can be seen in the Down Survey.[80] While Petty was long familiar with the practical mathematical arts, learning navigation as a sailor and geography at the Jesuit college at Caen,[81] he seems to have focused more on training and management than on day-to-day operations, the details of which he left to his undersurveyors. He provided guidelines on surveying to his undersurveyors on how to manage the project and to check the quality of the tasks performed by soldiers. This strategy of employing subcontractors was key to his taking control of the survey from the surveyor-general Benjamin Worsley.

Petty's simple compass and line methodology, determining the surround of a property, was extremely important in colonial contexts. The greater accuracy of a theodolite was not an advantage over the circumferentor when slow surveys held up possession and exploitation

[78] Petty, *Down Survey* (note 7), p. 1.

[79] Frances Harris, "Ireland as a Laboratory: The Archive of Sir William Petty," in Michael Hunter (ed.), *Archives of the Scientific Revolution* (Woodbridge, Eng., 1998).

[80] Petty, *Down Survey*, p. 1.

[81] BL Add. MS 72851, fols. 3–6; Petty, *Petty Papers* (note 39), II:193–98. Upon applying to Caen, Petty requested specifically to enroll in "ye first classe of your renowned College of Geography" (Sharp, *Sir William Petty* [note 38], p. 5). Sharp describes how the Jesuit colleges at La Fleche and Caen promoted a broad understanding of geography, which "reflected the vigorous and widespread missionary work of the order and the requirements of an education designed to produce military officers, surveyors, and skilled recruits for the country's growing maritime interests." (p. 5)

of land. For example, Hughes has shown the impatience of settlers in Virginia to have government surveyors measure their lands to obtain title: "Speed was more necessary than accuracy and simple instruments were more practical for the men who tramped for miles into virgin forests." Like the Irish case, large tracts of land could be claimed at little cost since the native rights to the land were rejected. Investors in the Virginia Company of London received land in return, again like the adventurer scheme in Ireland; half chose to settle.[82]

Trade and speculation in debentures to land in Ireland were matched by a similar trade in headrights, grants made to all heads of households willing to settle in Virginia. The result was increasing concentration of large properties. To get title in Virginia, a survey would need to be carried out within a certain period of time, and surveyors became key political figures as a result. Like Petty, many surveyors were able to identify key properties they would like to own and command fees in land that established themselves as large landowners. Whereas English surveyors were typically members of the middling classes, surveying the land without owning any themselves, Petty pointed the way for the colonial surveyor/gentleman (like George Washington).

This enhancing of the status of surveying was accompanied by a deskilling of actual surveying tasks as the function of the surveyor began to shift from the practitioner of a specialized skill to a administrator of others. To be sure, most Virginia surveyor-generals and later county surveyors continued to carry out surveys themselves, in between farming and political life. But surveying techniques needed to be quick and easy for that reason as well. In New England, by contrast, land was typically set out at once through a collective survey, with smaller plots of land and less commodification, in part due to the greater Puritan character of the communities there.

The two models came together following the U.S. Land Ordinance of 1785, when government surveys of the Northwest Territory divided up land in massive surveys to settle on individuals, an organizational pattern closer to the Down Survey.[83] Early experiments in execution took place in Ohio, but eventually the bureaucratic structure and technical practice was standardized, plotting six-mile townships on a

[82] Hughes, *Surveyors and Statesmen* (note 24), pp. 3–4.
[83] *Ibid.*

grid system before allocating land. Property maps for all sales were filed with the local, regional, and national land offices.[84]

> By the time the survey reached states such as Kansas and Nebraska (1854–85), a well-defined system of surveying and mapping with its own bureaucracy of contract deputy surveyors working under surveyors-general and supported by register clerks and draftsmen based in a network of local land offices had been established. In 1855 a manual entitled *Instructions to the Surveyors General of Public Lands of the United States . . . Containing also a Manual of Instructions to Regulate the Field Operations of Deputy Surveyors, Illustrated by Diagrams* was produced by the General Land Office as a codification of practices enshrined in earlier manuals relating to particular states and territories.[85]

Surveying had become a much more professionalized activity and had sought improvements in technique. Without something like the Down Survey emphasis on a quick and dirty approach, with its training of a wider pool of soldiers in surveying techniques, the execution of large surveys necessary for westward expansion would have been inhibited. Consequently, more attention should be focused on colonial and military surveys than traditional historiographical emphasis on cutting-edge, mathematical techniques that are prominent among mathematical writers looking to enhance the intellectual reputation of surveying.[86]

The increase in commodification of land under colonial situations was connected to this ability to settle land upon new claimants quickly. In the sixteenth and seventeenth centuries, the enclosures of commons and the displacement of peasants who previously had rights to land use and greater wealth in the late medieval period, left many peasants with a desire for land. At the same time, it encouraged landlords to treat their property as alienable property, with the right to evict peasants, rent land to the richer peasants who would improve yield, and even sell their property. In an irony of history noted by Kulikoff, "[t]he same men who evicted peasants financed colonial

[84] Kain and Baigent, *Cadastral Map* (note 55), pp. 293, 297.

[85] *Ibid.*, p. 293.

[86] On this point, see Bennett, "Geometry and Surveying" (note 22); Bennett and Brown, *Compleat Surveyor* (note 31). For the military context for some early mathematical practitioners, see Stephen Johnston, "Mathematical Practitioners and Instruments in Elizabethan England," *Annals of Science* 48 (1991): 319–344 at pp. 322–23.

ventures that promised land to former peasants." Though many immigrants to America had to work off their passage across the ocean by indentured servitude, two-thirds eventually acquired land themselves.[87] Later, during the American Revolution, the Continental Congress paid soldiers in future land rights, just as had occurred in Ireland.[88] The convertibility between land and money was more pronounced than in England. The alternating policies of colonial governments involving the granting of free land to farmers (making land cheap) and the encouragement of speculative markets in land (leading to higher prices) reflected the different interests in land of the peasant and the speculator. Both led to a need to survey the land and to treating land as something that can be bought and sold as needed. Markets in land evolved within the context of the legal requirement that lawful possession of land required a survey laying out its borders. Hence, surveying was crucial to the process of commodification of land and westward expansion, as speculation-driven land markets forced poorer farmers west in search of cheap land. From the context of the history of surveying, the connection between military conquest of Native Americans and the surveys of land made in order to grant land to new settlers is significant. For orderly settlement to take place, it was very important that settlers were able to identify the *borders* between properties; hence, traverse surveys carried out with circumferentors were important. Where disputes arose as the result of inaccurate or contradictory surveys, land owners also relied upon testimony of boundary trees and marks used as part of the initial survey.[89]

The greater tendency for colonial cadastral surveys to be executed as part of a territorial survey or shortly after land was claimed following conquest explains some of the differences between surveying in England and abroad. Petty's survey was unique in building a national map from traverse surveys of political boundaries and properties. Traverse property surveys remained important in many later surveys in America. John Love's 1688 textbook laid out the

[87] Alan Kulikoff, *From British Peasants to Colonial American Farmers* (Chapel Hill, N.C., 2000), pp. 2–3.

[88] D.W. Meinig, *The Shaping of America*, vol. I: *Atlantic America, 1492–1800* (New Haven, Conn., 1986), p. 355.

[89] Kulikoff, *From British Peasants*, pp. 111–112.

circumferentor and chain method (with angles measured against magnetic north) for surveyors in America faced with long and narrow plots that did not respond well to triangulation.[90]

In New England, land would typically be surveyed before granting land, while south of Pennsylvania, this pattern was reversed. In the case of New England, a greater amount of control over land was exercised and commodification was minimized by rules inhibiting the selling of land. In Virginia and other southern colonies, a great deal of land speculation took place and settlers could establish claims to large areas of land; to establish title, they needed to have Surveyors-General or county surveyors measure the land. Commodification, speculation, and corruption were much more prominent in this system, both among ordinary settlers and among surveyors. If the collective survey-and-distribute pattern in New England resembled the Down Survey's, another side of the Down Survey was reflected in the opportunities in the South for large-scale land grabs and the enhancement of the status of surveyor, where they could collect large fees and become significant landowners themselves. The colonization of the Northwest Territory after the revolutionary war opened up a fusion of the two perspectives in large scale government surveys that distributed land to settlers following survey as occurred in the Down Survey, although the surveys imposed a grid system, rather than following pre-existing political and natural boundaries.[91] Note, however, that this involved essentially creating territorial boundaries in the process of property surveying.

Kain and Baigent point out that some states settled arrears for soldiers with western lands still claimed by the states, while most land was given to the U.S. government, which led the federal government to address the limitations of the Virginia system's settle-first policy. The 1784 *Ordinance for Establishing a Land Office for the United States* and the *Land Ordinance of 1785* developed a system whereby government surveys of land on a grid system would lead to auctions of land to pay off government debt. Sales were slow since the land had not been cleared of Native Americans; like the Down Survey, native attacks on surveyors were common. Following the Battle of Fallen Timbers, the *Act Providing for the Sale of Lands of the United States*

[90] Kain and Baigent, *Cadastral Map* (note 55), p. 268; Love, *Geodaesia* (note 24), preface.

[91] Hughes, *Surveyors and Statesmen* (note 24); Kain and Baigent, *Cadastral Map*, ch. 8.

in the Territory Northwest of 1796 set out the Federal Land Survey, proceeding from east to west plotting rectangular surveys, which facilitated the commodification of land by standardizing plots. Between 1787 and 1837, 4.5 million people were moved into lands west of the Appalachians following U.S. Land Office surveys.[92]

Conclusion

When we think of military surveying, we usually consider maps as a direct tool of war. When we consider the power of maps in promoting colonization, we often focus on the rhetorical power of maps in claiming ownership of indigenous land. Lost in the shuffle is the role of maps in gaining effective control of land by ensuring that political and legal systems are effectively exported from the mother country. The class mobility allowed by westward expansion (beginning in Ireland) was accompanied by the relative stability of a class system based on land in Ireland and early America. The importance of surveying in the settlement of the American continent and its connection with military science at this time is demonstrated by the establishment of a military academy at West Point.[93] The Corps of Topographical Engineers was a branch of the army between 1813 and 1863, manned by officers trained at West Point, which played a crucial role in exploring and mapping the West.[94] The surveys of land carried out by the Federal Land Survey, building upon technical and organizational patterns introduced by the Down Survey, were essential prerequisites for the sale of land, which in turn ensured the financial solvency of the early Republic.[95]

At the same time, the commodification of land was facilitated by the substitution of land for wages (as when Petty asked to be paid in land at "Act [of settlement] rates").[96] This blurred the boundaries

---

[92] Kain and Baigent, pp. 289–97.

[93] Sidney Forman, "Why the United States Military Academy Was Established in 1802," *Military Affairs* 29.1 (1965): 16–28 at pp. 23–24.

[94] Richard A. Bartlett, *Great Surveys of the American West* (Norman, Okla., 1962), pp. 332–33; William H. Goetzmann, *Army Exploration in the American West, 1803–1863* (New Haven, Conn., 1959).

[95] Joseph W. Ernst, *With Compass and Chain: Federal Land Surveyors in the Old Northwest, 1785–1816* (New York, 1979); William D. Pattison, *Beginnings of the American Rectangular Survey System, 1784–1800* (New York, 1970).

[96] McKenny, "Seventeenth-Century Land Settlement" (note 21), p. 198.

between the landed wealth of free gentleman and the wealth based on wages or trade of professionals (including surveyors and physicians) or merchants. However, since newly wealthy professionals and merchants could attain large estates in lands outside England, class conflict within England was minimized. The Restoration settlement did take back some of the lands granted under Cromwell to pay back the King's allies, but those displaced were generally reimbursed with yet more land.[97] The ability of cash-strapped governments to settle land upon new claimants after military victories was instrumental in ensuring popular participation in colonization and attenuating internal conflict.

Acknowledgements

I would like to thank Steven Walton, Deborah Warner, and an anonymous referee for helpful comments on drafts of this paper. The research for this paper was supported by a sabbatical from Wayne State University and a fellowship from the Max Planck Institute for the History of Science in Berlin. I am grateful to Otto Sibum and the participants in his research group at the Max Planck Institute for their advice.

[97] BL Add. MS 72878.

CHAPTER THREE

LIKE CLOCKWORK? CLAUSEWITZIAN FRICTION AND THE SCIENTIFIC SIEGE IN THE AGE OF VAUBAN

Jamel Ostwald

The late seventeenth century was pivotal to the early modern Scientific Revolution. Its culmination is epitomized by Isaac Newton, who explained and predicted the workings of Nature in terms of mathematics. Synthesizing the observations and calculations of Copernicus, Brahe and Galileo, he developed a coherent, mechanistic explanation of planetary motion that appealed to the European fascination with the machine. Similarly, the late seventeenth century was the age of another 'scientific' achievement, this in the field of warfare. The French engineer Sébastien le Prestre de Vauban incorporated the practices and theories of previous military engineers into a coherent system of siegecraft, drawing on the period's *esprit géometrique* to create a scientific instrument capable of reducing even the strongest of fortresses in short order.[1] Military historians have drawn a parallel between the metaphor of Newton's clockwork universe, a system of planets and stars moving together in a mechanical, predictable way, and the metaphor of Vauban's siege, a formalized attack that captured fortresses in a similarly predictable and deterministic fashion. Further strengthening the connection between these two contemporaries, Vauban's technique of besieging fortresses is said to have ruled over generations of military engineers as completely as Newton's laws of motion

[1] Vauban has been the subject of numerous biographies in the last century. Paul Lazard, *Vauban, 1633–1707* (Paris, 1934); Reginald Blomfield, *Sebastien Le Prestre de Vauban, 1633–1707* (1938; New York, 1971); Michel Parent and Jacques Verroust, *Vauban* (Paris, 1971); F.J. Hebbert and George Rothrock, *Soldier of France: Sebastien Le Prestre de Vauban, 1633–1707* (New York, 1989); Bernard Pujo, *Vauban* (Paris, 1991); and Anne Blanchard, *Vauban* (Paris, 1996). The most direct connection between the Scientific Revolution and Vauban has been advanced by Henry Guerlac, "Vauban: The Impact of Science on War," in Peter Paret (ed.), *Makers of Modern Strategy from Machiavelli to the Modern Age* (1943; Princeton, 1986).

did over physicists.[2] A closer look at the validity of this metaphorical parallel will illuminate the ways in which this instrument of siegecraft was and was not 'scientific.'

Views of the Clockwork Siege

Historians have consistently described the Vaubanian siege attack in systematic, scientific terms. David Chandler explained that "Vauban had virtually imposed a series of standards on both attack and defense" by "perfect[ing] the techniques of the siege itself—with the laborious but almost mathematically certain 'sapping forward' by means of approach and parallel trenches, and the clever siting of batteries."[3] Martha Pollak stressed the mechanical nature of Vaubanian siegecraft:

> Vauban was considered a theoretical, systematic and machinating genius . . . His tables of calculations gave the impression of strategic unassailability since he calculated not only the dimensions of every element of the fortification, but also the length of time it would take the enemy to gain individual layers of the fortification, every stage of the siege could be predicted in advance. . . . The accountability of the smallest part of the defense, fortification and provisioning in Vauban reflect the earlier attempts by military theorists to set up a machine which can be expected to operate by itself, but which results—both in Vauban and his predecessors—in an obsession with the smallest detail.[4]

Other historians have also associated his achievement with the mechanical precision of clockwork (*e.g.*, *une sorte de mécanique guerrière*), placing it squarely within the scientific revolutionary context of seventeenth century France.[5] Christopher Duffy, the unparalleled historian of early

---

[2] Guerlac, "Vauban: The Impact of Science on War," pp. 72–73.

[3] David Chandler, *Marlborough as Military Commander* (1973; New York, 2000), p. 81 and Chandler, *Military Memoirs of Marlborough's Campaigns, 1702–1712* (Hamden, Conn., 1998), p. 234.

[4] Martha Pollak, *Military Architecture, Cartography and the Representation of the Early Modern European City: a checklist of treatises on fortification in the Newberry Library* (Chicago, 1991), p. xxxiv.

[5] Nicolas Faucherre and Philippe Prost, *Le triomphe de la methode: le traité de l'attaque des places* (Paris, 1992), p. 53. Michèle Virol discusses the use of mechanical metaphors in Vauban's period in *Les Oisivetés de Monsieur de Vauban* (Villeneuve d'Ascq, 2000), pp. 182ff. We should note that the clockwork metaphor—the human body, world, government, or society seen as a clock—was in widespread use far before the age of Vauban and Newton, flourishing from the early seventeenth century onward. See Otto Mayr, *Authority, Liberty, and Automatic Machinery in Early Modern Europe* (Baltimore, 1986), esp. pp. 28ff.

modern siegecraft, similarly emphasized the regularity of this military apparatus, claiming that Vauban "established a nearly-infallible routine which was accessible to ordinary mortals who were willing to take the trouble to become versed in it."[6] John Lynn's most recent work is only the latest to highlight the "Newtonian, systematized simplicity" of Vauban's attack.[7]

For military historians, the clearest proof of siegecraft's scientific nature is its predictability, one of the hallmarks of modern science. With Vauban's geometrically-aligned trenches and fields of fire, the outcome of a siege could purportedly be known in advance, just as the result of a scientific experiment could be derived from observation of past experiments or the movement of a clock hand. Duffy wrote that "if the besiegers made a serious trench attack on the fortress. . . . the progress of the rest of the siege could be predicted with reasonable confidence."[8] Other commentators have applied the clockwork metaphor more literally, arguing that siegeworks in themselves became comparable to a timepiece. In this interpretation, not only could the day of a town's fall be calculated, but some even assert that "Vauban claimed to be able to predict [a siege's] course on a daily timetable."[9] His timetable, in this view, served as a diagnostic instrument that could calculate a siege's length, even its day-to-day progress, with exact precision. It promised mechanical predictability in the most unpredictable of human endeavors, combat.

[6] Christopher Duffy, *The Fortress in the Age of Vauban and Frederick the Great 1660–1789* (London, 1985), p. 96. For other examples, see M.S. Anderson, *War and Society in Europe of the Old Regime 1618–1789* (New York, 1988), p. 88; Jean-Pierre Bois, "Armes, tactiques et batailles d'Azincourt à Fontenoy" in *Armes et Alliances en Europe* (Université de Nantes, 1992), p. 50; and Bois, *Maurice de Saxe* (Paris, 1992), pp. 222–226, where it is noted that de Saxe saw little room for improvement in Vauban's attack.

[7] John Lynn, *Battle: a history of combat and culture* (Boulder, Colorado, 2003), p. 119.

[8] Christopher Duffy, *The Military Experience in the Age of Reason* (New York, 1988), p. 294; see also H.T. Dickinson, "The Richards Brothers: Exponents of the Military Arts of Vauban," *The Journal of the Society for Army Historical Research* 46.186 (1986): 77.

[9] Frank Tallett, *War and Society in Early Modern Europe, 1495–1715* (London, 1992), p. 51; see also John Lynn, "Vauban" in Marsha and Linda Frey (eds.), *The Treaties of the War of the Spanish Succession: an historical and critical dictionary* (Westport, CT, 1995), p. 459; and Donald Neill, "Ancestral Voices: The Influence of the Ancients on the Military Thought of the Seventeenth and Eighteenth Centuries," *Journal of Military History* 62 (1998): 509.

Sieges: An Unpredictable Instrument

In reality, Vauban's siege instrument was far from predictable. Just as introductory physics courses ignore important real-world complications such as air resistance and speak of frictionless pulleys and massless ropes, so too do military historians describe an idealized, even stereotyped, siege attack. However, unlike modern physics, the expert literature on early modern siegecraft has remained at this level of generality, content to emulate the pedagogical structure used by contemporary siege manuals.[10] Instead of analyzing the details of actual sieges on their own terms, historians remain content to accept Vauban's theoretical siege as real and buttress this claim with isolated anecdotes from various sieges.[11] The result of this historiographical practice: the centrality of Vauban's perfect system is assumed *a priori* rather than tested. A closer reading of Vauban's own writings, as well as an analysis of the conduct of his mature attack at sieges directed in his last decades of life, provide a far different view of these 'clockwork' sieges. Upon closer examination, we find that the Vaubanian siege described above is only a theoretical construct, its implementation deviating dramatically from the predicted regularity.

In one respect at least, the stereotype of a predictable siege accurately reflects the reality of the period—by the end of Vauban's life fortresses consistently fell to the besieging force. The contemporary adage that Vauban took every fortress he besieged holds true for many of his contemporaries as well, and highlights a pronounced

[10] Within the past century only a handful of sieges have merited serious study, and even fewer of these case studies have placed their topic in the broader context of siege warfare. For these few case studies, see Roger Rapaille, *Le siège de Mons par Louis XIV en 1691: Etude du siège d'une ville des Pays-Bas pendant la guerre de la Ligue d'Augsbourg* (Mons, 1992); Françoise and Philippe Jacquet-Ladrier, *Assiégeants et assiégés au coeur de l'Europe, Namur 1688–1698* (Brussels, 1991); and Olaf van Nimwegen, *'Dien fatalen dag': Het beleg van Bergen op Zoom in 1747* (Gent, 1997). The most comprehensive remains the older Maurice Sautai, *Le siège de la ville et de la citadelle de Lille en 1708* (Lille, 1899).

[11] For examples, see: David Chandler, *The Art of Warfare in the Age of Marlborough* (1976; New York, 1995), pp. 240ff; John Childs, *The Nine Years' War and the British Army, 1688–1697: the operations in the Low Countries* (Manchester, 1991), pp. 92–96; and Jean-François Pernot, "Vauban, le siège devenu réglé ou l'économie des vies militaires," in J. Jacquart (ed.), *Les malheurs de la guerre I: De la guerre à l'ancienne à la guerre réglée* (Paris, 1996), pp. 256 and 258ff.

[11] Childs, *Nine Years War*, p. 93; John Lynn, *Giant of the Grand Siècle: the French army, 1610–1715* (Cambridge, 1996), p. 571.

imbalance between the attack and defense. Of the 115 sieges conducted in the War of the Spanish Succession (1701–1714), the first conflict to witness Vauban's perfected attack, the attackers were successful 85% of the time. Most garrisons simply could not expect to fight off their assaults without outside assistance and the likelihood of such relief diminished as the seventeenth century wore on. Nevertheless, until a relief effort was attempted and foiled, the ultimate fate of the siege remained in doubt. With hindsight eventual success may have been a statistical near-certainty, but the outcome of any particular siege was still too uncertain during its early phases for anyone to predict the results with much confidence.[12]

An analysis of French sieges conducted by both Vauban and his apprentices illustrates that the regular progression of the approach trenches remained a theoretical ideal, and it is exactly this aspect of the attack that has attracted the clockwork metaphor. The origin of the misconception that Vauban could accurately calculate siege lengths and even stages, it must be said, lies in his own words. In a 1672 treatise written for the Secretary of State for War, François-Michel Le Tellier, the marquis de Louvois, Vauban estimated that a typical siege would require 41 days. He prefaced his calculation by stating that it was only an "instructive calculation and not a hard, fast rule," yet his timetable has been taken much more literally than he ever intended.[13] The context of the schedule itself gives plenty of reasons for caution in accepting his figures as typical. The estimate's specific purpose was to calculate the supplies a garrison would need for a siege, since previous sieges had been "shorter than we might have hoped because we had failed to prepare the fortresses adequately to withstand a long siege."[14] Since his goal was to supply a fortress with enough supplies to preclude a premature surrender, he made a number of significant assumptions. First, he assumed that the fortress would have an adequately-sized garrison and defend itself competently.

[12] The French king reflected the popular wisdom when he noted that the tenacity of a garrison's defense was difficult to measure before their first obstacle, the covered way, was tested. Service Historique de l'Armée de Terre, Archives de Guerre [hereafter AG], série A[1] volume 2215, #148, Louis XIV to the French marshal Villars.

[13] This work was published only after Vauban's death, as *Mémoire pour servir d'instruction dans la conduite des siéges et dans la défense des places* (Leiden, 1740). George Rothrock has translated it into English as *A Manual of Siegecraft and Fortification* (Ann Arbor, 1968), pp. 140–141. In the text, a total of 43 days is stated, but the days for each stage add up to only 41. Quote on p. 140.

[14] Rothrock, *A Manual*, p. 138.

Second, he assumed that the besieger would attack the strongest sector of the defenses, further prolonging the siege and increasing the demands on the garrison's supplies. Third, the hypothetical fortress he based these estimates on was a regular (*i.e.*, symmetrically-shaped), six-bastioned fortress with ditches and well-revetted demi-lunes (masonry-faced fortified 'islands' in the ditch shielding the bastions), a firing step (termed the "covered way") with good palisades, and no other outworks. None of these assumptions could be taken for granted during an actual siege, but they satisfied Vauban's need to plan for the garrison's logistical worst-case scenario. Elsewhere in this same treatise, however, he was quite explicit about expectations of predictable progress:

> You cannot estimate how long it will take to arrive at the outer edge of the ditch; it depends upon the distance from the point where the trench was begun, the vigor of the garrison, the nature of the terrain, the availability of materials, and the availability of good workmen. I have seen some sieges that advanced steadily at the same speed and others where you could not make fifty paces in a night once you were close to the fortress; there were even times in the siege of Montmédy when we could not make one hundred and twenty yards in a week.[15]

His end-of-life efforts to codify his techniques follow the same formula almost verbatim, with the exception of adding another week to the expected siege length.[16] However, Vauban never intended these oft-cited estimates as prescriptive rules to be followed blindly by the besiegers.

In the real world, Vauban also harbored little doubt about the unpredictable nature of siegecraft: expectations of clockwork attacks were asking the impossible. We find an example in his response to Louvois' repeated requests for an estimate of how much longer the 1684 siege of Luxembourg would last. Six days into the siege, he repeated his theoretical caveats:

> It is not possible to tell you when we will lodge ourselves on the grand counterscarp, because this depends on the difficulties and quality of the parallel and the resistance of the redoubts that we are attacking; when I see the day that I can predict the future with some appearance of certainty, I will be sure to tell you; but I am not an astrologer. There

[15] Rothrock, *A Manual*, p. 63.

[16] Vauban, *Oeuvres de M. de Vauban* (Amsterdam, 1771), II:63–65. Faucherre and Prost mention this 48-day figure in *Le triomphe de la méthode* (note 5), p. 53.

> are some events of which God alone knows the outcome and its time-frame; it is up to men to do all that they can to succeed, as I am doing, God willing.[17]

It was only twelve days later on 26 May that he felt confident enough to inform Louvois that "the time when this place will be captured is not something that a man of good sense would dare to guess at; but as far as I can conjecture, I expect that it will last seven or eight days longer, give or take two days."[18] The town fell eight days later on 3 April, within the rather wide latitude of five to ten days (in a 26 day siege) he had given himself. Here we see the master himself refusing to attempt the impossible, putting off the Secretary of State for War's demands for prognostication. When Vauban did finally provide an estimate, he refused to pin down a narrow date range. In both the study and the siege camp, Vauban recognized the less-than perfect correspondence between theory and practice.

Vauban was wise to demur. The difficulty of forecasting events is illustrated by the great engineer's 1691 estimate of how long the recently-conquered fortress of Mons might hold out in a future siege. Not only was he unable to predict which particular sector of the fortifications the Allies would attack in 1709, but his estimate of 78 days was far longer than the 30 days the attack actually required.[19] Successive French engineers, lacking Vauban's status, were less fortunate since impatient commanders forced them to predict the unknowable on almost a daily basis—demanding much greater precision, it should be noted, than Vauban had allowed even himself. Even after such lesser engineers had reconnoitered the works and opened the trenches, their predictions were often still little more than guesses, as is indicated by the frequency with which they had to revise them.[20]

---

[17] Vauban to Louvois, Luxembourg, 14 May 1684 [Albert de Rochas d'Aiglun (ed.), *Vauban, sa famille et ses écrits, ses oisivetés et sa correspondance* (Paris, 1910), II:234–235]. On the Louvois-Vauban relationship, one grounded in frankness and mutual respect, see Blanchard, *Vauban* (note 1), pp. 136, 286; Hebbert and Rothrock, *Soldier of France* (note 1), pp. 29–30; and Lynn, *Giant of the Grand Siècle* (note 11), p. 560.

[18] Vauban to Louvois, Luxembourg, 26 May 1684 [Rochas d'Aiglun, *Vauban*, II:239].

[19] Bruno Van Mol, "Supputation par Vauban de la Durée apparente d'un Futur Siège de Mons," in *Cercle Archéologique de Mons, Actes du Colloque du 16 Mars 1991 sur le Tricentenaire du Siège de Mons par Louis XIV* (Mons, 1992), pp. 173–174. Van Mol states that the siege of 1709 lasted 45 days. However, contemporary sources indicate that the town was invested on 20 September, the trenches opened five days later, while the town finally capitulated on 20 October.

[20] Sieges of a month or more might require ten or more estimates of when it would fall, each projection giving a different answer.

On the offensive in Flanders for the first time since 1702, the French in 1712 persistently underestimated the resistance of the fortress of Douai by a week or more, much as the Allies had done with their own siege of the town two years earlier. At their next fortified target, Le Quesnoy, their initial calculations overestimated the length of the 16-day defense by a week, up to half of the entire duration of the siege. At Bouchain, the besiegers' early estimates inflated its 10-day resistance by sixty percent. Not surprisingly, as the besiegers' works approached the final defensive obstacles, the engineers usually improved their accuracy to within a day or two—validating Vauban's refusal to predict Luxembourg's fall until late in the siege.[21]

Nor is the purported regularity of sieges reinforced by the unpredictable changes in estimates of when a town would fall. The direction of the misestimates (*i.e.*, whether the defense was being overestimated or underestimated) might remain relatively constant, but at the least, the projected capitulation date fluctuated wildly from day to day. In the case of Douai, the besieger's first underestimate was off by seven days, a day later the margin of error surged to nineteen days, then fell back in the next to eight days; a week later it was down to three to five days earlier than the town's actual surrender. Another week saw the estimate only one day off, but then four days before the surrender the attackers veered the other way by *over*estimating the garrison's defense by four days. Little surprise that commanders would throw up their hands and admit to their superiors that they had no idea when some of these towns would fall. Even after a town's capture, more time would be required to put the battered fortress into a state of defense before the besiegers could leave—the exact number of days was unknowable until the engineers could view the works from the inside, and here too estimates might have to be revised. Despite a theoretical façade of clock-like regularity, projecting when a siege army would be available for other service was a fickle and erratic business.

[21] For estimates of Bouchain's length, consult Alègre to Maine, 6 October 1712, AG série A[4] Carton 8, folder 2; and AG série A[1] vol. 2386 #31. For Douai, AG A[1] vol. 2382 #7; #30; #132; Alègre to Maine, 2 September 1712, AG A[4] Carton 8 folder 2; and Alègre to Maine, 4 September 1712. On Le Quesnoy: Alègre to Maine, 12 September 1712, AG A[4] Carton 8 folder 2, as well as 18 and 23 September; and AG A[1] vol. 2385 #2.

War as Science?

Why, then, were engineers—including the great Vauban—unable to predict the progress of their sieges with the accuracy expected of them? The mechanical clock metaphor, as flawed as it is, provides an answer if it is extended further. The gulf between the theory and the reality of the clockwork siege is best explained by expanding the analogy to encompass another instrumental scientific metaphor commonly used in warfare, Carl von Clausewitz's concept of friction.[22] The renowned author of *On War* described his understanding of military friction as follows:

> Everything in war is very simple, but the simplest thing is difficult. The difficulties accumulate and end by producing a kind of friction. . . . Friction is the only concept that more or less corresponds to the factors that distinguish real war from war on paper. The military machine—the army and everything relating to it—is basically very simple and therefore seems easy to manage. But we should bear in mind that none of its components is of one piece: each part is composed of individuals, every one of whom retains his potential of friction.[23]

A closer look at Clausewitz's use of friction and the many points of friction in the trenches will correct the overly predictable view of the clockwork siege.[24]

Clausewitz's choice of mechanical friction is superficially at odds with his reaction against earlier Enlightened efforts to develop mechanistic, scientific, and hence universalistic rules of war.[25] He insisted

---

[22] Carl von Clausewitz, *On War*, ed. Michael Howard and Peter Paret (Princeton, 1984). Barry D. Watts, *Clausewitzian Friction and Future War* (Washington, D.C., 1996) discusses friction in chapter 2, available online at <www.ndu.edu/McNair/mcnair52/m52cont.html>. The topic is more briefly covered in Azar Gat, *The Origins of Military Thought: from the Enlightenment to Clausewitz* (Oxford, 1989), pp. 185–186 and in a nonlinear vein by Alan Beyerchen, "Clausewitz, Nonlinearity and the Unpredictability of War," *International Security* 17.3 (Winter 1992/1993): 75–77.

[23] Clausewitz, *On War*, p. 146.

[24] Clausewitz's first embryonic mention of the concept of "*Friktion*" (in an 1806 letter) dealt more narrowly with the internal issue of conflicting tensions between commanders and the "opinions of others" in the army. By 1812 Clausewitz had expanded this original, limited idea of 'internal' friction to a more overarching concept of 'general friction,' including various forms of 'external' friction such as unpredictable weather, faulty intelligence, the reactions of a thinking enemy, and most important of all, chance. Watts, *Clausewitzian Friction and Future War*, chapter 2.

[25] On Clausewitz as a Military Romantic (and Counter-Enlightenment) thinker, see Gat, *The Origins of Military Thought*, chapters 5 *passim*.

time and again that the "apparatus of scientific formulas and mechanics" could not discern the true essence of war.[26] His use of scientific metaphors is easily reconciled, however, when we recognize that they were primarily used to illustrate how *un*scientific real war was, at least in terms of the scientific models available at the turn of the nineteenth century. Right after introducing the analogy of resistance he clarified its obfuscating properties: "This tremendous friction, which cannot, as in mechanics, be reduced to a few points, is everywhere in contact with chance, and brings about the effects that cannot be measured." Elsewhere he declared that in theory "military action ought to run its course steadily like a wound-up clock. But no matter how savage the nature of war, it is fettered by human weaknesses."[27] With scientific examples drawn from fields as disparate as dynamics, optics, and metallurgy, the Prussian theorist used scientific metaphors as foils—by pointing out the inappropriateness of linear scientific comparisons, he contrasted the certainty of scientific knowledge with the uncertainty of real war.[28]

The choice of friction was therefore a strikingly appropriate response to Enlightenment military thought because it illustrated for Clausewitz what war *was*, rather than what it was not. Clausewitz considered such scientific metaphors productive only insofar as they accepted the organic whole of warfare (*i.e.*, psychological factors inseparable from physical ones) and accepted dynamic change and variable interaction as the only constants.[29] For decades military historians have

[26] Clausewitz, *On War*, p. 146. Gat cites a letter of Clausewitz's from 1809 in which he assaulted the eighteenth century's proclivity to transform warfare "into an artificial machine, in which the moral forces were subordinated to the mechanical" (quoted in Gat, *The Origins of Military Thought*, p. 184).

[27] Clausewitz, *On War*, p. 120 for the non-quantifiable nature of friction and pp. 216–218 on war's failure to follow a clock's steady course. Here, however, Clausewitz finds the clockwork metaphor useful in a more limited sense when explaining why wars do not degenerate into inactivity, and he actually uses an algebraic equation to express this. On p. 115 he argued that this friction itself was elastic, thus making "the degree of its friction exceedingly hard to gauge."

[28] For other scientific metaphors found lacking, see: p. 184 where the metallurgical metaphor of war as alloy fails; p. 113 on the "light of reason" refracting differently in combat; p. 597 (and pp. 220 and 232) on the disproportionality of military effort; p. 77 on the limited predictive power of numerical superiority in determining victory; p. 112 on the *intuitive* rather than mathematical genius of Newton and Euler; and p. 86 on the failure of absolute factors which makes wars resemble a game of cards or commerce rather than science. Beyerchen discusses two other examples in Beyerchen, "Clausewitz, Nonlinearity and the Unpredictability of War" (note 22), pp. 82–84.

[29] Clausewitz, *On War*, p. on 120 he equates waging war to moving in a resistant

embraced Clausewitz's concept of resistance in general terms, concurring that warfare is inherently unpredictable and full of surprises. Nor are recent scholars the first to exempt Vaubanian siegecraft from such wartime friction; Clausewitz himself did much the same. Even though he recognized that methods of war varied according to their political and technological contexts, in his discussions specifically focused on sieges he nonetheless viewed them as matters of purely material objects (*e.g.*, trenchworks and batteries) rather than the "actual conduct of war" dictated by moral forces. Siege warfare was therefore "intellectually uncreative," its success a matter of "mechanically" flattening the defenders with artillery and therefore governed by geometry rather than moral effort. As a result, siege theory was not much more than a "refined mechanical art" that assumed (as did Enlightened tactical theorists) that human beings were automata, "pieces of clockwork set off by a mere word of command."[30] Siege warfare in the real world was, theorists argued, no different from siege warfare on paper.

A Siege's Internal Friction

The assumption of a siege's frictionless operation was predicated on the careful planning and execution of the attacks according to the designs of skilled experts, but several sources of friction steadily eroded the engineers' ability to conduct the siege according to such strict standards. From early in his career Vauban recognized, as we have already seen, the many factors that played havoc with length estimates.[31] His many tactical innovations and his emphasis on planning

element; pp. 221–222 on the use of tension and equilibrium; p. 597 on delay's unequal advantages; pp. 596–597 on the center of gravity; p. 89 on the three magnets metaphor—Beyerchen highlights the nonlinear nature of this example in "Clausewitz, Nonlinearity and the Unpredictability of War," pp. 69–71.

[30] Clausewitz, *On War*, p. 113 on the material products of siegecraft and its limited intellectual effort; p. 393 the mechanical use of artillery; p. 214 for sieges governed by geometry; and pp. 133–134 for automata and clockwork. The conflation of siegecraft with Enlightened battlefield theorists is justified by the fact that eighteenth century writers sought to replicate Vauban's 'predictable' successes in the trenches on the battlefield. Gat, *The Origins of Military Thought* (note 22), p. 35.

[31] At his most 'scientific' and anti-Clausewitzian, Vauban attempted to account for such friction by providing a 'fudge factor.' He included in his estimate of the typical siege's length an extra four days "for the enemy's mistakes and negligence of preparations." Rothrock, *A Manual* (note 13), p. 141.

systematized the siege attack in order to minimize such external forms of friction—most famously the use of trench parallels to neutralize garrison sorties, elevated battery platforms (*cavaliers de tranchée*) to fire down into the covered way, and ricochet fire to strike at targets sheltering behind breastworks. Yet the most significant and persistent threats to the clockwork siege remained internal, 'command and control' friction that manifested itself as disputes over how the siege was to be conducted. At almost every link in the chain of command, many non-engineers were in contact with the siegeworks and provided points of potential friction. Some of these individuals pressed the siege forward while others applied a braking force in the opposite direction; the amount of force applied by each of these sources of friction could also vary throughout the course of a siege. As a result of these countervailing sources of push and pull, a siege's outcome was less certain and its precise progress impossible to predict, much as Newton's new mechanics failed to predict precisely the future location of three or more celestial bodies interacting with one another (the three-body problem), and much as clocks before 1650 lacked the mechanical precision and accuracy we associate with them today.[32] The unpredictable combination of these competing forces necessarily stressed the clockwork mechanism and made a siege's future progress erratic and unpredictable.

The history of the scientific understanding of friction also suggests how our understanding of siegecraft needs to progress. The effects of friction were known several millennia before the Age of Reason, but the first public scientific explorations of the phenomenon only occurred at the *Académie royale des sciences* in 1699.[33] Seventeenth and eighteenth century analyses of friction, culminating in the Amontons-Coulomb law, describe idealized friction as $F = \mu N$, where the degree of friction is dependent not on surface area but on the force of grav-

[32] For the inaccuracy of early clocks, see Silvio Bedini, "The Mechanical Clock and the Scientific Revolution," in Klaus Maurice and Otto Mayr (eds.), *The Clockwork Universe: German clocks and automata, 1550–1650* (Washington, DC, 1980), pp. 21–22.

[33] Leonardo DaVinci's now-famous experiments with friction and inclined planes were not fully revealed until the 1960s. Thus, eighteenth century Europe attributed the first scientific experimentation with friction to the Frenchman Guillaume Amontons. His priority would later be superceded by his compatriot Charles Augustin Coulomb, a member of the French *Corps du Génie* and author of many works on friction and magnetism. For a brief history of the scientific understanding of friction, see Greg Hahner and Nicholas Spencer, "Rubbing and Scrubbing," *Physics Today* 51.9 (1998): 22.

ity exerted on the object.[34] More recent analyses of friction (including microscopic examinations) have complicated the picture immensely, distinguishing different types of friction, reintroducing the importance of surface area and roughness on an atomic level, and making it impossible to generalize about friction in absolute terms. In contrast, the study of siegecraft has remained in its infancy, or perhaps it might be more accurate to say it has been infantilized in this century by the historiographical obsession with field battles, where historians simplistically treat the Vaubanian ideal as real. When we examine the siege attack 'microscopically' by tracing the progress of individual sieges and their conduct, we discover the many areas of contact otherwise ignored by superficial analyses. Friction was in fact a pervasive force that resisted the smooth operation of the clockwork siege.

The Ticking Campaign Clock

The first source of internal friction came from disputes between the chief engineer and those responsible for the overall direction of the war effort. Vauban's control over his sieges was not absolute, for the Sun King's *gloire* was reflected not only in the towns captured by his armies, but ultimately by whether he won his wars or not. Louis made sure to memorialize his capture of enemy fortresses in paintings, tapestries, and print, but he also wanted to dictate the peace (what better measure of *gloire*?) and this often called for pushing sieges faster than his expert would have liked. Vauban sought to conduct the most efficient siege possible, finding the optimal combination of low cost, low casualties and the fewest delays. For politicians and many military commanders, however, time was usually the most important variable by far: with campaign seasons averaging only 180 days out of the year and slow transportation technologies limiting rates of advance, decisive victories that shortened a war (saving both lives and money) were difficult to achieve, and almost impossible when fortresses blocked the way forward and could delay an advancing

[34] The force of friction (F) equaling the normal or perpendicular force (N) of an object multiplied by the coefficient of friction (μ), a value that can only be discovered experimentally. To complicate matters, for any given situation, there are separate coefficients of friction depending on whether the contact is static or sliding (kinetic), the former always being greater than the latter ($\mu_s > \mu_k$), which is why objects are easier to push once you get them moving.

army for weeks if not months, hopefully until bad weather forced the attacker into winter quarters. As a result of these seasonal constraints and with almost every path of advance blocked by fortresses, Louis accepted sieges but preferred that Vauban abandon efficiency for effectiveness: he pressed his chief engineer to push ahead even faster. The King wrote from Versailles of Vauban's model siege of Ath in 1697 that "It does not appear that, [given] the conduct of the governor of Ath, [the town] merits an attack in strict form. This is why I am persuaded that you can without concern go faster than you believe; it is important to finish it promptly and to not lose any time."[35] We also find, during Vauban's final siege of (Alt) Breisach, the King willing the siegeworks forward yet again. Louis' grandson Louis Bourbon, the Duke of Burgundy wrote to the Secretary of State for War Michel Chamillart from the siege: "I have received another letter whereby the King orders me to press the siege; I can assure you that we are doing all that we can and we are even going much faster than M. the *maréchal* Vauban would like."[36] The engineer confirmed these pressures, complaining that:

> It appears that His Majesty is resolutely determined to besiege Fribourg; he even orders Monseigneur [Burgundy] to press the conclusion of this siege; on this point I must tell you that we opened the trenches five or six days before the siege lines were finished against custom and only with a view toward accelerating it, and that it was pushed with so much speed that the batteries did not have enough time to have their effect with all the success we should expect.[37]

Once the French returned to the offensive in Flanders in 1712 without Vauban, they abandoned their defensive strategy of delays and again sought to shorten the lengths of their sieges against the wishes of their engineers.

As the example of Breisach makes clear, delays enraged politicians and commanders because they interfered with future operations, and

[35] Louis to Vauban, Versailles, 22 May 1697 [Rochas d'Aiglun, *Vauban* (note 17), II:453].

[36] Burgundy to Chamillart, Breisach, 3 September 1703 [Alfred Baudrillart and Léon Lecestre (eds.), *Lettres du duc de Bourgogne au roi d'Espagne Philippe V* (Paris, 1912) I:276, no. 166]. See also Chamillart's letter to the *maréchal* Tallard (Versailles, 30 August 1703) recommending he have Burgundy press Vauban to accelerate the siege [Vault and Pelet, *Mémoires militaires relatifs à la succession d'Espagne sous Louis XIV, extraits de la correspondance de la cour et des généraux par le lieutenant général de Vault, directeur du Dépot de la Guerre, mort en 1790* . . . (Paris, 1836–1842), III:442].

[37] Louis Bourbon to Chamillart, Biesheim, 5 September 1703 [Rochas d'Aiglun, *Vauban*, II:527].

herein lay the attractiveness of clockwork sieges—their predictability. The strategic planning needed to coordinate armies and fleets across several theaters had to be decided upon many months in advance, while unforeseen delays might throw grand strategic plans into disarray, throwing off coordination between theaters and allies, perhaps even forcing the cancellation of military operations altogether.[38] The Court, always anxious about when the army could start the campaign, similarly enquired when the army would finish up its current operation so as to be available for other tasks—troop reinforcements might need to be sent to shore up another theater, or quick action might be needed before an approaching enemy corps could prevent the investment of another town or eliminate the possibility of a battle with numerical superiority, or peace negotiations might require a kick-start from a military victory. Belligerents desired predictability regardless of whether they were on the defensive or the offensive—defensive planners needed to know when they could reinforce another position or would face the threat of a field battle (*i.e.*, once the enemy's forces were no longer divided between the besieging and covering armies). Besiegers also needed to know when they could try to force a field battle or whether or not they could invest another fortress without being hindered by a fully-assembled enemy army. The beginning of the 1710 campaign was particularly harried, the Allies beginning the season early with an uncontested investment of the northern French town of Douai in late April. The French command rushed to enter the field before this town surrendered, constantly pestering the garrison commander about his expectations for the expected length of the defense. Anticipated to last only a few weeks, the fortress's unexpectedly stubborn defense gave them an extra month to rush their army into the field and consume all the fodder around Arras before the enemy could capture Douai and move on to attack Arras.[39] Allied generals were furious that their

---

[38] Scholars have examined the tyranny of distance's effect on early modern governments in a number of works, most memorably in Fernand Braudel's *The Structures of Everyday Life: the limits of the possible*, trans. Siân Reynolds (Berkeley, 1992), pp. 415ff. In early modern European military history, the best recent example is Geoffrey Parker, *The Grand Strategy of Philip II* (New Haven, 1998), chapter 2.

[39] For a sampling of the French side's race against the clock, see: Sailly, 16 June 1710, AG A[1] vol. 2220, #123; Villars to Louis, Peronne, 16 May 1710, vol. 2215 #64; Albergotti, Douai, 7 June 1710, vol. 2225 #223; Bernières to Voysin, 16 June 1710, vol. 2225 #233; Voysin to Villars, 21 June 1710, vol. 2297–2 §2 #32.

plans had been upended, the benefits of their unopposed campaign opening squandered. A quick capture might have led them quickly on to Arras and would undoubtedly have influenced the peace negotiations currently underway at Geertruijdenberg. Instead they had to scramble to find fodder to finish off Douai since their forage magazines would run out a week before the town would fall.[40] As 1710 suggests, both sides had to continuously recalibrate their expectations and plans according to the varied pace of sieges. Hence we see the perpetual discussion of siege timetables and the demands for predictions of when the current siege would finally be finished—strategists were impatient to know when they could set future operations in motion.[41]

Such capitulation estimates were particularly important if one hoped to squeeze in yet another late-season siege, for the supplies needed to capture a town required weeks to collect and transport and it would be wasteful to pay for these preparations and then have to defer the siege due to inclement weather. To this preparation time planners had to add an estimate of how long it would take to capture the next target. The results of such guesswork helped them decide whether a future siege was even practical or not—could a besieging army capture the next fortress before bad weather forced it into winter quarters? This was a difficult calculation in itself, since the length of a future siege and the date of winter quarters could both deviate significantly in either direction from expectations. We have already seen the difficulty siege participants had predicting when an ongoing siege would end, so it is not hard to imagine the difficulties of predicting the resistance of a fortress that had not even been invested yet.[42]

---

[40] For an example of Marlborough's frustrations over the lost opportunity, see Marlborough to Godolphin in Henry Leonard Snyder (ed.), *The Marlborough-Godolphin Correspondence* (Oxford, 1975), III:1490, no. 1528. The length of the Douai siege had also forced the Allies to consume all the dry fodder in their magazines, thus making the grass surrounding Arras critical to any siege.

[41] The 1710 siege of Douai is also another telling example of the unpredictability of siege lengths: most Allied participants expected the town to fall within two weeks, but it defended itself for 63 days (52 of open trenches) and more than 25 length estimates were made over the course of these two months. Adding even more confusion, the garrison commander at first alarmed Versailles by promising no more than a brief resistance; only later did he regain his composure and provide the Court with more accurate predictions. See for example Voysin to Montesquiou, 15 May 1710, AG A[1] vol. 2215, #60.

[42] Vauban emphasized the importance of carefully reconnoitering an invested fortress, to make sure that the reality of its works confirmed to the expectations derived from previous intelligence and public-domain maps.

The end of the campaign was slightly more predictable, though a margin of error of a month was still prudent. In Flanders during the War of the Spanish Succession campaigns usually ended between early October and late November, though an exceptional campaign could extend through December. Given all the fluctuating variables in warfare, contemporaries clung to any certainty that might make the future more predictable.

'Command and Control' Friction

The genesis of Clausewitz's idea of friction came from the struggles he witnessed during the Prussian campaign against Napoleon in 1806, when the army had to answer to three different commanders at the same time, a situation further complicated by the advice they received from the Prussian general staff back in the capital. Though such divided commands were familiar to Louis XIV's reign, a more fundamental structural weakness made command particularly convoluted in the case of sieges. Vauban's ability to direct sieges derived not from his position as head of the engineers, but from the support he received from the king and his ministers.[43] With his rapid sieges in the War of Devolution (1667–1668), he gained Louis' trust, and with it, the command authority to implement his siege attacks. This could not last. In the latter part of his reign, Louis gave up campaigning, losing this immediate contact with a newer generation of engineers.[44] This took a toll not only on the authority of the engineers at sieges, but also resulted in a more general loss of the prestige previously associated with the engineering corps when it had been directly under the king's gaze.[45] Without Vauban present in person and without the king's active attention, the engineers lost much of the command authority at sieges that they had enjoyed previously.

[43] Charles Albert Samuel Lecomte, "Du service des ingénieurs militaires en France pendant le règne de Louis XIV," *Revue du génie militaire* 25 (1877): 122. John B. Wolf, *Louis XIV* (New York, 1968), pp. 232, 531–532. See also John Lynn, *The Wars of Louis XIV, 1667–1714* (New York, 1999), p. 120. Vauban did not even hold the top administrative position of *directeur-général des fortifications*, a post held first by the Secretary of War Louvois and then, after Louvois' death in 1691, by Michel Le Peletier de Souzy.

[44] Charles Albert Samuel Lecomte, *Les ingénieurs militaires en France pendant le règne de Louis XIV* (Paris, 1904), pp. 122–123.

[45] Wolf, *Louis XIV*, p. 199 notes that already at the 1688 siege of Philippsbourg, only Louis' royal presence ensured that Vauban's recommendations were followed.

The loss of Louis' personal endorsement and the eventual disappearance of Vauban would not have been so dangerous to the clockwork siege if the engineering corps had been integrated into the army chain of command. Unfortunately, assuring that an engineer's precise directions were followed was often difficult because engineering posts were not the same as regular army commissions, thus these technical experts served in a strictly advisory role even in their *métier* of siegecraft.[46] The engineering ranks of director-general, chief engineer, director of approaches, first, second, and third engineers, down to the extraordinary engineers were all outside of the normal military chain of command (infantry or cavalry). As a result, some held an engineer brevet at the same time as they held a commission in the foot, while the lower ranking pluralists among them might at times be expected to perform the duties required of both branches, even if these were in different theaters![47] The distinction between the services was even observed at the highest level of military service. As Vauban himself admitted, the baton symbolizing his rank as one of the elite *maréchaux de France* was different from his peers' since his expertise and experience involved the design of fortifications and the conduct of sieges, rather than the maneuvering of armies in the field or on the battle plain.[48] Unused to taking orders from engineers in the field, many field officers were hesitant to blindly accept their recommendations even in sieges. As a result, variable pressure from Court interacted with pressure from those officers in the trenches.

In this context, Vauban's reputation, as impressive as it appears today, did not impress the most renowned general officers of his day. Vauban may have had the king's ear, but such influence did not convince French generals to blindly implement the projects he sent them. The commanders had many opportunities to push the progress of sieges, for the overall responsibility for a town's capture was entrusted not to the chief engineer, but to the highest-ranking general (usually of the infantry).[49] Ideally the commander would solicit advice from the engineers on the attacks and follow their recom-

[46] Lecomte, *Les ingénieurs militaires*, pp. 113–118; Chandler, *The Art of Warfare in the Age of Marlborough* (note 11), pp. 219–220; Anne Blanchard, *Les ingénieurs du "Roy" de Louis XIV à Louis XVI: Etude du corps des fortifications* (Montpellier, 1979), p. 292.

[47] Blanchard, *Les ingénieurs du Roy*, pp. 106–108.

[48] Vauban to Puyzieulx, Saint-Malo [Hyrvoix de Landosle, *Vauban: Lettres intimes (inédites) addressées au marquis de Puyzieulx (1699–1705)* (Paris, 1924), p. 83].

[49] Lecomte, "Du service des ingénieurs militaires" (note 43), pp. 27ff.

mendations, but the decisions were ultimately his to make. The subordination of the chief engineers to the commanding general was taken for granted by Vauban. His first treatise made the engineer's consultative role quite clear:

> [The engineer] should sketch the design; he should inspect the camp with the general and show him the layout he has made and explain his reasons for it. Since the general must give the necessary orders, the engineer should show him alternative plans of the works, explaining the strengths and weaknesses of each and offering estimates of the construction time needed in each case.[50]

Several pages earlier he had highlighted the necessity of reconciling the engineer's plan with the means at hand, explaining how they needed to rely on the lieutenant-generals for accurate accounts of the men and resources available for the siege.[51] Expertise was not the same as authority: from the very beginning of a siege the engineers depended on the general officers for command authority.

Making contact with the siege machine at several points, army commanders applied their own pressures, usually laboring to push the clockwork faster than the engineers intended. Such pressures rarely compromised the overall success of the siege, but they did make the instrument's precise progress uncertain. One of the most important decisions in a siege was determining where to attack a town, and here the engineer's consultative role is quite evident. Fortification manuals might create for pedagogical purposes an imaginary regular fortress with each side sporting the same defenses in a flat plain, but few fortresses ever reached this ideal. An engineer's detailed understanding of the strengths and weaknesses of fortifications was needed to weigh the pros and cons of each front, but early in Louis' reign Vauban acknowledged that, ultimately, it was the commanding general who decided where to open the trenches.[52] In Germany in 1703, the French general Louis-Hector, duc de Villars ignored Vauban's proposal when attacking Fort Kehl and succeeded, against the advice of the majority of the engineers present.[53] Encamped

---

[50] Rothrock, *A Manual* (note 13), p. 29; also 27.
[51] *Ibid.*, p. 23.
[52] *Ibid.*, p. 39.
[53] Melchior, marquis de Vogüé (ed.), *Mémoires du Maréchal de Villars publié d'après le manuscrit original* (Paris, 1889), II:49–61. Villars reproduces his justificatory letter to Chamillart in II:67. In the secondary literature, see François Ziegler, *Villars: Le centurion de Louis XIV* (Paris, 1996), pp. 107–108.

before Landau later that year, Camille d'Hostun, *maréchal* Tallard chose an artillery officer's proposal over that of Vauban and was similarly rewarded with success.[54] At Nice in 1705, yet another French general, James Fitz-James Stuart, Duke of Berwick, also refused Vauban's siege plan and went on to capture the town by a different approach.[55] Pressured to accept Vauban's recommendations, head-strong Louis d'Aubusson, duc de La Feuillade, commander in Italy and son-in-law to Chamillart, made a point of reminding his father-in-law that Vauban considered Nice unassailable where Berwick attacked it, but the town only held out for 27 days.[56] The blue-blood continued to insist on his independence when besieging Turin: he flatly rejected Vauban's proposals and refused to relent even when the great engineer publicly criticized its conduct.[57] Berwick justified their explicit rejection of such 'meddling' by arguing that "those who see from close up are to be believed over those who see from far away."[58]

Nonetheless, commanders usually followed their experts' advice on the approaches to make. Commanders and generals often acquiesced only begrudgingly, though, for they often complained of their engineers' choices at the same time as they allowed them to determine the attacks. When Villars was forced to pick between two projected attacks, he settled on his least-favorite choice (proposed by the chief engineer) in order to avoid potential obstructionism: "if I were of a different opinion from [the engineers], I would still go along with their ideas anyway because it is too dangerous to make them do something against their wishes. . . . [Brigadier of infantry Erasme] Contades says that the engineers are like wet nurses [*nourrices*] which cannot be denied anything they want."[59] Later, he had to quell

[54] Hebbert and Rothrock, *Soldier of France* (note 1), p. 204; Pierre Alexandre Joseph Allent, *Histoire du corps impérial du génie . . . depuis l'origine de la fortification moderne jusqu'à la fin du règne de Louis XIV* (Paris, 1805), pp. 418–431.

[55] Berwick to Vauban's superior, the *directeur-général des fortifications* Michel Le Peletier de Souzy [Rochas d'Aiglun (ed.), *Vauban* (note 17), II:566]; also Jacques Fitz-James Stewart, duc de Berwick, *Mémoires du Maréchal de Berwick écrits par lui-même* (Switzerland, 1778), I:190 and 194. In fact, the chief engineer and head gunner at the siege both agreed with Berwick's attack: see the engineer Paratte to Chamillart, Villefranche [Gustave René Esnault (ed.), *Chamillart, Correspondance et papiers inédits recueillis et publiés par l'abbé G. Esnault* (Geneva, 1970), II: 55].

[56] Louis d'Aubusson, duc de La Feuillade to Chamillart, Casale [Esnault (ed.), *Chamillart*, II:91, no. 189].

[57] Vauban to Chamillart, Dunkirk [Rochas d'Aiglun (ed.), *Vauban* (note 17), II:582].

[58] Berwick to Le Peletier de Souzy [*Ibid.*, II:566].

[59] Villars to Voysin, AG A[1] vol. 2384, #1. See also, Jean-Robert LeFebvre d'Orval to Voysin, AG A[1] vol. 2382, #101.

rumors that they were attacking Le Quesnoy at its strongest side rather than its weakest, and reaffirmed his trust in the engineers on their choice of approach.[60] In 1713 at Freiburg near the Rhine, competing projects surfaced once again. Villars personally agreed with one sieur de la Battue, who had previously commanded in the château, but he allowed the chief engineer Charles-Guy Valory to conduct his attack against a different sector nonetheless: "Thus it is that one is often forced to give in to the reasoning of those who are directly charged with the attack, because if you force them to do otherwise [*quand on leur fait violence*], they are more than happy to make sure that nothing succeeds."[61] Villars' language evokes the image of commanders trying to push the clockwork siege as fast as possible without breaking the mechanism itself.

If commanders and generals were rarely willing to overrule the choice of approaches, they soon enough tired of the measured pace of the Vaubanian siege machine and forced the clockwork to run faster than intended once the trenches had been opened. These impatient officers resorted to a variety of techniques, believing that they would save precious weeks of the campaign season otherwise absorbed by less-than-decisive sieges. Vauban's advice for siege commanders suggests that he was already well aware of the friction their presence would engender:

> it is very important that the general-in-chief visits the trenches, but only occasionally and not every day; because his visits necessarily being long, they will cause too many distractions and delay the siege. . . . He only needs to visit from time to time with only a small retinue, to be personally informed of what is being done[62]

Vauban pleaded for a hands-off commander who implemented the chief engineer's plan of attack, yet commanders, Villars in particular, felt confident enough to actively intervene in the details of siegecraft in order to overcome the 'lethargy' of their engineers. In his memoirs, the impatient Villars used every opportunity to highlight the contrast between the correctness of his own decisive judgment and

---

[60] Villars to Voysin, AG A[1] vol. 2384, #106. The language used in his memoirs made sure to indicate the central role he played in deciding on the attacks with the engineers. Vogüé (ed.), *Mémoires du Maréchal de Villars* (note 53), III:192 and 217.

[61] *Ibid.*, III:217.

[62] Sébastien le Prestre de Vauban, *De l'attaque et de la défense des places*, edited by Antoine-Marie Augoyat (Paris, 1828), p. 234 [hereafter: Vauban, *Traité de l'attaque*].

the overly cautious attitude of his engineers.[63] Villars wrote of pressing Valory and the Florentine lieutenant-general François-Zénobie-Philippe, comte de Albergotti (who had defended the town during its 1710 siege) to accelerate their attack at Douai in 1712, accepting their excuses only begrudgingly.[64] Recalling his successful rebellion against Vauban's advice at the siege of Fort Kehl back in 1703, he notified the new Secretary of State for War, Daniel-François Voysin that as the attack on Douai was going slowly, he would give them two more days and then "conduct the siege according to my taste."[65] Later in life he recalled Valory's expectation of a fifty-day defense: "That was not my calculation, and I was used to leading the engineers a little bit faster than was their rule."[66] The cavalry brigadier François-Marie comte de Broglie, from his perspective, also considered their attacks far too cautious given the undersized garrison.[67] The town quickly fell and a detachment was immediately sent to invest the next target, Le Quesnoy. With the engineers still deemed "out of practice," Villars interjected himself yet again: "As we have not conducted sieges in a long time, my involvement was necessary to speed its pace."[68] Made acutely aware of the time constraints, Valory promised to quicken his attack to please both commander and Court.[69] The 1713 campaign in Germany saw the *maréchal* exercizing his freedom of action yet again. Describing his own behavior in the third person, he reiterated his impatience with the measured rhythm of the siege machine:

> He [Villars] forgot nothing in order to accelerate the preparations for the siege of Landau. The place was excellent, defended by a very strong garrison composed of the best Imperial troops. Thus people who like precautions had good reasons to object to the briskness of the marshal Villars, but as he followed the principal that vivacity is almost always necessary when it is not done carelessly, he only considered those precautions that were absolutely necessary.

[63] See, for example, a reproach of Albergotti on the capture of Fort Manheim in Vogüé (ed.), *Mémoires du Maréchal de Villars*, III:194.
[64] Villars to Voysin, Douai, AG A[1] vol. 2382, #148.
[65] Villars to Voysin, AG A[1] vol. 2382, #88. He repeated this threat in #124.
[66] In Anquetil, *Suite des mémoires rédigée par Anquetil* (Paris, 1828), II:383.
[67] Broglie to Voysin, Douai, AG A[1] vol. 2382, #196.
[68] Anquetil, *Suite des mémoires*, II:389.
[69] Valory, AG A[1] vol. 2384, #22.

After nine days of open trenches, he forced the commanding general to accelerate the siege, as not even an isolated redoubt well beyond the outworks had been captured. Lecturing the chief engineer, he recalled admonishing Valory: "one must not underestimate the enemy nor overestimate their strength, and, in judging its defense, even the enemy's greatest courage and skill is not enough to deter us from going faster."[70] Villars was only the most successful of the many generals who refused to watch the mechanism of the clockwork siege methodically unwind at its theoretically-prescribed pace.

The commanding general was not the only source of friction in the trenches, for lieutenant-generals provided just as much resistance to an orderly attack.[71] Once the chief engineer and his associates had drawn up the plan of approach and had it approved by the commander, its implementation would be entrusted to each attack's commander (usually the next-most senior general), each of whom would have a director of the approaches as his engineer assistant. In Vauban's later work he enumerated how the lieutenant-generals and their subordinates were to post the troops, regulate the detachments, supervise their service in the trenches, and furnish the necessary number of workmen.[72] Daily siege dispositions drawn up by the engineers illustrate this dependence, as they provided the infantry generals with lists of the number and types of soldiers to be provided when and where and for what service. The details of each day's work in the trenches would be overseen by the general of the day (usually a lieutenant-general), seconded by their majors of the trench (*majors de la tranchée*). The workmen and guards in the trenches rotated daily and were drawn from the siege army regiments, commanded by their regimental officers. A handful of engineers were charged with marking out the locations of the trenches and batteries, and supervising the construction of the breastworks.

Vauban's repeated emphasis on the need for generals to follow the engineers' plans, however, shows that they created their own friction as well. He had early on recognized how the noble officers' independent mindset exerted variable amounts of force on the siege mechanism:

---

[70] Vogüé (ed.), *Mémoires du Maréchal de Villars*, III:190–191 on Villars' vivacity; III:198 for his lecture to Valory.

[71] On the long tradition lieutenant-generals had of challenged orders from superiors, see Lynn, *Giant of the Grand Siècle* (note 11), p. 296.

[72] Vauban, *Traité de l'attaque*, p. 233.

> On their own authority they direct the line of the trenches as they please, breaking the design at every turn and negating all the precautions that the engineer has taken. Then, far from being able to observe an orderly conduct which might lead to a successful conclusion, the engineer finds himself forced to serve as an instrument of their various caprices. I say various because of the practice of alternating commanders, so that one will command one day in one fashion and another who relieves him tomorrow in another; as they are not always endowed with the greatest capacity for this sort of thing, God only knows the omissions and the wasteful expenditures they occasion and the amount of blood they spill needlessly by slowing the siege.[73]

With lieutenant-generals rotating service every day, each general complicated the smooth running of the siege instrument. The engineer's repeated appeal at the end of his life reflects this inevitable source of internal friction and the confusion it could cause:

> The lieutenant-general of the day commands the cavalry, infantry and artillery, engineers and miners, and generally everything related to the security and progress of the attacks; but he must consult with the director of the trenches, and neither undertake nor decide to do anything without the engineer's participation; because this last is the soul and the prime mover [*la véritable mobile*] of the attacks.[74]

Flouting the plan of attack, general officers remained yet another source of resistance to the clockwork siege.

Internal friction did not end once the engineers cajoled the generals into following the course of the trenches as they had been laid out. One of the most time-consuming stages of a Vaubanian siege was the slow trench advance towards the fortifications—much of this delay could be avoided if the besiegers could jump forward to the next phase, the capture of the covered way. This obstacle was composed of a gently sloping *glacis*, a zone where interlocking fields of the garrison's fire converged against any attacker brave enough to charge up it, crowned with a double row of palisades at the crest of the covered way (or counterscarp). Once this work was in their control, the besiegers could establish their breaching batteries on its crest and begin the breaching task, firing directly at the base of the outwork or curtain wall. As Vauban had identified in his early treatise, capturing the covered way was the most important tactical challenge,

[73] Rothrock, *A Manual* (note 13), pp. 167–168.

[74] Vauban, *Traité de l'attaque*, p. 232. This is one of Vauban's few explicit references to a scientific metaphor in his siegecraft *oeuvre*.

since it usually accounted for three-quarters of a besieger's overall casualties.[75] Over his lifetime he had perfected a number of tactical techniques to improve the odds of seizing this work, making this task both easier and safer. The industrious use of trenches, elevated firing platforms, and bouncing ricochet fire was now to replace the brute force of a frontal assault whenever possible.[76] Here, where Vauban was most methodical, is exactly where he was most strenuously resisted.

Consistently, siege commanders chose the quickest, most direct route—foregoing the recommendation of Vauban and successive chief engineers to methodically sap up the glacis. Instead, they chose frontal assault, sending a 'forlorn hope' of grenadiers charging up the glacis, past the shattered remains of palisades into the covered way itself. Scores of fusiliers would follow, as well as a group of workers directed by an engineer to construct a secure lodgment. Additional waves of reserve troops would be thrown into the fray as needed, in all several thousand men dedicated to establishing posts on the work. Once the besiegers had constructed such lodgments (usually at the salient angles), they could then begin the task of sapping to the right and left, clearing the branches of the covered way and starting descents underneath it towards the ditch. Against major fortresses, the casualties resulting from this massive effort numbered a thousand men or more, yet in siege after siege the commanders chose this bloody option rather than accept the several-day delay that was required for the slower process of advancing *à la sap*.

Villars was particularly captivated with shock tactics, using the technique again and again in the 1712 Flanders campaign. The *ingénieur en chef* Valory resisted pressures for a premature storm of Douai, but the place was finally stormed on 7 September and the garrison asked to negotiate surrender terms the next day.[77] Bouchain and Le Quesnoy quickly followed suit. Nonetheless, Villars remained skeptical of his engineers and in his memoirs he reminisced that the next year at Landau he intended to break with convention:

> The sentiments of the *maréchal* Villars, in the first days of the siege, had been to attack the covered way from a little further away than is normally practiced. The reason was that if this covered way completely

[75] Rothrock, *A Manual*, p. 65.
[76] Vauban, *Traité de l'attaque*, pp. 130, 263, maxim XIV.
[77] Valory to Villars, Douai, AG A[1] vol. 2382, #89.

> countermined, it was certain that the very wet terrain would force the enemies to charge the mines three days before they expected it to be attacked. The engineers opposed this plan, thinking it rash and too perilous. However, the rest of the siege would show that beyond the loss of time, which is very precious in wartime because thirty days of open trenches were needed to capture the detached outworks, the loss of men was also very considerable during this period of time which a premature attack would not have cost, and it was recognized by the end that the *maréchal* Villars' opinions were proved correct.[78]

Again, rather than acknowledge the necessity of the sap's delays, Villars was willing to risk an assault on a covered way only briefly prepared by artillery fire.

Reining in the impulses of assertive commanders was not enough to end command friction, however, for insubordinate lieutenant-generals could, in rare cases, break the siege machine entirely on their own by forcing it forward too abruptly. One of the most (in)famous examples of such recklessness is the covered way storm on the Italian fortress of Coni in 1691. This unprepared assault was repulsed with heavy losses, and in a panic the elderly French siege commander abandoned the siege. It was rumored that Lieutenant-General Antoine de Pas, the marquis de Feuquières, was responsible for this unauthorized storm and it is perhaps an indication of his reflection on Coni's failure that his later memoirs include the conviction that Vauban's method of capturing the counterscarp by *industrie* rather than brute force was "certainly [the] best, and most effectual [method], at the same time that it is least fatal to the Men."[79] Such theoretical pronouncements did not end the practice though, as Joseph Sevin, the chevalier de Quincy, reminded his readers that this mindset remained widespread even after Vauban's forty-year tenure. After describing a similar failed attack at Freiburg in 1713, he concluded:

> It will not suffice to leave to the desires and caprice of the general officers of the day in the trenches to attack the covered way or other works that we besiege; because we can say, to the credit [*à la louange*] of the majority of these Messieurs (we have already remarked on it far too often), that they are extremely ignorant on the day they find

[78] Vogüé (ed.), *Mémoires du Maréchal de Villars* (note 53), III:199.

[79] On the siege of Coni, see Camille Rousset, *Histoire de Louvois et de son administration politique et militaire* (Paris, 1864), 4:492–497. For Feuquières' later disavowal of the technique he is accused of using, see: *Memoirs Historical and Military: Containing a Distinct View of all the considerable states of Europe* . . . (London, 1736), II:286 and 289.

> themselves in the trenches. . . . To remedy this abuse, it will be necessary that the chief engineer be called by the generals of the trenches to determine if the work that is to be attacked is indeed ready to be assaulted. Certainly, this precaution would save many men and we would no longer make such stupid mistakes [*et on ne feroit point de cacade*].[80]

With impetuous commanders and generals at the helm, pushing and overruling their engineers, it is no wonder that sieges failed to be as predictable as clockwork.

The engineers' lack of command authority also allowed sources of resistance to retard a siege's steady progress as much as accelerate it. Engineers were just as incapable of controlling this backwards momentum as they were to resist the forward impetus of their superior officers. While generals tended to push the siege machine faster than the engineers expected, the erratic behavior of artillery officers and the workmen in the trenches worked in the opposite direction.

Just as the engineers were an independent branch of the army, so too was the artillery.[81] With a separate chain-of-command and headed by the *grand-maître d'artillerie*, this autonomous arm guarded its privileges as closely as did the other branches.[82] Its mission also differed from the engineers and thus they held divergent interests. French artillery schools established under Louis included an extensive curriculum that focused on casting, firing, maintaining and transporting a wide variety of artillery pieces for both sieges and field battles, and in this institutionalized setting gunners developed a professional identity that stressed their autonomy from their sister service. Pierre Surirey de Saint-Rémy's textbook for aspiring gunners mentioned the critical topic of artillery-engineer coordination in sieges only long enough to discourage gunners from following engineers' orders to place their batteries in the trenches rather than construct separate batteries.[83] Not surprisingly then, relations between these two branches

---

[80] Léon Lecestre (ed.), *Mémoires du chevalier de Quincy* (Paris, 1898–1901), III: 256.

[81] On Louis XIV's artillery, see Susane, *Histoire de l'artillerie française* (Paris, 1874), ch. 4; Ernest Picard, *L'Artillerie française au XVIIIe siècle* (Paris, 1906); Michel Decker, "Louvois, l'artillerie et les sièges," *Histoire, économie et société* 15.1 (1996): 75–94; and most recently Frédéric Naulet, *L'artillerie française (1665–1765): Naissance d'une arme* (Paris, 2002).

[82] Blanchard, *Les ingénieurs du Roy* (note 46), p. 137. The Marine department had their own gunners, which on occasion might serve at a terrestrial siege. For an overview of the office of grand-master of artillery, see Lynn, *Giant of the Grand Siècle* (note 11), pp. 99–100.

[83] For the duties of artillery officers during sieges, see Surirey de Saint-Remy, *Mémoires d'artillerie* (The Hague, 1741), II:255–269.

were problematic, as gunners were particularly hesitant to surrender their initiative and follow the orders of engineers in the trenches.

Vauban would experience this conflict first hand. Before rising to prominence in the Sun King's armies, he had already discovered examples of the artillery's ignorance of siegecraft and its willingness to ignore well-planned attacks.[84] At a number of subsequent sieges he lamented his inability to convince the gunners of the error of their ways; his lack of authority over the artillery prevented the full systematization of the attack in spite of his friendship with several of their highest-ranking officers.[85] At the siege of Charleroi in 1693 he identified other weaknesses that eroded the siege's efficient operation:

> I have moreover suffered greatly from the ignorance of the cannoniers and bombardiers, who, with the exception of a small number, I cannot be pleased with. The proper use of artillery and bombs requires an exact art [*un art particulier*] with precise rules that not a single one of these people know. They do things as they have always done them, and have primarily their own interests in mind, which means that all the batteries are defective and imperfect. Furthermore, as none of them understand the principles of fortifications, they know neither where nor how to fire against the works. They are not familiar with their own guns, which does not surprise me, because they are only infantrymen that serve the cannon and who obey the artillery officers only when it pleases them; and since there is usually only one artillery officer for several pieces and because the officers cannot keep an eye on everything when they are aiming a piece, it is often the case that more than half or two-thirds of the rounds are squandered or miss the target completely.[86]

Apart from such technical matters, the artillery corps' independence from the engineers was asserted even more strongly once Louis-Auguste Bourbon, duc du Maine and an illegitimate son of the king, acquired the post of *grand-maître d'artillerie* in 1694.[87] One of Vauban's aide-de-camps, chronicling the 1697 siege of Ath, saw little improvement from what Vauban had described thirty years earlier:

---

[84] Rothrock, *A Manual* (note 13), p. 167.

[85] Vauban, for example, co-authored a proposal for artillery reform with the artillerists François Frézeau marquis de la Frézelière, lieutenant-general of French artillery during much of Louis XIV's reign, and Armand de Mormes, sieur de Saint-Hilaire. Printed in Vauban, *Traité de l'attaque*, "De l'artillerie."

[86] Vauban to Le Peletier de Souzy, Charleroi [Rochas d'Aiglun (ed.), *Vauban* (note 17), II:399]. Saint-Rémy confirmed the reliance on conscripted soldiers to serve the cannon, much as engineers were forced to oversee regular soldiers as sappers. Saint-Rémy, *Mémoires* (note 83), vol. II, p. 265.

[87] Lecomte, "Du service des ingénieurs militaires" (note 43), pp. 123, 127ff.

> Everybody knows that the common practice of the officers of the train [of artillery] is to fire at all before them, with all the fury that is possible, ruining without distinction all objects that present themselves to view, whether works, batteries, towers, or cavaliers [*i.e.*, elevated firing platforms]; they even forget themselves so far as to fire upon any building which affords a fair mark, as gates, bridges, corps de gardes, sentry-boxes on the walls, sometimes at houses and steeples, for the mattrosses [gunners] only want to make havock appear, which was not Monsieur de Vauban's taste, who had so frequently seen the fruitlessness of it at so many sieges. . . . Bounce and clatter and readiness for action had hitherto composed the whole merit of the train at sieges.[88]

The same old complaints of poorly-constructed battery breastworks and poorly-served guns can also be found in his final siege, Breisach 1703, where he was driven to write:

> if the artillery had done its job, we would very soon be inside the town; but it is infinitely difficult to direct them. They are all men who have hardly ever seen a siege and who know only how to fire straight ahead; they do not even know how to construct proper ramparts.[89]

Just a few years before his death, he admitted that his revolutionary development of ricochet fire (*i.e.*, cannonballs fired with an undersized powder charge that skipped along the ground in order to smash defenders sheltered behind walls) was still used sparingly almost a decade after its introduction.[90] At the end of his long career Vauban identified yet one more source of internal friction he had been unable to eliminate.

Nor were such complaints limited to Vauban's tenure. Besieging the weak forts protecting Huy in 1705, the French commander noted that the gunners' profligate consumption of gunpowder was having very little effect; he concluded that further improvements were needed in the artillery schools.[91] Evidence of continued internal friction between engineers and artillerists comes from the Secretary of State

[88] Louis Goulon, *Memoirs of Monsieur Goulon, being a Treatise on the Attack and Defence of a Place. To which is added, a journal of the siege of Ath, in the year 1697, under the conduct of Monsieur de Vauban* (London, 1745), p. 103. Augoyat attributed this journal to Vauban's aide-de-camp Ferry.

[89] Vauban to Chamillart, Biesheim [Rochas d'Aiglun (ed.), *Vauban*, II:525]. From this general condemnation Vauban excluded their head, 31-year old Jean-Angélique Frézeau marquis de la Frézelière, brigadier of artillery, son of the late lieutenant-general of artillery François de la Frézelière.

[90] Vauban, *Traité de l'attaque*, p. 116.

[91] Henri duc d'Harcourt to Chamillart, AG A[1] vol. 1835, #269.

for War's reiteration to an artillery officer of the rather fine division of labor between the two branches: "All your capacity must be confined to promptly constructing the batteries and serving them with as much dexterity as diligence; but it is up to the engineers to indicate where they want them to be placed."[92] Not surprisingly, such intimate coordination between the two services was "*assez rare*" according to *maréchal* Berwick.[93] Pierre d'Artaignan, comte de Montesquiou and newly-minted *maréchal de France*, noted how the return to bickering at Douai in 1712 hindered his efforts to propel the attack forward:

> This siege is going too slowly for my taste, there is too much division between the engineers and the artillery and as no one person is charged with overall command of the siege, the service is often delayed. To negotiate such dialogues a man who follows each step and who has overall command is needed, but as no one has overall authority each general officer of the trenches follows his own ideas.[94]

Given the interactions between all these conflicting sources, some pressing forward, some pulling backwards, and none with the same amount of force for the same amount of time, it is no surprise that the estimates of when exactly a town would fall could fluctuate so dramatically from day to day.

Less powerful sources of friction could also accumulate like sand in the siege machine, particularly the thousands of workmen engaged in the trenches. Getting the officers and gunners to accept the engineers' authority was not enough to assure that the real siege met the ideal, for the rank and file infantry did the digging of trenches and building of earthworks, just as they assisted the gunners with their pieces. Properly-constructed ramparts were only one cause for concern; flight was another. Soldiers on guard duty had their weapons to reassure them, whereas the workmen—motivated primarily by money and drink—were armed only with pick and shovel. Working at night and presented with the sudden threat of enemy matchlocks and bayonets, laboring soldiers often chaotically abandoned their works when sallying defenders approached, and sometimes even when

---

[92] Chamillart to d'Houville, quoted in Lecomte, "Du service des ingénieurs militaires," p. 131.
[93] *Ibid.*, p. 26.
[94] Montesquiou to Voysin, AG A[1] vol. 2382, #173.

they did not, for "nothing is more common than for the workmen to take to their heels," as Louis Goulon recalled.[95] Focusing the troops' attention on properly constructing the trenches to the exclusion of thoughts about their own personal safety was a challenge for both officers and engineers alike.

Vauban had recognized this critical weakness and constantly petitioned Louis for independent sapper brigades, soldiers who would become experts in the dangerous task of advancing the trenches under enemy fire. Just as importantly, they would be trained to abandon the advanced trenches for the safety of the rear parallel in an orderly fashion.[96] The best plans accounted for little without skilled workmen, and in 1672 he warned that his method for attacking a fortress

> demands intelligence both on the part of directors and workmen, I should add that we cannot hope to make the best use of it with those whom we usually employ in our sieges. They are naturally maladroit and surly; besides they are neither trained nor drilled to execute with precision the tasks to which they are assigned. It is, therefore, absolutely essential to form and train a special body of well-versed men, either drawn from several regiments or raised separately, as a corps of engineers.[97]

Refused this request, at Charleroi Vauban warned of the dire results if his petition continued to be rejected—sieges would continue to be more expensive in both time and money.[98] After complaining of a shortage of skilled miners at Breisach in 1703, he repeated at the end of his career the same request for three regiments of artillery and a separate company of sappers.[99] Turning to the younger generation, his 1704 treatise dedicated to Burgundy repeated his belief that sapper companies officered by engineers could quickly learn the trench skills that would speed up and regularize the siege attack and at the same time decrease the number of casualties among the engineers, the artillerists and the common soldiers alike.[100] His requests

[95] Goulon, *Memoirs of Monsieur Goulon* (note 88), p. 16.

[96] The issue of unreliable workmen is discussed in Lecomte, "Du service des ingénieurs militaires" (note 43), pp. 124–126.

[97] Rothrock, *A Manual* (note 13), p. 105.

[98] Vauban to Le Peletier de Souzy [Rochas d'Aiglun (ed.), *Vauban*, II:400].

[99] On Breisach at Biesheim, see *ibid.*, II:522. For his later call, see Vauban, *Traité de l'attaque*, pp. 269–295 for the artillery and 296–309 for the sappers.

[100] Vauban, *Traité de l'attaque*, pp. 306–307.

continued to be rejected for being too expensive. Here too Vauban was unable to eliminate the engineers' reliance on others, adding yet another source of internal friction that made precise progress impossible to predict.

Adding Friction to the Clockwork Siege

Sieges were indeed more predictable instruments than battles in the open field. Most were ultimately successful because of the variety of materiel and tactics brought to bear on an isolated and overpowered garrison. The overwhelming superiority of besiegers' firepower, its superiority in manpower, the semi-permanence of the trenches, and the isolation of the garrison all served as 'ratchet wheels,' acting as a brake on a siege machine's backwards motion. Unlike the ideal siege found in military manuals, however, real sieges were affected by both external and internal friction. If we wish to retain the clockwork metaphor, we must incorporate Clausewitzian friction as well, since final success is not the same as precise regularity. There were many points of friction in each siege, and the degree of friction could vary from siege to siege as the operational context varied from month to month. Even within an individual siege the rate of progress was influenced not only by changing external factors but also by the individual participants, whether it was the siege commander, the here-today-gone-tomorrow generals of the day, artillery officers, or the humble foot soldier. A siege's internal friction was also variable (though less so than on the battlefield), since the force that these different actors applied varied over the course of a siege. The unpredictable arrival of news of a distant defeat, a marching relief force, or an approaching detachment from another theater might suddenly send a siege lurching forward or even backwards (if the besieging army was stripped of troops to reinforce the threatened observation army), while the late autumn months often allowed a slower pace, since the loss of a week or more made little difference given the inevitable rains of late autumn. These conflicting internal pressures on the siege machine rarely broke the mechanism, but they did add a degree of uncertainty and made sieges much less predictable than historians have recognized.

Adding Clausewitzian friction to Vauban's metaphorical clockwork, we can conclude that Newton and Vauban did in fact share one common trait. Just as Newton recognized that his generalized laws of motion were only approximations of a much more complicated reality, so too did Vauban recognize the gulf between his perfect siege and the reality on the ground.[101] As the engineer explained: "The principles on which I have founded my method are drawn from those of fortification itself, which assumes a regular system as the most perfect, and all that one can do is strive to approximate this perfection as closely as the different situations allow."[102] In every stage of the siege, the engineers' influence was eroded by the internal friction of those struggling to control its conduct. Thus the timetable's predictive, instrumental power, only theoretical to begin with, vanished as a result.

[101] George Smith, "The Methodology of the *Principia*," in I.B. Cohen and G.E. Smith (eds.), *The Cambridge Companion to Newton* (Cambridge, 2002), pp. 152ff.
[102] Vauban, *Traité de l'attaque*, p. 202.

CHAPTER FOUR

CALORIMETERS AND CRUSHERS: THE DEVELOPMENT OF INSTRUMENTS FOR MEASURING THE BEHAVIOR OF MILITARY POWDER[1]

Seymour H. Mauskopf

How good is the gunpowder? This has been the persistent question facing gunpowder makers and ordnance authorities since gunpowder's introduction into European warfare more than six centuries ago. To answer this question, one doesn't have to be a general to recognize some obvious relevant characteristics: the powder has to be a reliably consistent product with strictly limited variation in characteristics between batches. Moreover, the powder has to maintain its original characteristics over time so that it will have them when used. And the most significant of its characteristics is ballistic performance; gunpowder needs to be forceful enough to enable the gunner to carry out his strategic aims with his particular firearms. At the same time, the force had to be controlled so that the setting off of the powder charge doesn't blow up the gun (and gunner).

The reliability of gunpowder was a perennial problem that, arguably, was not fully under control even at the end of the black powder era in the late nineteenth century. Nor was the smokeless successor to black powder free from this problem. Generally speaking, at the manufacturing site, consistency and reliability were (and are) addressed in the same way they were for any manufactured product: by insuring, as best they could, the uniformity of material constituents and of the manufacturing processes.

But the ultimate test of reliability was ballistic performance. It was, of course, of crucial importance to determine this for munitions prior to their actual use on the battlefield. To meet this challenge in the

[1] I want to thank Dr. Patrice Bret and Professor Alex Roland for their careful, critical reading of a draft of this paper. They are, of course, not responsible for any stylistic infelicities or errors of fact and interpretation.

eighteenth century and nineteenth century, scientific principles and procedures were deployed in increasingly complex and sophisticated instruments to provide quantitative, "objective" and (hopefully) useful measurements of ballistic performance. The focus of this paper will be on two laboratory instruments developed in France in the last quarter of the nineteenth century by the military engineer, Paul Vieille, for testing aspects of "ballistic force." They are the *bombe calorimétrique* and the *manomètre enregistreur*. The first was used primarily for determining the heat produced by a weighted sample of an explosive substance. With the second instrument, Vieille studied the way in which an explosive combusts, particularly, whether or not a powder burns regularly through parallel material layers.[2]

These two instruments are emblematic of the imposition of scientific theory (in this case, thermochemistry) and laboratory practice on the determination of "how good is gunpowder." The bomb calorimeter became a standard instrument to study the heats of reactions of a wide variety of substances. Vieille's studies of the combustion processes of explosives via the *manomètre enregistreur* was the experimental context for his development of the first gelatinized smokeless military powder, *Poudre B*.

Moving explosives testing from the field gun into the laboratory was not a straightforward process. During the eighteenth and nineteenth centuries, a variety of instruments were developed that depended upon scientific principles and new science-based technologies, but were not in any usual sense laboratory instruments. Most depended upon measuring the mechanical behavior of a projectile actually shot from some kind of gun.[3] Some, like the "mortar-eprouvette," measured how far a fixed amount of powder could project a cannon ball of fixed size and weight from a small mortar. Others, like the

[2] These were not the only military-scientific instruments developed by Vieille; others included the "bombe à grain d'érosion" and the "tube à ondes explosives." See René Amiable, "Les instruments scientifiques de Paul Vieille au 19ème siècle," in *Troisièmes journées scientifiques Paul Vieille: Instrumentation, expérimentation et expertise des matériaux énergétiques (poudres, explosifs et pyrotechnie) du XVIe siècle à nos jours*. Centre de recherche en histoire des sciences et des techniques, Cité des sciences & des industrie, Paris, October 19–21, 2000 (Paris, n.d.), pp. 97–106. Amiable terms the *manomètre enregistreur* the "bombe à manomètre enregisteur" but I am using Vielle's original name.

[3] The following discussion is taken from my article, "Bridging Chemistry and Physics in the Experimental Study of Gunpowder," in Frederic L. Holmes and Trevor H. Levere (ed.), *Instruments and Experimentation in the History of Chemistry*, Dibner Institute Studies in the History of Science and Technology (Cambridge, Mass., 2000), pp. 335–365.

ballistic pendulum, measured the impact of a projectile on a pendulum target at a standard distance from the gun. From fairly elementary dynamics, one could compute the velocity of the projectile as it struck the pendulum; from assumptions about the effect of air resistance on the projectile as it traveled from gun to pendulum, one could extrapolate back to the muzzle velocity of the projectile as it left the gun. In the nineteenth century, a refinement was made so that the gun itself became the pendulum; this enabled projectile force and muzzle velocity to be more directly computable from the recoil of the gun. By mid-century, developments of electromagnets led to appearance of "electro-ballistic chronographs," which measured the velocity of the projectile by successive ruptures of electrical circuits.[4]

Laboratory Explorations

In all of these instruments, testing of gunpowder took place in real guns, approximating its military use. However, by the middle of the nineteenth century, the development of thermodynamics and thermochemistry introduced a new approach to ballistic research derived more from chemistry than from mechanics, and one that was centered in the laboratory, where gunpowder was treated as a normal laboratory substance (if a rather dangerous one).[5] This chemical/laboratory testing of the gunpowder ballistics originated in research carried out by the German chemist, Robert Bunsen (of eponymous burner fame) and his Russian student, the artillerist and chemist Leon Schischkoff, and was published as "A Chemical Theory of Gunpowder" in 1857.[6]

One of the components of their research was an analysis of the gunpowder combustion reaction in unprecedented detail. This was closely connected to the thermochemical approach of Bunsen and

---

[4] See my article, "Explosives, Instruments to Test the Ballistic Force of," in Robert Bud and Deborah Jean Warner (eds.), *Instruments of Science: an historical encyclopedia*, Garland Encyclopedias in the History of Science 2 (New York, 1998), pp. 234–236.

[5] At the end of the eighteenth century, a chemical/laboratory approach had been utilized by the chemist, Joseph-Louis Proust but had not led to a sustained research tradition. See S.H. Mauskopf, "Chemistry and Cannon: J.-L. Proust and Gunpowder Analysis," *Technology & Culture* 31 (1990): 398–426.

[6] "Chemische Theorie des Schiesspulvers," *Annalen der Physik und Chemie* 2, Folge 102 (1857): 321–353.

Schischkoff in which they employed a *calorimeter* to measure the heat of reaction of gunpowder explosion. The calorimeter itself was introduced by Antoine-Laurent Lavoisier in the early 1780s in research on heats of reactions carried out in collaboration with Pierre-Simon Laplace. Although gunpowder explosion was one of the reactions studied by them,[7] there was no immediate sequel to their thermochemical study of this reaction. But by the 1850s, as the fundamental laws of thermodynamics were being enunciated and developed, there was an uptake in thermochemistry, most notably in the calorimetric studies of heats of reactions by Pierre Antoine Favre and Johann Theobald Silbermann. Favre had briefly examined the heat of reaction of gunpowder explosion[8] but with nothing like the scope of Bunsen and Schischkoff.

They used their calorimetric results to compute the actual temperature of the explosion by dividing the determined heat of reaction by the specific heats of each of the products of the reaction. Hence, their need for detailed chemical analysis of these products. From the computed temperature, they derived the pressure of the gunpowder explosion. It will be noted that, in this laboratory research, the principal parameter of the ballistic testing tradition, velocity, was not even considered. It was the laboratory study of gunpowder explosion itself that was the object of study, and not its use in a weapon for propelling a projectile.

Nevertheless, their 1857 paper was very influential in the ensuing decades despite (or perhaps *because*) its results were soon contested. It gave rise to several attempts to extend the results from the sporting powder that Bunsen and Schischkoff had used to military powder.[9] More significantly, the research approach and results of Bunsen

---

[7] S.H. Mauskopf, "Gunpowder and the Chemical Revolution," in Arthur Donovan (ed.), *The Chemical Revolution: essays in reinterpretation* [*Osiris*, 2nd series 4] (1988): 111–115 for an overview of the Lavoisier-Laplace thermochemical study gunpowder explosion.

[8] P.-A. Favre, "Recherches thermo-chemiques sur les combinaisons formées en proportions multiples," *Journal de pharmacie et de chimie*, 3ième série 24 (1853): 338–339.

[9] *E.g.*, L. v. Karolyi, "On the Products of the Combustion of Gun-cotton and Gunpowder, under circumstances analogous to those which occur in practice," trans. Dr. Atkinson, *The London, Edinburgh, and Dublin Philosophical Magazine and Journal of Science*, 4th series 26 (1863): 266–280. Karolyi's experiment differed from Bunsen's and Schischkoff in (a) testing military rather than sporting powder (and guncotton); (b) using an apparatus that mimicked what took place in a gun and (c) concentration on analyses of the products of explosion and not on the thermochemistry;

and Schischkoff served as points of departure for two ambitious research programs in the 1870s to study the physical and chemical states-of-affairs of explosive reactions.

One was British, and was carried out by the artillery officer and ordnance expert Andrew Noble with the military chemist Frederick Abel. Unlike Bunsen and Schischkoff's laboratory study, this one had military as well as scientific utility as a goal, and consequently aimed at testing gunpowder for its ballistic behavior as well as the laboratory chemical reaction itself. For example, fairly sizeable charges of gunpowder were exploded in an instrument mimicking the conditions in a gun: an "explosion-apparatus," described as "a mild steel vessel of great strength, carefully tempered in oil."[10] Moreover, actual field data was also collected and evaluated.

Yet the influence of Bunsen and Schishkoff is evident not only in the references to "A Chemical Theory of Gunpowder" in Noble and Abel's work but also in their employment of a calorimeter containing water of a standard weight and temperature, into which the test apparatus was placed immediately after explosion. Although Noble and Abel made changes in the algorithms used to compute items like temperature of explosion, they basically followed the thermochemical approach of Bunsen and Schischkoff.

Although Bunsen and Schischkoff's approach was an important component of Noble and Abel's theoretical and experimental structure, it was not the only one. Thus, variations in the physical characteristics of powders—*e.g.*, density and grain size—as well as variations in the pressure of firing were tested comparatively for the effects on the nature of the chemical products of explosion. And the ultimate goal of the massive experimental project was to produce ballistic tables of muzzle velocities for any standard projectile in any standard British large gun for every description of suitable gunpowder from the thermodynamic data.[11]

---

hence, no employment of a calorimeter. For other examples and their tabulation, see: Andrew Noble and Frederick Abel, "Researches on Explosives—Part I (1875), in Noble, *Artillery and Explosives* (London, 1906), pp. 108–111 and 130–131 (table).

[10] Noble and Abel, "Researches on Explosives—Part I," p. 114. There were vessels of two sizes, one of which could hold a charge of about 2.25 lbs. (1 kg) and the other about half that charge.

[11] Velocity was derived from the total work per pound of powder (computed maximum work multiplied by a "factor of effect" for each gun and charge). *Ibid.*, pp. 204–205, footnote.

For the argument of this paper, it is important to mention one other "physical" feature of the experimental setup. Unlike Bunsen and Schischkoff, Noble and Abel were not content to deduce theoretical pressure from the thermodynamic data, but also measured it directly *via* a "crusher gauge." This instrument was Noble's refinement of the "indenting apparatus" invented by the American ordnance officer, T.J. Rodman, to measure the pressure of gunpowder explosion at various points along the bore of the gun. It consisted of what Rodman called an "indenting tool," a shank or piston carrying a knife edge and a "copper specimen," a thick disk of soft copper, both screwed into a cylindrical housing fitted into the wall of the gun bore. Upon firing, the gaseous products of explosion pressure exerted on the base of the indenting tool caused its knife edge to cut into the copper specimen; the depth of the cut being correlated to the degree of pressure exerted on the indenting tool. Noble's refinement was to replace the knife-edge and copper disk with a piston, a copper cylinder, and a rigid anvil at the opposite end of the cylinder from the piston. The gas pressure from the explosion pushed the piston against the copper cylinder, which, in turn, was compressed against the anvil. The magnitude of the gas pressure was measured by the degree of compression of the copper cylinder.[12] The crusher gauge was placed directly into the explosion-apparatus.

Contemporaneous with the British research of Noble and Abel, a French group was also taking up the new thermochemical approach. The chief formulator of this approach was the chemist, Marcellin Berthelot. Unlike Abel, Berthelot came to the chemistry of explosives relatively late in his career. His interest in research on explosives and munitions developed through his activities in the Franco-Prussian war. After the defeat of Napoleon III at Sedan, with Paris facing siege, Berthelot was appointed head of a seven member "Comite Scientifique pour la Defense de Paris" on 2 September 1870.[13] Berthelot himself took up the study of explosives with expedition: his lecture course at the Collège de France during the siege of Paris was on "the force of powder and explosive materials" and he addressed

[12] For details, see Andrew Noble, "On Methods that have been Adopted for Measuring the Pressures in the Bores of Guns," in Noble, *Artillery and Explosives*, pp. 487–493.

[13] The other members included: d'Almeida, Bregnuet, Fremy, Jamin, Ruggieri and Schutzenberger.

the topic in three lengthy reports to the Académie des Sciences in November 1870.[14] By 1871 he produced the first of three successively longer versions of his thermochemical analysis of explosives and munitions: *Sur la force de la poudre et des matières explosives*.[15]

Like Noble and Abel, Berthelot acknowledged the precedent of Bunsen and Schischkoff, if somewhat critically. But there were considerable differences in scientific "styles" between the English and French research programs, due to the different professional loci of the researchers and perhaps also to differences in national styles. As I have noted, Andrew Noble was an artillery officer and a gun manufacturer; Frederick Abel was (unusual for the nineteenth century) a military chemist through his whole career. They therefore naturally started with the concrete situation of military propellants and their functions in guns. Although both Noble and Abel were conversant with—and utilized—the latest theoretical and experimental developments in physics and chemistry (including, of course, Bunsen and Schischkoff's work), they were not particularly interested in developing elaborate theoretical structures from which their experimental consequences could be deduced.

Berthelot, by contrast, had come to explosives research from very different research domains of organic chemistry and, more recently, thermochemistry. Moreover, despite being a "positivist" regarding the existence of theoretical entities such as atoms, he was also a theory builder, in the French tradition of theoretically-based rational science. As he made clear in the final paragraph of the first edition of *Sur la force de la poudre*, his vision was to make the study of explosives a deductive branch of his thermochemistry:

> In summary, up to now, the force and the mechanical properties of diverse explosive substances have only been compared by empirical means. I have tried to establish this comparison upon theoretical notions, and one should see that the deductions so obtained generally accord with experience; one might therefore take them [theoretical notions]

[14] Louis Médard and Henri Tachoire, *Histoire de la thermochimie: Prélude à la thermodynamique chimique* (Publications de Université de Provence, 1994), p. 198. Médard and Tachoire express doubts about the value of the work of the "Comite Scientifique pour la Defense de Paris" (p. 195).

[15] Paris, 1871. The successive editions were 1872 and 1883 (the latter, 2 vols. with the title changed to: *Sur la forces des matières explosives* ***d'après la thermochimie*** [my emphasis]).

> as a guide, for obtaining the maximum effect from known explosives, for associating these with other substance, or for discovering new explosive compounds which have properties already determined in advance.[16]

Moreover, the military engineers who became Berthelot's collaborators were all alumni of the *École polytechnique*. There they had imbibed a research orientation that shared the theoretical and rational characteristics of Berthelot's science.

By the time of the third edition, Berthelot had developed his thermochemistry fully and consequently was able to refer the conditions of explosion back to the governing principles of his thermochemistry.[17] In the first two editions, the deductive path was not yet so rigorous or comprehensive, but the general model remained the same for all three editions.

The "force d'une matière explosive" was defined in terms of two fundamental data: the *volume of gases produced*, and the *quantity of heat liberated* in an explosive reaction. Berthelot defined (and determined) the released quantity of heat as equal to the difference between the heats of formation of the explosion products and the heats of formation of the explosive's chemical components.[18] From his general theory, Berthelot developed a research program comprised of three components: (1) The determination of the heats of formation of the chemical components of explosives and of their reaction products. (2) The study of gaseous dissociation during explosion. Gaseous dissociation under high temperature and pressure affected the gaseous volume and the temperature of the reaction (since dissociations were endothermic)[19] and, hence, had to be thoroughly studied for a proper understanding of explosion. These two components were the essential ones for determining the explosive force according to Berthelot's theory. (3) The study of how shocks (either mechanical or chemical, *e.g.*, a primer) could cause an explosive to be set off.[20]

---

[16] *Sur la Force* (1871), pp. 38–39. He ended the 1872 edition with exactly the same words. *Sur la Force* (1872), p. 191.

[17] These were (1) The principle of molecular work; (2) The principal of the calorific equivalence of chemical transformations (the principle of the initial and final states); (3) The principle of maximum work. C. Napier Hake and William Macnab (trans.), *Explosives and their Power* (London, 1892) pp. 114ff. This was an English translation and abridgment of *Sur la Force* (1883).

[18] *Sur la Force* (1871), p. 9. Discussion of what comprised the "force d'une matière explosve is on pp. 6–9.

[19] Hake and Macnab, *Explosives and their Power*, p. 9.

[20] *Sur la Force* (1871), pp. 30–31. These had been previously studied by Abel, but Berthelot thought that they could be explained by "thermodynamical theories."

Berthelot himself began to carry out research on nitro-carbons in 1871, using a calorimeter as a primary instrument. He justified it in terms of his theoretical model of explosion:

> The explosive force of nitro-carbon compounds results from a kind of internal combustion, analogous to that of guncotton. . . . That force is the stronger insofar as the compound under consideration generates more gas and heat. But the heat released by the combustion will be the greater, all things being equal, as the preliminary union of nitric acid with the organic constituent itself releases less heat. That is, the energy of the system formed by the combusting acid and the combustible will be less diminished in the act of combination.[21]

In subsequent studies, Berthelot reached back to the heats of formation of the ultimate constituents of these nitro-carbons: the compounds between nitrogen and oxygen. In one of the first of these,[22] he established an ambitious four-part research objective to study heats released (or taken up) in the transformation of nitrites into nitrates, in the formation of nitrites and nitrates from their elements, in the formation of all the gaseous oxides of nitrogen and, lastly, the application of the above research to the study of various reactions but including that of gunpowder and other explosive substances.[23] Berthelot concentrated on the first objective, specifically, in determining the heat of formation of nitrous acid (endothermic) by analyzing the decomposition of ammonium nitrite.[24]

Berthelot was not very forthcoming about what kind of calorimeter he had used.[25] At any rate, Berthelot did not return to the subject for six years, during which time it was taken up by others. Indeed, the next paper on the subject to be published contained research that had been carried out just prior the Franco-Prussian War, antedating

---

[21] Berthelot, "Sur la formation des composés organiques qui dérivent de l'acide nitrique," *Comptes Rendus de l'Académie des Sciences* (henceforth *CRAS*) 73 (1871): 260.

[22] Berthelot, "Sur la chaleur dégagé dans les combinaison de l'azote avec l'oxygène," *CRAS* 78 (1874): 99–106.

[23] *Ibid.*, p. 99. He wrote about this research that, although difficult, obscure and controversial, "Il offer cependant une grande importance, tant au point de vue de la mécanique chimique que des applications relatives aux substances explosives."

[24] He asserted that his thermal determination of the heat of formation of nitrous acid "agreed, as I shall show, with the determination of the heat of combustion of gunpowder made recently by MM. Roux and Sarrau and by M. de Tromenec, who have given figures very much higher than the ancient ones of Bunsen and Schischkoff." *Ibid.*, p. 106.

[25] Médard and Tachoire were not very impressed by Berthelot's accuracy (*Histoire de la thermochimie* [note 14], p. 200).

Berthelot's own work.[26] The author was Louis-Francois le Bihannic de Tromenec, an *École polytechnique*-trained artillerist and professor of applied science at the *École d'artillerie* at Rennes.[27] Nevertheless, Berthelot helped to see the paper into publication.[28]

The alumni of the military engineering training of the *École polytechnique* were conversant with both the practical and theoretical side of military science but showed a leaning (or yearning) towards the theoretical. This was evident in the very first lines of Tromenec's paper, where he made a sharp distinction between physical/field methods of testing gunpowder and the newer chemical (or thermochemical)/laboratory method developed by Bunsen and Schishkoff. He characterized the methods by which gunpowder was normally tested, using such instruments as the ballistic pendulum and mortar eprouvette, as "generally judged insufficient" because "they can only compare powders that differ little between themselves by their *physical* characteristics." These resulted in different ballistic results in different guns.[29]

Instead, Tromenec proposed a method "already employed by MM. Bunsen and Schishkoff" that would give an *absolute value*, independent of the weapon in which it was fired.[30] By "absolute value" Tromenec meant "absolute force," which he defined as the work that gunpowder explosion could perform, measured

> By this principle of thermodynamics: When a body detonates without producing a dynamical effect, the disposable force is transformed into heat. It thus suffices to explode the powder in a closed vessel and measure the heat produced.[31]

[26] A date of 2 July 1870 is given for the calorimeter experiment. De Tromenenc, "Sur un moyen de comparer les poudres entre elles," *CRAS* 77 (1873): 128.

[27] Information on Tromenec from Médard and Tachoire, *Histoire de la thermochimie*, pp. 190–191.

[28] De Tromenenc, "Sur un moyen de comparer les poudres entre elles," p. 126, n. 2 [Note de M. Berthelot]. Berthelot wrote that Tromenec had sent the paper to him in 1872 but not for immediate publication because he wanted to complete his analysis of the products of gunpowder combustion. Berthelot continued: "Mais, ayant, appris que d'autres savants s'occupaient de la même question, je crois devoir faire connaître ce travail, tel qu'il m'a été communique, afin de réserver les droits de l'auteur."

[29] *Ibid.*, pp. 126–127, my italics. The only example he gave here of physical characteristic was powder grain size. "Il est évident, par-exmple, que, si l'on tire dans le mortier éprouvette une poudre très-fine, puis une poudre très-grosse, les résultats que l'on obtient ne peuvent servir de termes de comparaison entre les deux poudres."

[30] *Ibid.*, p. 127, my italics.

[31] *Ibid.*

This was done, of course, in a calorimeter, of which Tromenec gave a description. It was a cylindrical vessel made of cast steel with an interior capacity of about one half liter with walls of 3 to 4 centimeters in thickness. It was "hermetically sealed" by a screw stopper through which electrical wires passed which served to spark 5 grams of powder. The vessel was placed in a calorimeter made of sheet metal and filled with 1,500 grams of water;[32] this, in turn, was placed in a tub filled with cotton to contain the heat. The vessel was kept immobile by a binding screw supported by the screw stopper. The calorimeter was equipped with a thermometer graduated to almost one-hundredth of a degree and an agitator. The heat produced from the explosion of five-gram samples of three kinds of powder were reported.[33]

Although Médard and Tachoire found Tromenec's determinations to be too high by about 10%, they lauded his work as the first that allowed "the measure of a heat of reaction in a closed vessel at an elevated pressure. The bomb that he [Tromenec] had put together can be considered to be the ancester to our bomb calorimeters."[34]

However, this was Tromenec's last appearance in print on the subject. But, as Berthelot had noted, others—also *polytechneciens*—had taken up this approach, notably the munitions engineer, Jacques-Ferdinand-Émile Sarrau and his superior, Louis Roux, director of the *Dépot Central des poudres et salpetres.* Sarrau had taken part in the establishment of a powder mill during the siege of Paris and had subsequently been given a place at the *Dépot Central.* There he worked closely with Roux. In 1873, Roux and Sarrau published two short studies of the forces of different explosives. In the second, the impact of Berthelot's thermochemistry (as well as the origins of this approach in Bunsen's and Schischkoff's work) was heralded in the opening statement:

> Berthelot's research on the theoretical calculation of the force of powder and explosive materials brought to light the importance, from this point of view, of measuring the quantities of heat released by the combustion of those substances. The only experimental result of this type

[32] The rest of the apparatus was calculated to weigh 526 grams. *Ibid.*, p. 128.

[33] *Ibid. Poudre à canon de Bouchet (1861): 840 calories; poudre de mine: 729 calories; poudre de contrabande, d'origine anglaise: 891 calories.*

[34] Médard and Tachoire, *Histoire de la thermochimie* (note 14), p. 191. They attribute his errors in estimation to an overestimation of the water in his calorimeter.

that has been published up to now, as far as we know, is that put forth by MM. Bunsen and Schischkoff for a powder similar to our sporting powder. We thought that it would be of interest for this purpose to set up a simple and inexpensive apparatus, of reasonably sure and rapid operation, to make a practical complement to the proofs to which the various military and industrial explosives are submitted to the Depot central des Manufactures de l'Etat.[35]

I have given an account of the investigations of Roux and Sarrau elsewhere.[36] Suffice it here to reproduce their account of the instruments they employed for carrying out the explosion and measuring the heat given off:

> Explosion is produced in cylindrical bombs made of cast iron, of a thickness of 6 millimeters and having an interior capacity of between 270 and 280 cubic millimeters. The bombs are closed by a bronze screw cap traversed by an insulated wire. The passage of a current through this heats a thin wire placed in the interior to red and thus inflames the test substance. They [the bombs] are placed in a red copper vessel of a diameter of 0.140 meters and a height of 0.160 meters, containing 1.830 kilograms of water. The temperature of this water bath is measured with the aid of a thermometer graduated in tenths of degrees and affording a reading to a hundredth of a degree. In order to measure the released heat it suffices to equilibrate the temperature of the bath to that of the surroundings, produce the explosion and, agitating the water, observe the change in the temperature of the bath. In designating this change by Δ and total weight *of the water* of the calorimeter by P, the released heat is equal to PΔ.[37]

This apparatus was first used to measure the heat released in five types of black powder to a high degree of accuracy.[38] The follow-

[35] Roux and Sarrau, "Sur la chaleur de combustion des matières explosives," *CRAS* 77 (1873) [Séance of 14 July 1873]. Pagination is from publication in *Les Explosives Modernes: Mémoires par MM. A. Nobel, L. Roux, Sarrau etc.* (Paris: Société générale poud la fabrication de la dynamite[Typographie Lahure], 1876), pp. 171–172. Médard and Tachoire suggest that Tromenec's bomb and calorimeter may have been the example Roux and Sarrau were following. Médard and Tachoire, *Histoire de la thermochimie*, p. 192.

[36] S.H. Mauskopf, "Introduction: The Experimental Study of Munitions: Scientific and Military Traditions," *Troisièmes journées scientifiques Paul Vieille* (note 2), 19ff.

[37] Roux and Sarrau, "Sur la chaleur de combustion des matières explosives," p. 172. Italics in text. They noted that corrections had to be made for two causes of error: (1) heat radiation from the calorimeter (which can normally be neglected) and (2) the difference ["généralement très-faible"] between the interior and exterior temperature of the bomb after the explosion and when the temperature of the water bath his attained its maximum (pp. 173–176).

[38] Médard and Tachoire, *Histoire de la thermochimie*, p. 192, who determined that the mean error did not exceed 0.5%.

ing month, using a similar apparatus but including a mercury manometer to measure gas pressure, the two experimenters widened their tests to include not only black powder but also a variety of high explosives. They made use of the gas law to convert the determinations of pressure into volumetric determinations.[39] An assortment of thermodynamical and thermochemical data were extrapolated up from the one-gram test weight to one kilogram, and tabulated for the test explosives.[40] The resultant high temperatures and pressures presumably mimicked those found in gun barrels when large charges of powder were actually exploded.[41] The following year, these studies were extended yet further to other high explosives.[42]

Vieille, Craftsman of the Bomb Calorimeter

In 1877, Sarrau succeeded Roux as director of the *Dépot Central des poudres et salpetres* and, in the same year, established a close relationship with Berthelot through their work on a commission investigating means of preventing firedamp explosions. Their association was enhanced in 1878 when they both became members of the eight-member *Commission Scientifique des Substances Explosives* constituted by the War Minister (Berthelot was named President). Through these associations, Berthelot came to know a young protégé of Sarrau

[39] In using the gas law, they departed from Berthelot's procedure. Roux and Sarrau, "Recherches experimentales sur les matieres explosives," *CRAS* 77 (1873) [Séance of 18 August 1873]. *Les Explosives Modernes*, p. 181. The manometer was constructed by A. Clair. The explosives used included dynamite, guncotton, and picrate and chlorate of potassium.

[40] *E.g.*, heats and temperatures of explosion, volumes and pressures of permanent gases produced, maximum work. *Ibid.*, pp. 182–185.

[41] The computed explosion temperatures for the five types of black powder ranged from 3372°C to 4654°C; the pressures from 3792 atmospheres to 4339 atmospheres. *Ibid.*, p. 182.

[42] Roux and Sarrau, "Recherches expérimentales sur les substances explosives," *CRAS* 79 (1874) [Séance of 5 October 1874]. *Les Explosives Modernes*, pp. 186–192. The explosives tested were: fulminate of mercury, nytroglycerine, guncotton (pyroxyle), picric acid and potassium, barium, strontium and lead picrate, with a military powder as a baseline comparison. Médard and Tachoire find the results in these experiments less exact than those of the previous year (*e.g.*, the heats of explosion are too high by 6–9%). They are unsure as to the cause but suggest, in some cases (*e.g.*, fulminate of mercury) the air in the bomb could have produced supplementary oxidation of the products and hence additional heat. Médard and Tachoire, *Histoire de la thermochimie*, p. 192.

who was working at the *Dépot Central des poudres et salpetres*: Paul Vielle. Born in Paris in 1854, Vielle entered the *École polytechnique* in 1873 and upon graduation in 1875 went directly into the newly reorganized (1874) *Corps d'ingenieurs des poudres et salpetres.* Vieille attracted the attention of Sarrau, who secured for him a position at the *Dépot Central.* From 1877 until Sarrau's retirement in 1897, Sarrau and Vieille were to work closely together first at the *Dépot Central* and then from 1887 at the *Laboratoire Central des poudres et salpetres.*[43]

One of the first tasks that Sarrau set for Vieille was to improve "the simple and inexpensive" bomb calorimeter that he and Roux had earlier employed. Working with the principal mechanic of the *Dépot Centr*al, named Bianchi, Vieille produced two of what he initially called an *éprouvette calorimétrique.* (figure 1) In comparison to its predecessors, this instrument was not so much innovative as it was much more carefully and precisely crafted.[44] It consisted of a cylinder made of very tough hammered wrought iron of a wall thickness of 7 mm and a capacity of 305 cm^2.[45] One end had a spherical cap; on the other end was a flat-bottomed tray tightly joined to the wall of the cylinder by conic surfaces fitted to the two pieces and ground to each other. This part of the apparatus was tightened by means of a screw nut. Appended to the flat-bottomed basin was an apparatus for sparking the test material by an electric filament and a stopcock for enabling the gaseous products to be collected.[46] One of the two original calorimeters was lined with platinum foil; this became standard when test materials were exploded in an atmosphere of excess oxygen.

To determine heats of reaction, the bomb was immersed in a calorimeter made of copper and surrounded by a double water envelope.

---

[43] Information on Vieille from Amiable, "Les instruments scientifiques de Paul Vieille au 19ème siècle" (note 2), p. 97 and Henri Tachoire, "Les techniques de mesure à la bombe calorimétrique. Des travaux de Paul Vieille aux méthodes de micro-mesure et à la bombe rotative," *Troisièmes journées scientifiques Paul Vieille* (note 2), p. 109.

[44] Amiable, "Les instruments scientifiques de Paul Vieille," p. 98.

[45] *Ibid.*, 99; Tachoire, "Les techniques de mesure à la bombe calorimétrique," p. 110. Médard says that the bomb was made of hammared steel [*acier embouti*]. Louis Médard, "L'oeuvre scientifique de Paul Vieille," *Revue d'histoire des sciences* 48 (1995): 390. The wall thickness is given here. Médard also states that it could support a pressure of more than 250 atmospheres (p. 392).

[46] In subsequent versions of the bomb, the gas collection stopcock was placed on the other end from the sparking apparatus. Émile Sarrau, *Théorie des explosifs* (Paris, 1895), pp. 11–13.

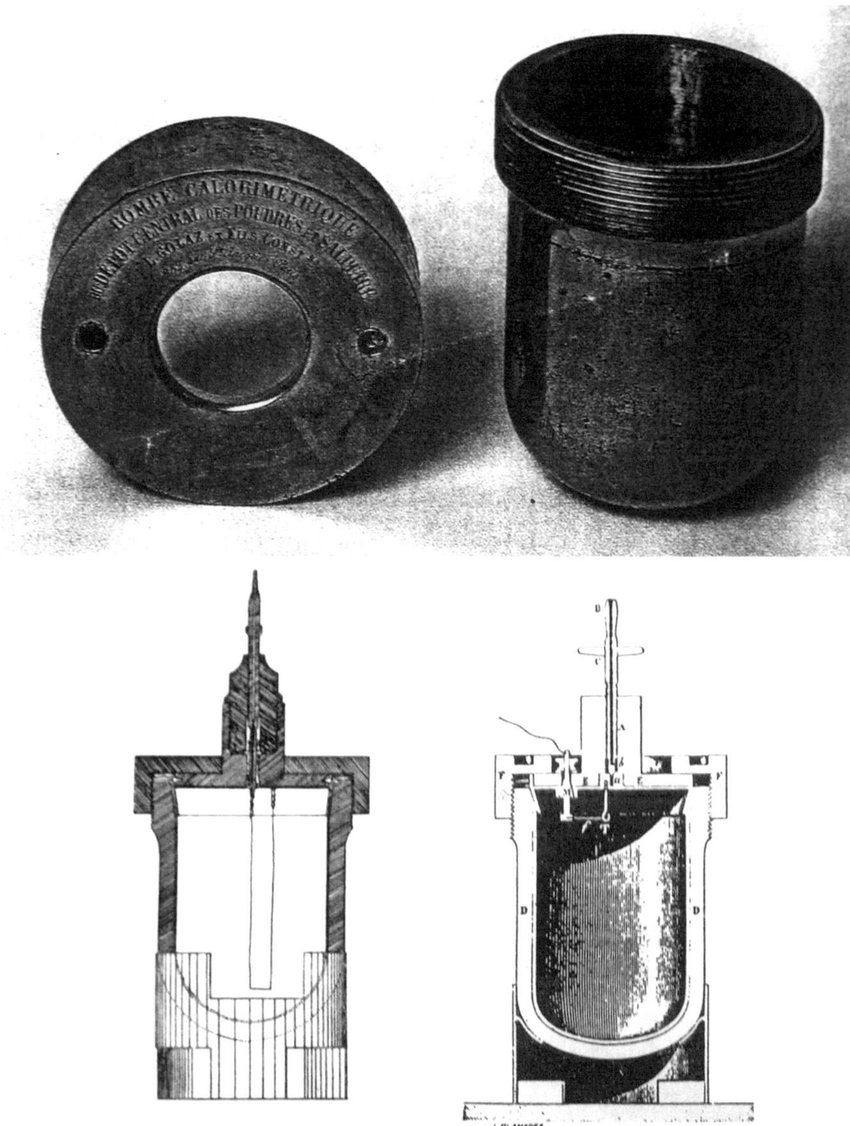

Figure 1. Paul Vieille's bomb calorimeters. Top: bomb constructed by Golaz (1885). Bottom: first model constructed by Bianci (1879, left) and second model by Bianci (right). From Médard & Tachoire [1994], pp. 202, 204.

The calorimeter was a cylindrical vessel 140 cm in diameter and 180 cm in height, holding 1.8 kg of water. The temperature was determined by a Baudin thermometer graduated to one fiftieth of a degree and allowing evaluation down to one two-hundredths of a degree.[47]

Vieille, in collaboration with Sarrau and Berthelot, carried out extensive heat of reaction determinations though the mid-1880s. Their first publication in which the new *éprouvette calorimétrique* was reported concerned guncotton made at the powder-works at Moulin-Blanc.[48] Sarrau and Vieille, as military engineers, were also interested in measuring directly the pressure of explosion of guncotton. For this, they exploded the guncotton in an *éprouvette cylindrique* that used a crusher gauge of Andrew Noble's design.[49]

Berthelot himself was led to return to the experimental study of heats of reaction by means of Vieille's *éprouvette calorimétrique*. On first viewing this instrument, Berthelot immediately comprehended the revolutionary effect that it would have on thermochemical experimentation[50] and, by the end of 1879, he had had two similar models (but slightly smaller capacity than Vieille's) constructed by a mechanic named Golaz. In fact, it was Berthelot who gave the first printed description of the instrument, which he called a "petite *bombe* ou *détonateur calorimétrique*."[51] His name, the bomb calorimeter, became

[47] This account has been put together from a number of different sources: Amiable, "Les instruments scientifiques de Paul Vieille" (note 2), pp. 98–99; Tachoire, "Les techniques de mesure à la bombe calorimétrique," pp. 109–110; Médard and Tachoire, *Histoire de la thermochimie* (note 14), pp. 205–207; Sarrau, *Théorie des explosifs*, pp. 11–13.

[48] Sarrau and Vieille, "Recherches expérimentales sur la décomposition du cotonpoudre en vas clos," *CRAS* 89 (1879): 165–166. Other data given included the temperature of the calorimetric water bath (18°C), the mean density of the products (0.023), and the pressure (250 kg/cc^2). The reaction was carried out in an atmosphere of nitrogen. A test weight of seven grams yielded a reaction heat of 1045 cal/g. According the Médard and Tachoire, the published value for the heat of reaction was only 0.5% different from the later, standard value. *Histoire de la thermochimie*, p. 206, n. 15.

[49] *Ibid.*, p. 165. A version of the instrument is described in Hake and Macnab, *Explosives and their Power* (note 17), pp. 21–22 (with picture). The eprouvette was strengthened "according to Schultz's system" by steel wire of 0.8 mm in diameter wound fifteen times around on the cylindrical tube at a tension of 35 kg.

[50] Recollection of Henry Le Chatelier cited in Médard, "L'oeuvre scientifique de Paul Vieille" (note 45), p. 391.

[51] Berthelot, "Sur la chaleur de formation des oxydes de l'azote," *CRAS* 90 (1880): 780, italics in original. It had a capacity of 200 cc. A detailed account (with pictures) was given by Berthelot in "Appareils pour mesurer la chaleur de combustion de gas par détonation," *CRAS* 91 (1880): 188–191. The original apparently suffered interior degradation and leaked gas, and was consequently refined by Golaz in May, 1880. Médard, "L'oeuvre scientifique de Paul Vieille," pp. 392–393. Regarding the

standard and Berthelot himself subsequently received credit for Vieille's (and Bianchi's) conception.

As the title of the paper in which Berthelot described his instrument indicates, Berthelot had resumed the research on the heats of formation of compounds of nitrogen and oxygen that he had started (and abandoned) in 1874, specifically using the thermal differences of the detonation of cyanogen by oxygen and by nitrogen dioxide [AzO_2 for Berthelot; NO in modern notation] to determine the heat of formation of this latter as well as several other nitrogen-oxygen compounds.[52] In the tabulation at the end of his paper, Berthelot fulfilled his earlier program by presenting the heats of formation for all the oxides of nitrogen, the nitrates (including nitroglycerine, nitrobenzene, and dinitrobenzene) as well as various ammoniacal salts.[53]

Berthelot's program of determining heats of formation was taken up by Sarrau and Vieille. Their reliance on Berthelot's thermochemical framework was emphasized at the start of their first paper:

> The researches of M. Berthelot have highlighted the importance that the determination of the heat released by the decomposition of an explosive has for the theoretical evaluation of its force. It is also this element what enables the determination of the maximum work that the explosive seems theoretically capable of furnishing.[54]

Their thermochemical model of the explosion reaction was exactly the same as Berthelot's. Hence they, too, prioritized the determination of heats of formation of explosives and of their products.[55]

"legend" of Berthelot's invention of the bomb calorimeter, see Médard and Tachoire, *Histoire de la thermochimie*, pp. 209–211.

[52] Combustion of 26g of cyanogen by oxygen yielded a means of 130.9 cal for 3 readings (constant volume); by nitrogen dioxide [NO_2], 174.6 cal (same parameters). The difference, 43.7 cal, represented the heat *released* by the decomposition of two units of nitrogen dioxide [NO_2]. Therefore, the heat *absorbed* in the formation of one unit of nitrogen dioxide was −21.8 cal. For the proto-oxyde [AzO for Berthelot; N_2O in modern notation], Berthelot obtained a value of −10.3 cal by combustion with CO (oxyde de carbone) [corrected by doubling to −20.6 cal]. Emile Sarrau, *Théorie des explosifs* (Paris, 1895), p. 84. Berthelot, "Sur la chaleur de formation des oxides de l'azote," pp. 781–783.

[53] *Ibid.*, pp. 783–785.

[54] Sarrau and Vieille, "Sur la chaleur de formation des explosifs," *CRAS* 93 (1881): 213.

[55] "En effet, quand un explosif se decompose, la chaleur degagee est egale, d'apres un principe fondamental de Thermochimie, a l'exces de la chaleur de formation des produits sur la chaleur de formation de l'explosif." Or: $q = q_1 - q_o$, therefore $q_o = q_1 - q$, where q = heat released in explosion; q_1 = heats of formation of products; q_o = heat of formation of explosive. *Ibid.*, pp. 213–214.

However, they introduced a major change of method in order to avoid "uncertain results, either of the analysis of products or the determination of their heats of formation": they combusted the elements of a substance, in producing its decomposition, in an excess of oxygen.[56]

Their method of determination captured the thermochemical theoretical position. A number of high explosives (nitroglycerine, nitro mannitol [*nitromannite*], and guncotton) were combusted with excess oxygen in a calorimeter whose container they styled as "identical to that which has been adopted and described by M. Berthelot."[57] Their method was to determine the heat released in each combustion, and subtract that figure from that of the calculated heats of formation of the products of the explosion (previously ascertained).[58] Working with Vieille, Berthelot, in turn, soon adopted the use of combustion with oxygen to determine heats of formation.[59] Lastly, they refined this method of combusting solid hydrocarbons with oxygen under pressure (of about seven atmospheres).[60]

The tradition of bomb calorimetry that I have recounted parallels the experiments of Bunsen and Schischkoff in being largely thermochemical laboratory investigations. But the apparatus, particularly as developed by Vieille and Bianchi, became much more refined and precise, and much better able to cope adequately with powerful explosives, high temperatures, and high pressures than had the apparatus of Bunsen and Schischkiff. The bomb calorimeter became an important instrument of laboratory and industrial chemistry in the twentieth century.[61]

---

[56] *Ibid.*, p. 214. Berthelot himself had hitherto used an atmosphere of nitrogen, *e.g.*, in one of the early studies he carried our with Vieille: "Etude des propriétés explosives du fulminate de mercure." *CRAS* 90 (1880): 948.

[57] They gave one of the first reasonably complete descriptions of Vieille's *éprouvette* here. Sarrau and Vieille, "Sur la chaleur de formation des explosifs," *CRAS* 93 (1881): 214.

[58] Values of heats of formation for 1 equivalent: nitroglycerine (227 grams): 94.0 cal; nitromannite (452 grams): 161.5 cal; guncotton (1143 grams): 639.5 cal; potassium picrate (267 grams): 117.5 cal; ammonium picrate (246 grams): 80.1 cal; Sarrau and Vieille, "Chaleur de formation des explosives; donnees numeriques," *CRAS* 93 (1881): 269–270.

[59] Berthelot and Vieille, "Sur la chaleur de formation de perchlorate de potasse," *CRAS* 93 (1881): 289–291.

[60] Berthelot and Vieille, "Nouvelle methode pour la mesure de la chaleur de combustion de charbon et des composes organiques," *CRAS* 99 (1884): 1097–1103. This was later raised to 24 atmospheres. Berthelot and Recoura, "Sur la bombe calorimétrique et la mesure des chaleurs de combustion," *CRAS* 104 (1887): 875.

[61] Eugene S. Domalski, "Calorimeter, Bomb," in Bud and Warner (eds.), *Instruments of Science* (note 4), pp. 82–84.

How Powder Really Works

Research with the bomb calorimeter was not very helpful in elucidating the military performance of explosives, particularly as propellants in weapons. What this research did *not* reveal was the crucial data of how explosions proceeded over time, and how explosive material actually combusted under the conditions that obtained in the barrel of a gun. To get at these issues, Vieille embarked on a research program complementary, as it were, to the thermochemical research of the bomb calorimeter. This one was focused in another instrument he devised: the *manomètre enregistreur*, developed by 1881, no doubt as an elaboration of the *éprouvette cylindrique* he and Sarrau used two years earlier. The perspectives of the laboratory and of the testing field were melded in its use as in no other previous instrument for the study of explosives. Vieille himself gave cogent voice to the position between industry, field and laboratory that his instrument held in his great retrospective narrative, published in 1893:

> Between the fabrication, properly speaking, of explosives, that is, their industrial production, and their utilization in weapons of war, that is, the determination of their ballistic effects, it seems indispensable to make a place for what one might call *physiological study of explosives*, laboratory investigations in which their intimate mode of function could perhaps be analyzed without field test with the aid of simple apparatus, whose handling is silent [and] not dangerous for the environment, if not in an absolute fashion for the operator.[62]

The particular "physiological study" that Vieille took up was the question of how gunpowder actually burned in a gun. The accepted view when Vieille began his research was that of the premier ordnance researcher of mid-century: Guillaume Piobert. Piobert had studied the dynamics of gunpowder explosion: how combustion took place, and how its speed, pressure, and other factors were correlated with physical characteristics of powder, such as size, shape and density of powder grains. Like Sarrau and Vieille after him, Piobert brought the experimental orientation of the *polytechnecien* to his study:

> It is . . . necessary to bring to the examination of these phenomena an analytical method, which proceeds slowly and step by step to reach the truth more surely. In following this way, one must begin with the results

[62] Vieille, "Étude sur le mode de combustion des matières explosives," *Mémorial des poudres et salpetres* 6 (1893): 256–257, my italics.

> of experience, isolate, as best one can, particular facts without denaturing them, study them separately, then endeavor to connect them by utilizing all that the physical sciences have taught us concerning the nature and properties of the material which one has under consideration.[63]

His objective was to arrive at simple but quantitative "laws" governing the way gunpowder burned, but he recognized that observing the actual burning of real military powder could at best yield a sketch of what actually took place because of the small size of gunpowder grains and speed and complexity of gunpowder explosion. Therefore it was necessary to construct an experimental situation that would yield more details about the process of explosion—detail from which one could infer what happened with actual powder grains.

He did this by "scaling up" the powder grains to parallelepiped shaped powder cakes, whose lateral sides and bottom were protected by coats of tallow or lard and whose lower end was placed in water. Carrying out the combustion in air, Piobert found that the "free" surface of these cakes burned slowly enough and the inflammation was passed to the next layers with sufficient regularity as to gain data to arrive at some laws of inflammation. The first such law was that of the "uniformity" of burn: "The speed of communication of the fire to the interior of the same type of powder remains invariant, even if the surface area is greater or less." But secondly, the speed of propagation of combustion was inversely related to the density of the powder. It therefore, "results from the preceding law that the quantity of matter burned in equal times is the same, for equal surfaces, for each type of composition, regardless of the density."[64]

Having established "laws" of burning for the same type of powder, Piobert then compared the speeds of burn of different types of powder, those fabricated by edge-runners, stamping mills or steel balls of trituration, and composed of either black or red (distilled) charcoal.[65] Piobert inferred that the law of uniformity of burn applied to individual powder grains as well as to larger cakes; taking the simplest case of spherical grains, he postulated a uniform burn from

[63] G. Piobert, *Traité d'artillerie théorique et pratique: Partie théorique et expérimentale* (Paris, 1847), p. 103. He concluded the *Traité* by claiming that his "theory" of the effects of powder as "a theory founded uniquely upon experience and disengaged from all hypothesis." (p. 382).

[64] *Ibid.*, pp. 108–117. Quotations from 111 and 117.

[65] *Ibid.*, p. 135.

circumference to center. Cannon powder grains were sufficiently close to spherical grains that the same law should apply to them.[66]

"Following the method of Piobert," Vieille proposed to study directly the influences that the various components of the process of powder-making had on the mode of combustion of powder but, "in working these combustions, not in the open air, as did Piobert, but under pressures comparable to those found in actual firearms."[67] The apparatus Vieille had designed to replicate in the laboratory the conditions in a gun, the *manomètre enregistreur*, consisted of two parts: (a) an *éprouvette* and (b) a rotating cylinder. (figure 2) The *éprouvette*, where the actual testing of the explosive took place, was a cylindrical tube of variable dimensions, depending on the nature of the powder being tested. The tube was threaded at the extremities, each of which held a steel plug. One served as the firing device and the other was connected to a crusher gauge of the type developed by Andrew Noble.[68]

The crusher gauge was connected to the second part of the apparatus, the rotating cylinder, by an appendage attached to the head of the gauge's movable piston, which protruded beyond the *éprouvette*, and communicated to a steel foil nib. The nib was placed so it could make a trace on blackened paper affixed to a bronze cylinder of 14 cm in diameter turning on an axis parallel to the axis of the *éprouvette* and operated by an electric motor by means of a pulley and wire.[69] The linear speed of the rotating cylinder could be varied between 0.50 meters and 10 meters per second. The apparatus operated as follows:

> Before the explosion of the charge, the nib traces a circle on the rotating cylinder corresponding to the initial position of the piston [of the crusher gauge]. When the charge is set off, the piston moves, crushing the cylinder [the copper cylinder of the crusher gauge] and the nib traces a curve on the rotating cylinder. Then, with the combustion ended, the piston stops moving and the nib traces a second circle corresponding to the final position of the crusher piston.

[66] *Ibid.*, p. 155.
[67] Vieille, "Étude sur le mode de combustion des matières explosives," p. 259.
[68] *Ibid.*, pp. 264–267. Details of how the firing device and crusher gauge were connected to the central cylinder were given in the text. For powder of small grain size, a cylinder of capacity of 22 cc was used; for large-grain powder and colloidal powders, cylinders of 75 cc, 150 cc and 350 cc capacity were used.
[69] A Marcel Deprez electric motor or a "moteur à poids" was used.

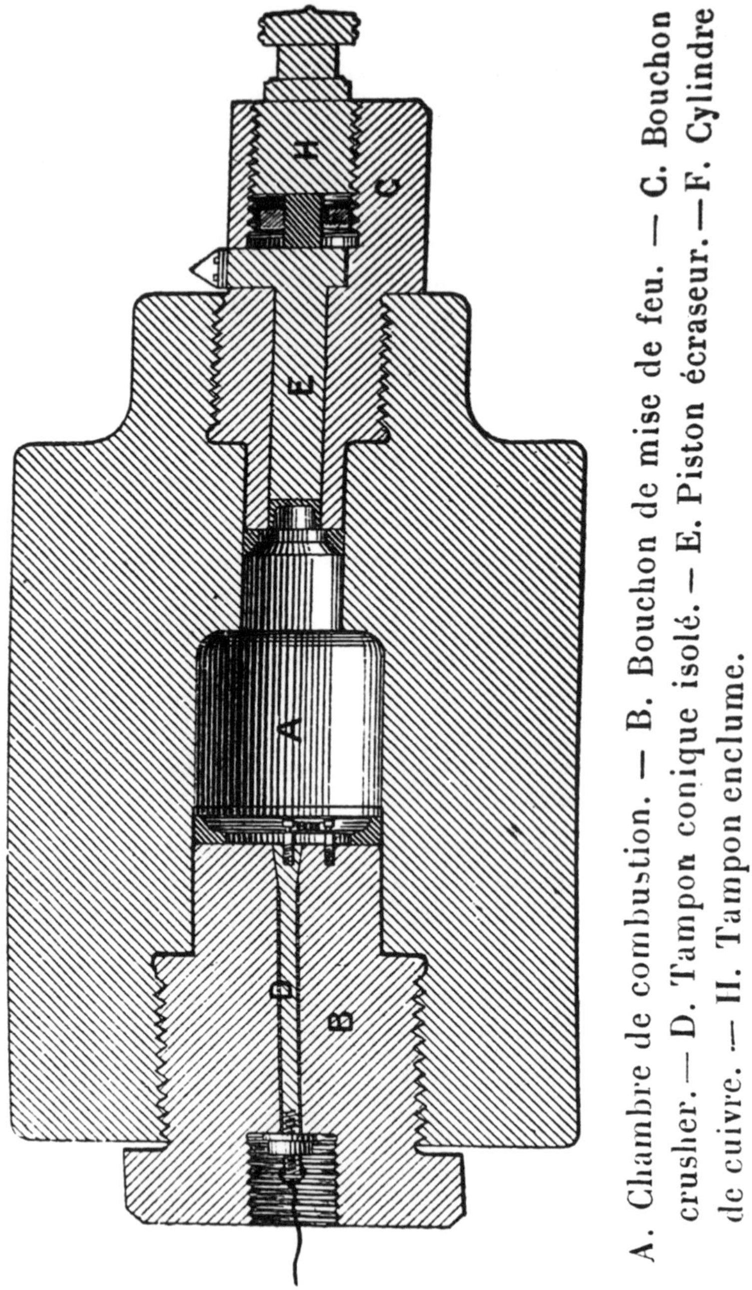

Figure 2. Paul Vieille's *manomètre enregistreur* (1893), designed to record the performance of powder under conditions analogous to being fired in a gun. F, unlabeled, is the darker, diagonally-hatched part between E and H. From Vielle [1893], p. 265.

> The distance between the initial and final circle, measured by a marker of the rotating cylinder is equal to the amount of crushing undergone by the copper cylinder.[70]

The time of the tracing (and therefore of the motion of the piston in the crusher gauge) was calibrated by another tracing by a nib connected to an electrically operated tuning fork. This second tracing was of sinusoidal waves, each of which rigorously corresponded to a pre-determined time interval.[71]

Copper obturator rings sealed the steel plugs of the cylinder and one in the form of a cap sealed the movable piston and prevented the escape of gas. The whole apparatus was placed in a cage of thick sheet metal; this was sealed, in turn, in a strong housing. (figure 3) At the moment of firing, the cage could be shut by iron flaps supported by solid latches. The temporal scale of the sinusoidal waves was determined with a micrometer; the crusher gauge curve was established along two axes by microscopic inspection.

This instrument was the climax of the attempts to gain experimental control over phenomena whose characteristics—very short duration, very high temperatures and very intense pressures—presented the greatest challenges to the scientific researcher. As Louis Médard, one of the scholars of Vieille has put it, "At first glance, it would seem chimerical . . . to record correctly by mechanical means a movement whose amplitude is of the order of 5 mm and whose duration lasts from 3/10000 to 2 or 3 hundredths of a second."[72]

The principal research program for which Vieille used the *manomètre* was to re-examine Piobert's law of "uniformity of burn" for explosion of gunpowder in a weapon. Vieille comparatively studied an extensive array of powders of many types and countries, systematically varying and testing a wide variety of characteristics and manufacturing techniques. These included powder density, grain size, mode of trituration, type of charcoal used, glazing of the powder grains. These were similar to the range of characteristics that Piobert had investigated decades before, using much less refined means of testing rate and pressure of burn of the powder.

---

[70] Vieille, "Étude sur le mode de combustion des matières explosives" (note 62), p. 266.

[71] Vieille favored a tuning fork giving 500 double vibrations/second.

[72] Attributed to Louis Médard in Amiable, "Les instruments scientifiques de Paul Vieille" (note 2), p. 101.

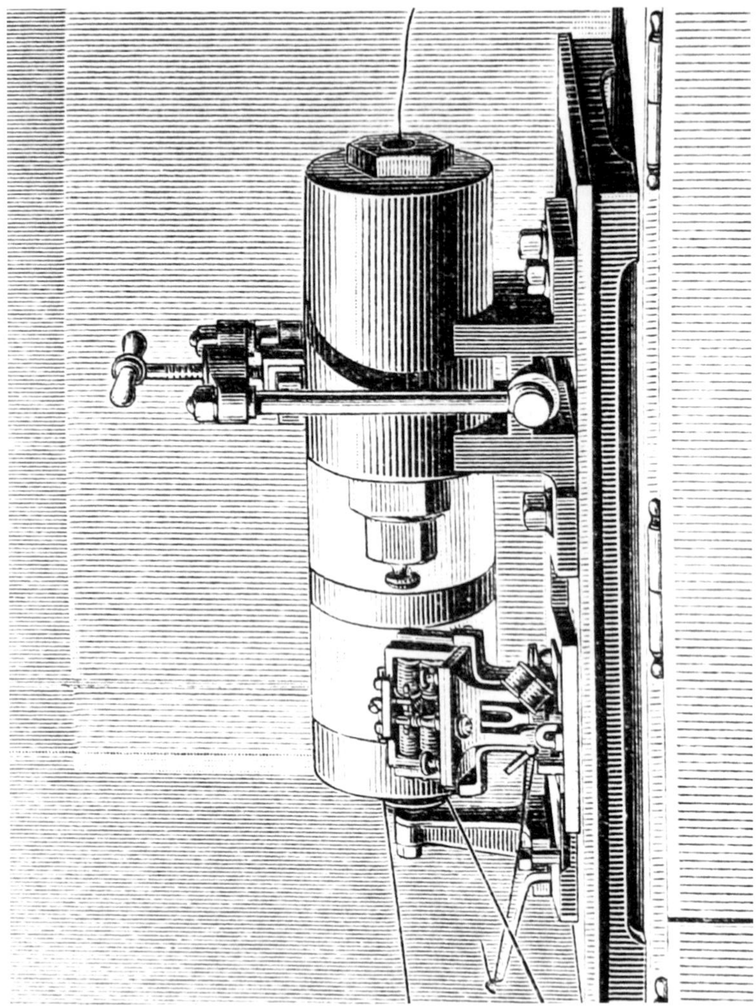

Figure 3. Vieille's overall experimental setup for his recording manometer, showing (from left to right): the carbon-black paper covered rotating cylinder with tuning fork recording head, the manometer itself, and electrical detonation leads. From Médard & Tachoire [1994], p. 256.

Uniformity of burn was defined by Vieille as burning of gunpowder by "concentric layers".[73] Operationally, this meant that the period of burn of a particular powder (as determined by the above graphical procedure) was proportional to the thickness of the test samples. Vieille discovered (contrary to Piobert) that burning by concentric layers was not at all the case with military powders:

> The fundamental fact that emerges from these comparisons is that the period of combustion of the materials used in the fabrication of black or brown powders is actually markedly independent of their thickness and, as a result, *these materials do not burn by concentric layers.*[74]

Moreover, the external shape and dimensions of the powder grains seemed to make very little difference regarding regularity of burn.

Vieille did note that for powders of very high density there was "a proportionality approaching that of times to thicknesses and, thereby, combustion by parallel layers."[75] After many more tests, Vieille enunciated a general schema and a theoretical explanation for what he had observed. Density (or, better, "compression") seemed to be the critical factor. He distinguished four stages or "periods" (in his terminology) of compression, corresponding to very different modes of combustion. For powders of very low compression (the first stage), the times of combustion were completely independent of the thicknesses of the test samples and, indeed, did not differ from those of their constituents. Powders of greater compression (the second stage) exhibited periods of burn four or five times slower than that of elementary grains of powder-dust but they were still independent of the thickness of the test samples. These powders corresponded to normal black or brown artillery powders. With powders of even greater compressibility (the third stage), "the influence of thickness [on duration of burn] is increasingly apparent" but these materials were very irregular in behavior. Finally, with the compacted material of the fourth

[73] His phrase was "couches concentriques", Vieille, "Étude sur le mode de combustion des matières explosives," p. 293. He also used "couches parallèles" (p. 291) and "surfaces parallèles" (p. 288).

[74] *Ibid.*, p. 293, italics are Vieille's. Although Vieille wrote of testing the "matières servant à la fabrication des poudres noires et brunes," he was, in fact, reporting comparative tests of actual powders, mostly prismatic, some of which were of very high density. Tables of results, pp. 292–293. On p. 296, Vieille reiterated his conclusion, adding "sous les pressions comparables à celles qui sont réalisées dans les bouches à feu."

[75] *Ibid.*, p. 299.

stage, the times of combustion slowly increase with the density and are proportional to the thicknesses of the test samples. This last period corresponded to a highly compacted material of "perfect continuity."[76]

Vieille's theoretical explanation for why powders of his first three stages did not burn in time periods proportional to the thickness of their test samples focused on their material discontinuity. For the least compressed powders of the first stage, for example:

> There exists a system of interstices in the charge that assures the inflammation of all the grains without any appreciable delay or, at the least, in a negligible time period in comparison with the characteristic time of combustion of the grains.[77]

For more compressed powders of the second stage, the inter-grain interstices were largely eliminated by "a kind of entanglement of the grains" but they were replaced by a new set of channels of dimensions similar to the original ones but now between "polyhedral nuclei formed by the partial agglomeration of several grains."[78] For Vieille, this scenario explained both how second stage powders could have slower periods of combustion which was, nevertheless, still independent of test sample thickness.

With the third stage, the growing agglomerated powder nuclei gradually eliminated the interstitial canals and the proportionality between duration of burn and thickness of sample began to be apparent; yet lack of complete continuity gave rise to irregularities in behavior of the powder. These disappear with the fourth stage of compression, where the powder approached "the limit corresponding to perfect continuity" and the duration of combustion, changing only very slowly with increased compression, became proportional to sample thickness. Regularity of burn marked this stage.

As this model makes clear, degree of compressibility, or density, was the critical factor in the way powder combusted for Vieille. As a result, he discounted completely a factor that had been taken as the prime one in understanding (and predicting) the powder would burn in a weapon since the dissemination in the 1860s of T.J. Rodman's research: powder grain size (and shape).[79] "Thus," he concluded,

---

[76] *Ibid.*, pp. 321–323.

[77] *Ibid.*, p. 322.

[78] *Ibid.*, p. 323.

[79] First clearly set forth in T.J. Rodman, *Reports of Experiments on the Properties of Metals for Cannon, and the Qualities of Cannon Powder, with an Account of the Fabrication and Trial of a 15-Inch Gun* (Boston, 1861).

> the dimensions of [powder] grains, which powder makers consider to be the principal variable that one must be able to manage to produce slow burning powders, only plays a secondary role to all intents and purposes in the function of the resultant products. The improvements achieved in this process result from minor modifications dependent on the increase in density and on the size of the elementary powder grains in the largest [grained] powders.[80]

As an example, Vieille cited the case of prismatic powders, the large grained powders dominant in the 1880s and derived from Rodman's own "perforated cake cartridges." To Vieille, their success was due to the "cessation of a useless search for an increase in the slowness of powder in the dimensions of grains" and to its replacement by "development of the true variables in the formation of slow [acting] materials, whose preponderant role we have recognized." Vieille cited changes such as the reduction of sulfur content, the use of charcoal "peu carbonisé," ("lightly carbonized"—presumably Vieille was referring to brown [cocoa] prismatic powder), and the production of a regular elementary grain not intermixed with powder dust as the prime examples of his "true variables."[81]

Laboratory Research and the Invention of Smokeless Powder

By the time Vieille published this study on the manner of combustion of explosives, the issue of the role of size of powder grain versus other factors in the efficacy of prismatic powder, or, for that matter, all traditional military gunpowder, was rapidly becoming academic. They were being superceded by a new kind of powder, one that was perfectly fitted to Vieille's fourth stage of regularity of burn and "perfect continuity." These were the "colloidal powders," composed of guncotton "gelatinized by the action of appropriate solvents," subsequently removed by volatilization.[82] The final research

[80] Vieille, "Étude sur le mode de combustion des matières explosives," p. 356.

[81] *Ibid.*, pp. 356–357. The role of the new charcoal was to implement slowness because of "une remarquable propriété des charbons faiblement carbonisés. Ces charbons conservent, en effet, une rétractilité qui permet aux poudres d'augmenter de densité au course du séchage." The regularity of elementary grain and lack of powder dust resulted from "la facilité de chargement des moules prismatiques."

[82] *Ibid.*, p. 357. This was true of so-called single base powders. In others, termed double base (ballistite, cordite), the solvent, nitroglycerine, remained a component of the powder.

section of Vieille's study was devoted to showing that these powders did indeed fulfill his criterion for regularity of burn: the duration of combustion was proportional to the thickness of test samples.[83] As Vieille somewhat ingenuously noted at the start of his discussion of these powders, "this type of explosive had been introduced in France at the end of 1884, and its use has been progressively generalized in every European country since then."[84] Indeed, it was Vieille himself who developed the first of these, the famous single base powder made of gelatinized nitrocellulose, "Poudre B".[85]

It has been asserted for a long time that Vieille's studies of the mode of combustion of explosives utilizing the *manomètre enregistreur* was the essential context for his development of the first successful military colloidal powder.[86] The logic of the "Étude sur le mode de combustion des matières explosives," as recounted here, would appear to support this. Vieille had demonstrated, after all, that colloidal explosives were at the positive end of his spectrum of explosive materials regarding regularity of burning. Therefore, the "speed" of these powders could be controlled by regulating the shape and thickness of the powder charge. This was made all the more easy for colloidal materials since they were horn-like, flexible and easily worked to any desired shape.[87]

Veille began his 1893 "Étude" by stating that he had begun the research recounted in it "about ten years ago" and "most of the results reported here were obtained in the years 1884 and 1885."[88] As has also been noted, Vieille gave the date of "the end of 1884" as the time when colloidal powder was "introduced in France," which we can take to mean: successfully developed by him. In fact, labo-

[83] *Ibid.*, pp. 357–364.

[84] *Ibid.*, p. 357.

[85] Originally called "Poudre V" (after Vieille). J. Challeat, *Histoire technique de l'artillerie de terre en France pendant un siècle (1816–1914). Tome second (1880–1914)* (Paris, 1935), p. 215. Challeat referred the reader to Vieille's "Étude sur le mode de combustion des matières explosives," as an account of the discovery of Poudre B (pp. 210, 212). The exact chemical formula and the mode of fabrication of this powder was a state secret for many years, the first such state secret that I know of regarding powder fabrication. Marshall gave its analysis as: 68.2% insoluble nitrocellulose, 29.8% soluble nitro-cellulose, 2.0% paraffin. Gelatinization was by means of acetic ether. Arthur Marshall, *Explosives*, 3 vols. (London, 1917–32), I:294–295.

[86] Challeat, *Histoire technique de l'artillerie de terre en France*, pp. 210–215.

[87] Vieille, "Étude sur le mode de combustion des matières explosives," pp. 365–366.

[88] *Ibid.*, p. 256.

ratory notebooks of Vieille's survive from this period.[89] The first ten cover the period July 1881–September 1885. Unfortunately, the ninth notebook, covering what appears to have been the crucial month of December, 1884, has been missing for many decades. Through the good office of M. René Amiable, I had the opportunity to examine the available notebooks a few years ago.

Although only examined briefly by me, the notebooks appear to bear out the general contention that Vieille's work with the *manomètre enregistreur* was occupying most of his research time in the early and mid-1880s. But they do not do so in quite the logical, linear fashion of the narrative of the "Étude sur le mode de combustion des matières explosives." Rather, they illustrate a characteristic of Vieille's research that has marked the research of many great laboratory scientists: he was engaged on a number of independent, if related, research projects, and his focus shifted back and forth from one to the other.[90]

In October, 1881, for instance, Vieille seems to have been working out the details of the *manomètre enregistreur* and beginning to measure the "pression & durée d'ecrasement [pressure and time of crusing]" of powder—but also of nitrocellulose.[91] During much of 1882, Vieille turned his focus directly to the study of different types of nitrocellulose, examining both chemical composition and chemical properties (such as solubility and gelatinization by ether-alcohol) and physical properties such as the pressures developed in closed chambers.[92] In this same period (June 1882), he also began investigating

[89] Carton "Manuscrits redigés par Paul Vieille (note de laboratoire)" in the archives of the Bibliotheque de l'Ecole Polytechnique (hereafter BEP). To the best of my knowledge, there is no further identification of these notebooks. Each notebook is denominated a "Carnet de laboratoire,"; they are sequentially numbered and dated.

[90] This characterization has been elucidated most notably for Claude Bernard and Antoine-Laurent Lavoisier by the late Frederic L. Holmes: *Claude Bernard and Animal Chemistry: the emergence of a scientist* (Cambridge, Mass., 1974) and *Lavoisier and the Chemistry of Life: an exploration of scientific creativity* (Madison, Wis., 1985). Vieille was no doubt also carrying out commissioned testing of various explosives.

[91] Vieille, Carnet de laboratoire: No. 1 (1881–1882) [BEP]. At the same time, he was still working out the mechanics of the piston and the details of his recording device. October, 1881 is a bit earlier than the date of inception given (or at least implied) in his 1893 article.

[92] Vieille, Carnet de laboratoire No. 2 (2 March 1882–10 July 1882) [BEP]; Carnet de laboratoire No. 3 (11 July 1882–1 December 1882) [BEP]. Starting on 18 October with the heading, "Tensions en Vase clos du Fulmicoton" Vieille tested a wide variety of nitrocelluloses for the pressures they developed in closed vessels.

the properties of one of the principal "proto" smokeless powder, Schultze's powder.[93]

These two research programs left published traces in the *Comptes Rendus* of the *Académie des Sciences* in 1882. In the 3 July 1882 session Vieille and Sarrau published an article analyzing the mechanics of how Noble's crusher gauge actually worked. The rhetoric of the introduction is imbued with the analytical, mathematical, and theoretical perspective of the *École polytechnique*:

> But the nature of the numerical evaluations which these indications [of the deformation of the cylinder] are capable of giving has not been sufficiently defined up to now, and the use of it which has been made for the measure of pressures has not been sufficiently justified.[94]

They followed this up two weeks later, at the 17 July session, with a generalized (and mathematical) analysis of how the *manomètre enregistreur* worked.[95]

Immediately after this second paper appeared, one by Vieille alone on the degrees and limits of nitrification of nitrocellulose. This research was the most ambitious survey of the chemistry of nitrocellulose that had yet appeared. Vieille had nitrated cotton under carefully controlled conditions with nitric acid of varying degrees of concentration, and with mixtures of sulfuric and nitric acid, tested for complete nitration with an iodine indicator, and estimated the degree of nitration by the volume of nitric oxide gas released from the nitrated products made from a standard weight of cotton.[96]

---

[93] Vieille, Carnet de laboratoire No. 2 [BEP]. Schultze's powder was made by nitrating wood grains and purifying the product in much the way that guncotton was purified. It was invented by a Prussian artillery officer, Major Schultze, and introduced around 1865. Its rate of burning was reduced through the addition of potassium and barium nitrate. In the early 1870s, further refinement was made by the holder of the Austrian patents (firm of Friedrich Volkmann): the nitrated wood grains were partially gelatinized by treatment with a mixture of ether and alcohol. This was marketed from 1872 to 1875 under the name Collodin. Although too violent for rifles, it was very effective in shotguns. See Marshall, *Explosives* (note 85), I:47–48 for details. "'Proto' smokeless powder" is my term.

[94] Sarrau and Vieille, "Recherches sur l'emploi des manomètres à écrasement pour la mesure des pressions par les substances explosives," *CRAS* 95 (1882): 26–29 at pp. 26–27. A comprehensive set of essays on the *École Polytechnique* was published on its bicentenary: *La formation polytechnicienne, 1794–1994* (Sous la direction de Bruno Belhoste, Amy Dahan Dalmedico, Antoine Picon) (Paris, 1994).

[95] Sarrau and Vieille, "Recherches sur l'emploi des manomètres à écrasement pour la mesure des pressions développées par les substances explosives," *CRAS* 95 (1882): 130–132.

[96] Vieille (presented by Berthelot), "Sur les degrés de nitrification limites de la

Vieille was particularly interested in the changes in the dissolution of nitrated cotton with degree of nitrification, noting how "narrow is the zone of nitro-cottons that can produce collodions and what small variations in the acid strength suffice to give products that are insoluble in ether." He noted that, in general, "the properties of nitrated celluloses vis-à-vis solvents depend only on their chemical composition."[97] In an elaboration of this paper published in 1883, Vieille also took up the research he had done on explosion pressure in closed chambers of different types of nitrocellulose; his conclusion was that his measurements "show that the force . . . diminishes with the percentage of nitration. . . . The percentage of nitrogen constitutes a true measure of the explosive qualities of a product."[98]

During 1883, Vieille continued his research on various types of nitrocellulose, no doubt for the expanded version of his article on nitration of cotton just cited.[99] But, in June, Vieille recorded testing another "proto" smokeless powder, "English military powder [*poudre de guerre*] called E.C." as he named it. In connection with this, there is a very suggestive reference in the notebook to an "infantry rifle [*fusil d'infanterie*]".[100]

In the first part of 1884, Vieille returned to focus on the experiments with various kinds of black powders (and others) recounted in the "Étude sur le mode de combustion des matières explosives." Indeed, it would seem that Vieille was still grappling with issues of the relationship duration of burn, thickness of test sample, and density of the powder in May and June 1884.[101]

---

cellulose," *CRAS* 95 (1882): 132–135. The iodine indicator was a solution of iodine in potassium iodide; the "method of Schloesing" was used to produce and estimate the volume of nitric oxide. He noted that the time necessary for nitration varied inversely with the concentration of the nitric acid used.

[97] *Ibid.*, pp. 133–134 for both quotations.

[98] Vieille, "Researches upon the Nitration of Cotton," trans. John B. Bernadou, in John B. Bernadou, *Smokeless Powder, Nitro-Cellulose, and Theory of the Cellulose Molecule* (New York, 1901), appendix I, p. 96. The article, dated September 1883, was originally published in the *Mémorial des poudres et salpetres*, vol. 2.

[99] Vieille, Carnet de laboratoire No. 5 (6 April—10 September 1883) [BEP].

[100] *Ibid.*, entry for 7 June. First entry was on 6 June 1883, where the reference to an infantry rifle was made. E.C. powder was not a military powder. E.C. powder (made by the Explosives Company at Stowmarket, England) was patented in 1882. Its nitro-cotton base was mixed with barium and potassium nitrate, coloring matter, and some other organic compounds. It was made up as grains, hardened by being partially gelatinized by ether-alcohol solvent. See Marshall, *Explosives* (note 85), I:48.

[101] Vieille: Carnet de laboratoire No. 7 (1 January 1884–25 June 1884) [BEP].

Further study of these notebooks will be needed to flesh out (and test) this account.[102] Although no dramatic breakthrough to Poudre B was recorded in these notebooks, what does emerge is a context for its invention. It concerns the interaction of the two general research programs on which Vieille was engaged in these years. The study of the nitration of cotton, particularly the gelatinization of certain levels of nitration in ether-alcohol solvent, interacted with the work on regularity of burn. This was perhaps as early as June 1883, when Vieille turned to E.C. powder, or perhaps not until he had sorted out issues of regularity of burn a year later. But it will take much more detailed study of the laboratory notebooks to sort out the chronology of Vieille's research.

Conclusion

The research programs developed to determine the force and behavior of munitions in the 1870s in France by Berthelot, Sarrau and Vieille and in England by Noble and Abel were the most scientifically ambitious devised up to that time. In both cases, a thermochemical orientation was combined with more direct physical measurements (for example, the crusher gauge for measuring explosion pressure). Moreover, the advancement of scientific knowledge was combined with the more practical goal of improving military munitions. The instruments were designed to mirror conditions that occured in a gun during explosion as closely as possible in a laboratory setting.

Although the two research programs had many similarities, there were also differences that mirrored national differences in research orientation. In particular, research characteristics of Vieille and some

As can be inferred from the numbering of the experiments cited in "Étude sur le mode de combustion des matières explosives," Vieille continued to work on this program throughout the 1880s.

[102] I have examined the 8th Carnet de laboratoire (October—[?] 1884) [BEP] and it has headings suggesting that Vieille was testing a nitrocellulose powder (*e.g.*, an entry for 21 October). A page from the now missing 9th Carnet de laboratoire, dated "19 Dec." is reproduced in Challeat's *Histoire technique de l'artillerie de terre en France* (note 85), p. 214, as an example of Vieille's testing of nitrocellulose-based smokeless powder, although the heading simply refers to "des Echantillons de Cot. aggloméré." In the 10th Carnet de laboratoire (January–July 1885) [BEP], "Poudre V" (Vieille original name for Poudre B) is mentioned as early as entry immediately following one dated "13 Janvier".

of his colleagues reflected long standing research traditions of the *École Polytechnique* and, perhaps, French science in general. Primary among these was a concern for rational, systematic research that was grounded in theory and explicitly was aimed at discovering—or better, proving—general natural laws. It is significant, I think, that Émile Sarrau published highly mathematical works on "theories of explosives".[103] The concern of Sarrau and Vieille with providing a theoretical basis for the operation of the crusher gauge is another example of this orientation.

The two instruments developed by Vieille that have been the focus of this paper, the bomb calorimeter and the *manomètre enregistreur* were both inscribed in general, theory-based research programs. If invented by Vieille and not by Marcellin Berthelot, the bomb calorimeter was utilized by Vieille in the service of Berthelot's program for comprehending the nature and behavior of explosives of all kinds under his general thermochemical theory. And the *manomètre enregistreur* was put to work to devise an updated general theory of how explosives burn under conditions approximating those actually obtained in a weapon at the moment of explosion. Vieille's study of nitrocellulose also illustrated the penchant for generalized, systematic studies.

The instruments themselves, although designed to enable the mimicking of extreme conditions, are also peculiarly French in their elegant, exquisite, and precise construction. If the comparison be permitted, they are quite different from the much more robust, no-nonsense "explosion apparatus" of Noble and Abel. In general, the rational, theory-based research conducted in the laboratory with these extraordinary instruments proved to be extraordinarily productive. It enabled an astute researcher like Vieille to arrive at conclusions—and a propellant—that might not have been possible using the more traditional and standard methods of testing the ballistic behavior of propellants on the military testing field.

In the twentieth century, the bomb calorimeter became more an instrument of chemical science and of industry than of war. The reason was as much theoretical as practical; namely, Berthelot's and his colleagues' thermochemical conception of the "force d'une matière explosive" proved, retrospectively, to have ignored a critical factor:

[103] *E.g., Recherches théoriques sur les effets de la poudre et des substances explosives: Force et travail des substances explosives* (Paris, 1874); Sarrau, *Théorie des explosifs* (note 52). These kinds of publications simply did not appear in England.

the rate of explosion. In the twentieth century, the application of a new scientific perspective, chemical kinetics, to the comprehension of the explosion reaction, brought this factor to prominence. But the principle behind Vieille's *manomètre enregistreur* remained important in the scientific study of military explosives. The technology of the instrument evolved in the second half of the century with the introduction of the piezoelectric transducer.[104]

I have described the English and French research programs as primarily scientific. But the French munitions engineers like Vieille were employed by the military arm of the government (as, I might add, was Abel in England). I would like to know more than I do now about how Vieille's research activities were affected by extra-scientific agents and objectives. Were his major projects of the early 1880s—the study of the manner in which munitions burned and the typological analysis of nitrocellulose—research programs of his own devising, or were they (partly or totally) set by outside military authorities? Was the development of a nitrocellulose based military powder part of his (or the military's) original plan, or did that emerge in the course of his more strictly scientific study of manner of burning? These are questions that have scarcely begun to be asked about the development of munitions before the twentieth century. The answers to them will give us a truly contextualized account of the development of some of the most sophisticated and important instruments for the study and development of nineteenth-century munitions.

[104] "Research Test Techniques Applied to Gun Interior Ballistics," in Herman Krier and Martin Summerfield (ed.), *Interior Ballistics of Guns* (New York, 1979), pp. 283–284; B. Baschung, D. Grune, "De la bombe manométrique de Vieille à la bombe plasma: un aperçu de l'évolution de la loi de combustion de Vieille," *Troisièmes journées scientifiques Paul Vieille* (note 2).

CHAPTER FIVE

TELEGRAPHING THE WEATHER: MILITARY METEOROLOGY, STRATEGY, AND 'HOMELAND SECURITY' ON THE AMERICAN FRONTIER IN THE 1870s

James R. Fleming

Introduction

To extend the metaphor implied by Tom Sandage's recent book, *The Victorian Internet*,[1] what if the U.S. Army Signal Office had a web site in the 1870s? It might resemble that of the current U.S. Department of Homeland Security:

> The tragic events at *Little Big Horn* show just how critical it is that America have a coordinated and comprehensive national strategy to help protect the United States against *domestic* threats or *Indian* attacks. . . .
>
> *General Albert J. Myer, head of the U.S. Army Signal Office* has been given authority to tackle this challenging mission. *General Myer* has answered the call to develop a new strategy to help our nation move forward after the tragedies of *June 25, 1876*. . . .
>
> *General Myer* will coordinate all federal government terrorist prevention and protection activities within the U.S., and will work with state and local governments on, among other things: detection, preparedness, prevention, protection, response and recovery, and incident management, *including telegraphing intelligence about storms, strikes, Indian uprisings, and other threats to domestic order.*[2]

Surprisingly, the modern language of homeland security captures a sense of anxiety and need for control applicable to the 1870s. I have merely made substitutions, where appropriate, for the terms "September 11," "Governor Tom Ridge," and "terrorism."

---

[1] Tom Standage, *The Victorian Internet: the remarkable story of the telegraph and the nineteenth century's on-line pioneers* (Berkeley, Cal., 1999).

[2] FAQ: Department of Homeland Security, <http://www.whitehouse.gov/response/faq-homeland.html> (accessed 2 April 2004). See also <http://www.dhs.gov>.

The story of the military's involvement in "telegraphing the weather," on the American frontier in the post-bellum era is part of a much larger story of modern meteorology. In 1831, at a meeting of the British Association for the Advancement of Science, James David Forbes bemoaned the state of meteorology and wondered if it could yet be called a science. He observed that it currently lacked sound theoretical grounding, standard methods of measurement and observation, and a worldwide system of data collection and communication. Over the course of the next four decades, national weather services and international cooperation gradually replaced earlier volunteer systems and ad hoc organizations. Experiments in weather telegraphy in the 1850s led to the establishment of storm warning services in the 1860s and 1870s. While it still could not be considered a coherent *science*, meteorology had found its niche, as a *service* provided by governments to the public through the medium of telegraphy. While the connection between the telegraphs and the railroads is well known, as is their connection to the military (at least during the Civil War), the conjunction of telegraphs, the military, and meteorology is not.

For the past two centuries the U.S. military has been an interested promoter of meteorology in general and instruments and measures to predict the weather in particular. During the War of 1812, Surgeon General of the Army James Tilton, motivated by prevailing environmental theories of disease linking illness and epidemics to weather and climate, issued a general order directing all hospital surgeons, mates, and post surgeons under his command to prepare quarterly reports as part of their official duties and to "keep a diary of the weather." For the next six decades, the Army Medical Department continued its support for meteorology by observing airs, waters, and places for the protection of the health of the troops.[3] In the 1830s the U.S. Navy also initiated a program to collect meteorological data at navy yards and on board its ships.

In Europe, a shipping accident during the Crimean War led to experimental efforts in telegraphic meteorology. In 1854 a destructive gale in the Crimean Sea near the port of Balaclava wrecked Anglo-French transport ships. Thinking that timely warnings might

[3] U.S. Army Medical Department, "Regulations for the Medical Department," in *Military Laws and Rules and Regulations for the Army of the United States* (Washington, D.C., 1814), pp. 227–28; for fuller accounts of military involvement see James R. Fleming, *Meteorology in America, 1800–1870* (Baltimore, 1990).

have averted the disaster, Urbain Le Verrier, director of the Paris Observatory, initiated a system of telegraphic weather reports in 1856 and began issuing weather warnings in 1863.[4] In America the U.S. Army Signal Office was established during the Civil War and reconstituted afterwards as a national weather service and telegraphic surveillance network.

Military meteorology reentered the picture in World War I when the Signal Corps and Weather Bureau (now part of the U.S. Department of Agriculture) coordinated their efforts to provide meteorological support for the war effort. In concert with their European colleagues, meteorologists supported new fields of study on "battlefield climatology," the effects of high-level winds on shell trajectories, and weather conditions near the ground conducive to launching or surviving poison gas attacks. Of course, meteorological support for aviation also made its debut in World War I. Hundreds of meteorological officers received their "aerological" training during the war; after their discharge, many went on to continue their careers in weather-related areas.

During World War II the U.S. Army Air Forces and the U.S. Navy trained approximately 8,000 weather officers. Personnel of the Army's Air Weather Service (AWS), an agency nonexistent in 1937, numbered 19,000 in 1945. About 4,500 of this total were officers. Even after demobilization the AWS averaged approximately 11,000 soldiers during the Cold War and Vietnam eras.[5] In 1954 a National Science Foundation survey of 5,273 professional meteorologists in America revealed that 43 percent of them were still in uniform on active duty, 25 percent held Air Force reserve commissions, and 12 percent were in the naval reserve. Thus almost a decade after the war, 80 percent of American meteorologists still had military ties. Postwar meteorology also benefited from new tools such as radar, electronic computers, and satellites provided by or pioneered by the military.[6]

In the Cold War era, techniques of modifying the weather and possibly the climate led the military to institute crash programs in cloud seeding in search of a weather weapon that could be used

[4] James R. Fleming, "Meteorological Observing Systems Before 1870 in England, France, Germany, Russia, and the USA: A Review and Comparison," *World Meteorological Organization Bulletin* 46 (1997): 249–58.

[5] Charles C. Bates and John F. Fuller, *America's Weather Warriors 1814–1985* (College Station, Tex., 1986).

[6] James R. Fleming, "Meteorology," in Brian S. Baigre (ed.), *A History of Modern Science and Mathematics* (New York, 2002), 3:184–217.

surreptitiously to release the violence of the atmosphere against an enemy, tame the winds in the service of an all-weather air force, or, on a larger scale, perhaps disrupt (or improve) the agricultural economy of nations and alter the global climate for strategic purposes.[7]

Finally, a recent study by the U.S. Air Force claims that by "2025, US aerospace forces can 'own the weather' by capitalizing on emerging technologies and focusing development of those technologies to war-fighting applications." In addition to traditional cloud seeding methods, the air force visionaries propose computer hacking to disrupt an enemy's weather monitors and models, and using "nanotechnology" to create clouds of microscopic computer particles that could block an enemy's optical sensors or guide smart weapons to their targets; the cost developing these clouds would be borne by the private sector. In a recurring theme, the military points out that weather modification, unlike other approaches, "makes what are otherwise the results of deliberate actions appear to be the consequences of natural weather phenomena."[8]

In a relationship forged over the course of centuries between science and its state patrons, it is safe to say that scientists seek support from the state and access to political *power*, while the state (especially the military) seeks *powers* over nature as promised (and sometimes delivered) by scientists. As geophysicists pursue the knowledge of the Earth they had long sought, they willingly focus their investigations on those areas of knowledge where military interests have made resources readily available. Today's military desires the capability to fight and win wars on a global battlefield that includes the atmosphere, the oceans, the polar caps, and near space environments. The strategic and tactical needs of an all-weather air force, a nuclear navy, and the targeting requirements of offensive and defensive missile systems all require detailed geophysical knowledge and sophisticated technical systems—especially reliable remote sensing instruments and global communications systems.[9] So the relationship was forged—

---

[7] James R. Fleming, "Fixing the Weather and Climate: Military and Civilian Schemes for Cloud Seeding and Climate Engineering," in Lisa Rosner (ed.). *The Technological Fix: how people use technology to create and solve problems*, Hagley Center Studies in the History of Business and Technology (New York, 2004), pp. 175–200.

[8] Col. Tamzy J. House, *et al.*, "Owning the Weather in 2025," <http://www.au.af.mil/au/2025/volume3/chap15/v3c15-1.htm> (accessed 1 June 2003).

[9] James R. Fleming (ed.), Special theme issue on geophysics and the military, *Historical Studies in the Physical and Biological Sciences* 30.2 (2000).

far beyond simple patronage, with roots much deeper than we might have imagined. A very relevant root involves instrumentation, strategy, and homeland security on the American frontier in the era following the Civil War.

Albert James Meyer and the Signal Corps

The early history of the Signal Corps is intimately intertwined with the biography of its founder, Albert James Myer. (Figure 1) Myer was born in Newburgh, New York, in 1828.[10] After graduation from Geneva College in 1847, he worked for telegraph offices in Buffalo and studied medicine, receiving his M.D. from Buffalo Medical College in 1851. He wrote his thesis on "A New Sign Language for Deaf Mutes" that was based on the Bain telegraphic alphabet. For this Geneva College also awarded him an M.A.

Myer passed the army medical board examination in 1854 and was appointed assistant surgeon, assigned to military outposts in Texas. In addition to his regular medical duties, he served as post treasurer and supervised the diet of the troops. Notably, one of his regular duties involved filing reports on weather observations with the Surgeon General. Myer was seriously ill in 1855, by his own diagnosis with remittent fever and scurvy, but recovered in several months. His fortunes improved dramatically in 1857 when he married Catherine Walden, daughter of judge Ebenezer Walden, Buffalo's first lawyer and a prominent real estate developer.

In 1859 Myer's proposal for a comprehensive system of military and naval signals was reviewed favorably by a panel of officers headed by Lieutenant Colonel Robert E. Lee and underwent successful field tests (An earlier proposal had been shelved by Secretary of War Jefferson Davis in 1856). Myer's signaling system, later known as "wigwag," employed flags by day and torches by night. In 1860, Congress created a Signal Corps in the U.S. Army and Myer was appointed its first officer, with the rank of major. He was assigned to the Navajo Expedition of 1860–61 in New Mexico under Major Edward Canby.

[10] This sketch of Myer's life is based on James R. Fleming, "Albert James Myer," in John A. Garraty and Mark C. Carnes (ed.), *American National Biography* (New York, 1998), 16:197–98; and on Paul J. Scheips, "Albert James Myer, founder of the Army signal corps: a biographical study," Ph.D. dissertation, American University, 1966.

Fig. 1. General Albert James Myer (1828–1880). Author's collection.

At the onset of the Civil War, Myer was ordered to establish a signal system for the Union Army. He served as an aide to General Irvin McDowell at the Battle of First Bull Run and then under General George McClellan as Chief Signal Officer (CSO) of the Army of the Potomac. He coordinated training in wigwag and telegraphy for officers and enlisted men detached to him from various departments of the Union Army. He also introduced the "Beardslee magnetoelectric tactical telegraph machine," the army's first electrical communications equipment suitable for field use.

Late in 1863 Myer, now a colonel, attempted to recruit skilled telegraphers for commissions in the Signal Corps and came into conflict with the United States Military Telegraph, which used civilian telegraphers to staff its lines. Secretary of War Edwin M. Stanton relieved Myer of his position, denied him access to the electric telegraph, and exiled him to the Memphis-Cairo area (western Tennessee). Further misfortune struck in 1864 when the Senate failed to reconfirm his now expired position as Chief Signal Officer. His appointment was formally revoked, his rank reverted to Major, Signal Corps, and he was placed on inactive duty.[11] He was befriended, however, by his old comrade General Canby who used him as signal officer for the Military Division of the West Mississippi. After a long campaign by Myer to restore his reputation, President Andrew Johnson ordered Stanton to reinstate Myer as Chief Signal Officer in 1866 with rank of Colonel. A year later U.S. Grant, now Secretary of War, recommended Myer for promotion to the ranks of Brevet Lieutenant Colonel and Colonel (retroactive to 1862), and Brevet Brigadier General (retroactive to 1865), "for distinguished service."[12]

With his position restored, his promotion in hand, and his reputation intact, Myer now faced the task of rebuilding the postwar Signal Corps and redefining their mission for peacetime service. He convinced the military academies at West Point and Annapolis to teach his system of military signaling and to adopt his *Manual of Signals*, first published in 1864 during his exile.[13] He also established his own

[11] Paul J. Scheips, "Union Signal Communications: Innovation and Conflict," *Civil War History* 9 (1963): 399–421.

[12] William H. Powell, *List of Officers of the Army of the United States from 1770 to 1900* (1900; Detroit, 1967), *s.v.* Myer, Albert J., p. 499.

[13] Myer's *Manual of Signals* was expanded and reissued by Van Nostrand (1866, 1868, 1872 and 1874) and in revised form by the US GPO (1877, 1879).

camp of instruction near Washington, D.C., first at Fort Greble and then at Fort Whipple and conducted other training courses for the Corps of Engineers. Myer, who was now solely responsible for electric telegraphy within the military, argued that even in peacetime the Signal Corps should be involved in surveillance of potential enemies. Still, there was no legislative authorization for a separate Signal Corps and his modest budget request for 1869 had been cut in half.

Meteorology, the Military and Modernity

The Civil War, as historian Alan Nevins noted, "transformed an inchoate nation, individualistic in temper and wedded to improvisation, into a shaped and disciplined nation, increasingly aware of the importance of plan and control."[14] Although the war was the first major conflict influenced by the industrial revolution, it drew upon existing technology rather than generating much that was new.[15] Nevertheless, as Peter Hall has observed, "the war changed everything"; it transformed and revitalized the elite "culture of organization," introduced new models of organization, and, through national networks: railroads, telegraph lines, and the banking system, infused new technologies into American society.[16] Looking back at the conflict on its fiftieth anniversary, President Woodrow Wilson pointed out what everyone already knew, that the Civil War had "created in this country what had never existed before—a national consciousness."[17]

In the more managerial era that dawned after the war the federal government established a massive and well-funded national weather service with direct links to the military. Although weather observations had been made in America since colonial times, the first national service providing daily weather reports and forecasts was established in 1870 in the War Department.[18] Before that, meteorological and

[14] Alan Nevins, *The War for the Union* (New York, 1959–60), I:v; quoted in George M. Fredrickson, *The Inner Civil War: northern intellectuals and the crisis of the Union* (New York, 1965), p. 111.

[15] Robert V. Bruce, *Lincoln and the Tools of War* (Indianapolis, 1956).

[16] Peter Dobkin Hall, *The Organization of American Culture, 1700–1900: private institutions, elites, and the origins of American nationality* (New York, 1982), pp. 242 and 227–39.

[17] Woodrow Wilson, Memorial Day address, May 31, 1915, Arlington National Cemetery, Virginia.

[18] Donald R. Whitnah, *A History of the United States Weather Bureau* (Urbana, Ill., 1961); Joseph M. Hawes, "The Signal Corps and Its Weather Service, 1870–1890," *Military Affairs* 30 (1966): 68–76; and Lewis J. Darter, Jr., *List of Climatological Records*

climatological observations were collected by volunteer observers and a number of federal agencies including the Army Medical Department, the General Land Office, the Navy, and the Smithsonian Institution.[19]

The U.S. Army Signal Service, highly decorated for its successes in monitoring Confederate troop movements during the Civil War, had been mustered out with the volunteer army of 1865, leaving Chief Signal Officer Myer a bureau chief without a bureau, in charge of one lieutenant and two clerks.[20] Myer, who was known as an innovator and aggressive administrator, now faced the task of rebuilding the Signal Corps and redefining its mission for peacetime service. He was grasping for a mission that would keep the corps alive when, in December 1869 a bill was introduced in Congress establishing a national storm warning service. Because of its prior involvement in meteorological research, some scientists considered the Smithsonian Institution as a likely candidate to organize the new system, but at age 72, Joseph Henry (the Secretary of the Smithsonian) was in no position to take on this responsibility. Moreover, the mission of the Smithsonian was "the increase and diffusion of knowledge", not the routine provision of public services. The final version of the bill named the Secretary of War as the responsible party.

The Chief Signal Officer immediately called on Wisconsin Congressman Halbert Paine, the sponsor of the bill, to stake his claim. Defining storms as the "enemy" of commerce, Myer argued that the Signal Corps could use telegraphy to track their movement and provide meteorological intelligence in advance of their arrival: "The telegraph can announce meteorological observations, statistics, and reports giving the presence, the course, and the extent of storms . . . and their probable approach, as it would, in time of war, those of an enemy."[21] As Paine recounted the events two decades later,

in the National Archives (Washington, D.C., 1942). See also Paul J. Scheips, "'Old Probabilities': A.J. Myer and the Signal Corps Weather Service," *Arlington Historical Magazine* 5 (1974): 29–43; George M. Kober, "General Albert J. Myer and the United States Weather Bureau," *Military Surgeon* 65 (1929): 65–83; and Rebecca Robbins Raines, *Getting the Message Through: a branch history of the U.S. Army Signal Corps* (Washington, D.C., 1996), especially chapter 2.

[19] Fleming, *Meteorology in America* (note 3).

[20] More complete documentation of this case study appears in James R. Fleming, "Historical Introduction: The Signal Office and the Bibliography of Meteorology," in Flemming and Roy E. Goodman (eds.), *International Bibliography of Meteorology: from the beginning of printing to 1889* (Upland, Penn., 1994).

[21] Albert J. Myer to Congressman Halbert F. Paine, January 18, 1870, "Letters . . . relative to storm telegraphy," U.S. House of Representatives, ex. doc. 10, pt. 2, 41st Congress, second session, p. 22.

> Before I finally obtained the unanimous consent of the House to the passage of the joint resolution, a stranger called upon me and introduced himself as Colonel Myer, of the Signal Corps. He exhibited the most intense anxiety for the success of the measure. It seemed to me that his zeal and enthusiasm marked him as the fit man to launch the new enterprise. As soon as the joint resolution passed, I called upon General [William] Belknap, then Secretary of War, and suggested that he should consider the propriety of assigning Colonel Myer to the duty. He said at one that he had already in his own mind fixed upon Colonel Myer as the officer to commence the work under the joint resolution.[22]

Congress was persuaded by Myer's zeal, signaling expertise, and the promise of military discipline in the system. The first national weather service was established in 1870 under the direction of the Chief Signal Officer. With generous support from Congress, the Signal Office budget soared from $5,000 in 1869 to $400,000 by 1874. During the same period Myer's command expanded from three enlisted men to over five hundred college-educated observers. The Signal Office had become a major military and scientific service, providing "telegrams and reports for the benefit of commerce" and weather predictions to the public, which was rapidly becoming accustomed to receiving daily weather forecasts.

Initially, the Signal Service depended on commercial telegraph lines, notably Western Union; observer-sergeants were located only in areas where such service was available. There was a great deal of contention, however, about the public responsibility of the telegraph companies. In 1866 Congress had granted several companies the right to construct lines on public lands using public resources. Each new telegraph station constructed under this act was allotted thirty acres of land. In return the US government and its agents were to have top priority over use of the lines at "rates to be fixed by the Postmaster General." After 1870, according to the telegraph companies, the heavy message traffic of the weather service threatened to overwhelm commercial uses and, because of loss of profits, limit further private expansion of the system. In response Myer suggested shortening government messages (the weather service used an elaborate code), detailing Signal officers to outlying stations to clear government traffic, and building military lines into remote and strategic areas

[22] *Ibid.*, Halbert E. Paine to H.L. Dawes, U.S. Senate, Sept. 18, 1888, *Congressional Record* 19, no. 224 (19 Sept. 1888), pp. 9564–65.

that would connect with the nearest commercial office. When the private companies resisted these proposals Congress reacted, imposing severe penalties, including fines of $100 to $1,000 per event, for any telegraph company refusing to or neglecting to transmit government communications.[23]

In the 1870s Myer received authorization to construct military telegraph lines to lighthouses and mountain stations, and into Indian Territory in the southwest and northwest frontiers. Signal Office wires along with those of a score of different telegraph companies all converged on the "Telegraph Room" of the Signal Office. There is a major and stunning archaeological discontinuity in the records of the U.S. Army Signal Office preserved in the National Archives. Before 1870, Myer's correspondence contains a modest number of letters, both typed and handwritten. After 1870, the files are dominated—overwhelmed—by telegrams rather than letters. For example, in fiscal year 1876–77, his office logged well over 700,000 telegraphic messages received and over 52,000 telegraphic communications sent. This placed Myer at the center of an electric intelligence network, a giant web spanning the nation.[24]

Surveillance

Although telegraphic reports "for the benefit of commerce and agriculture" formed the primary rationale for the weather service, the men of Myer's command served as both meteorological observers and at times as secret service agents reporting to him on domestic enemies such as striking workers in the rail strikes of 1877, Indian uprisings in the southwest, and natural hazards to commerce and agriculture. Signal service observers reported on the hatching and migration of locust swarms, on frost and drought in the cotton, corn and tobacco-growing regions, on hazards to shipping along the coast. Mercantile interests were advised of weather conditions affecting the packing and shipment of perishable goods such as oysters, pork, and ice.

[23] U.S. Statutes at Large, 42nd Congress, second session, ch. 415 (1872), pp. 366–67.

[24] Tom Standage's summary of the history of telegraphy likens Samuel F.B. Morse to "an immense spider in the center of a vast web he himself had woven"; Standage, *Victorian Internet* (note 1), p. 182.

Sailors received notice of fogs, storms, and fair winds. Insurance companies received data useful to them for setting rates for shipping. River reports warned of floods and low water conditions; railway reports announced heavy snows and track conditions; sanitary reports tracked the course of cholera and yellow fever epidemics in the interest of public health. All of these missions, including daily weather reports, involved potential threats to commerce, agriculture, and the domestic order.

The Great Railroad Strike of July and August 1877, the nation's first general labor strike and large-scale violent protest, was monitored throughout by the U.S. Army Signal Office, with telegraphic strike reports arriving from across the nation. On 16 July 1877 workers on the Baltimore & Ohio Railroad, who had just been notified of a ten-percent pay cut and a reduction in their working hours, decided to go out on strike and blockaded freight trains in Maryland and West Virginia. Employees of the Pennsylvania Railroad, in response to pay cuts and work speedups, also halted the movement of trains by seizing control of rail depots. By July 25th, the strike had spread to other companies and had reached all major cities, adversely affecting two-thirds of the nation's rail traffic. Major rail hubs including Baltimore, Pittsburgh, Chicago, Kansas City, and San Francisco were shut down. In St. Louis, city government was closed for two days, 100 people were killed in rioting, and an estimated $10 million in rail property was destroyed. Governors in Maryland, West Virginia, and Pennsylvania mobilized their state militias, while President Rutherford B. Hayes sent in federal troops to restore order.

The Marxist-influenced Workingmen's Party came to the support of the strikers, calling for government ownership of railroads and telegraphs, and an eight-hour day for workers. Concerning the strike, Karl Marx wrote to Friedrich Engels, "What do you think of the workers of the United States? This first explosion against the associated oligarchy of capital which has occurred since the Civil War will naturally again be suppressed, but can very well form the point of origin of an earnest workers' party. . . . A nice sauce is being stirred over there, and the transference of the center of the International to the United States may obtain a very remarkable *post festum* opportuneness."[25]

During the rail strikes of 1877 the Signal Office provided special

[25] Robert V. Bruce, *1877: year of violence* (Indianapolis, Ind., 1959), p. 276.

reports every three hours from across the nation to the War Department and, by direct telegraph line, to President Hayes at the White House. After completing their weather reporting duties, Signal Corps "observer sergeants" would dress in street clothes and attend strike rallies, then return to the weather service office to telegraph their findings to Myer. Some of the strike telegrams were actually recorded on the back of weather observation forms. The following four telegrams (for example) document the situation in New York City between 6:00 P.M. and 11:30 P.M. on 25 July 1877, a critical day of the strike:

> Private Pollak, New York to CSO, 6 p.m.—Trouble may follow tonight. Meeting at Thompkins Square though policemen well prepared. Will watch proceedings despite the demand of internationalists not to turn out. Private Jewell will attend meeting and report circumstantially.
>
> Thomas S. James, New York to CSO—At 9:15 p.m. meeting adjourned and square was cleared. The crowd assembled on 9th Street. Inspector Murray ordered them to disperse. They replied with a shower of brickbats and stones. Police immediately charged and after a sharp and decisive struggle they were completely routed. All is reported quiet now.
>
> Private Jewell, NY to CSO—At nine thirty the crowd attempted to form a procession and march up 8th Street, but the policemen would not allow it. The crowd fled rapidly before their charge.
>
> Myer, CSO to President Hayes—It is now half past eleven here. The New York International meeting has ended and is a failure. Albany, Buffalo, Cleveland, Indianapolis, Cincinnati, Pittsburgh, Philadelphia, Baltimore are quiet. Louisville is a little excited but well in hand by special police. Chicago is well controlled. St. Louis is not so well but no serious violence. San Francisco is alert and at 6 o'clock controlled. There seems no reason to fear really serious outbreak anywhere tonight.[26]

According to Robert Bruce, President Hayes used such detailed information in deciding not to escalate the federal response and the strike ended that summer with many of the railroad companies conceding to at least part of the striker's demands. On 5 August 1877 the Signal Corps issued its final strike report: "*Pax Semper Ubique*" [peace always everywhere].[27]

Because of Indian unrest and uprisings on the frontier, however Signal Service weather observers also reported from the northwest and southwest territories. A special set of telegraph lines was constructed

---

[26] *Ibid.* Telegrams were edited for clarity.
[27] *Ibid.*, pp. 73, 230, 271, 279–80, 291.

to military posts on or near reservations where any suspicious activity could be reported instantly. The federal government was determined to shatter the civilization of the Native Americans. The destruction of the bison is a well-known example. Between 1872 and 1874, by some estimates, over nine million bison were slaughtered, a mere 150,000 by Native Americans for their needs. General of the Army Philip Sheridan, attending a session of the Texas legislature to discuss this issue, voiced the majority opinion, "Let them kill, skin and sell until the buffalo is exterminated, as it is the only way to bring lasting peace and allow civilization to advance."[28] This was in line with the congressional testimony of Secretary of the Interior Columbus Delano in 1874, "The buffalo are disappearing rapidly, but not faster than I desire. I regard the destruction of such game as Indians subsist upon as facilitating the policy of the Government, of destroying their hunting habits, coercing them on reservations, and compelling them to begin to adopt the habits of civilization."[29] As historian William Leckie put it, "Disappearance of the buffalo would be synonymous with a disappearance of the Indians' way of life."[30] Out of an estimated population of fifteen million bison in 1865, barely one thousand remained by 1885—a four-fold "decimation" verging on the extermination of the species.

The Signal Office did its part as well to shatter the ecology of natural knowledge. Through the wires and communication devices of the Army Signal Office, science, technology and settlement made their incursion into the American West. According to journalist Mary Clemmer Ames, a period of unusually wet and stormy weather followed the establishment of a telegraphic weather station at Fort Gibson in Indian Territory. The natives, thinking the Signal Office observer was responsible, were dissuaded from tearing down the station only after being told by their agent that the observer only recorded the weather, he did not control it.[31]

[28] Sheridan quoted in Martin S. Garretson, *The American Bison: the story of its extermination as a wild species and its restoration under federal protection* (New York, 1938), p. 128.

[29] Secretary of the Interior Columbus Delano, testimony before Congress, 10 Jan. 1874, House Report no. 384, 43rd Congress, first session, p. 99, quoted in Robert Wooster, *The Military and United States Indian Policy 1865–1903* (New Haven, 1988), p. 171.

[30] William H. Leckie, *The Military Conquest of the Southern Plains* (Norman, Okla., 1963), p. 7.

[31] Mary Clemmer Ames, *Ten Years in Washington: or, inside life and scenes in our national capital as a woman sees them* (Hartford, 1882), p. 502.

The discovery of gold in the Black Hills in 1874 was accompanied by a furious campaign by the Signal Office to extend its telegraph lines into the northwest. Frontier expansion in the southwest also stimulated a program construction and operation of telegraph lines into the frontier "against Indian and Mexican depredations." By 1879 there were 5,000 miles of lines and 73 stations from the Dakotas to Washington Territory in the northwest and from Texas to California in the southwest. Signal office telegraphers may have filed regular weather reports, but they also filed more desperate calls for help:

> Telegram from Maricopa Wells, Arizona Territory to Lt. Reade, San Diego, June 1, 1877—A band of 50 Apache Indians committing depredations 3 miles from here, killed one child and mortally wounded one woman. Request protection by troops.

The telegraphic network was vulnerable to Apache raiding parties who pulled down the poles and cut the wires. An even more effective technique was to allow the poles to stand and join the cut wires with a thin strip of leather, making a break in the circuit extremely difficult to locate. Because of this, the Army had to detail repair teams and bring in signal officers trained to use the heliograph when the telegraph lines were down.

By 1877 the Signal Office had also constructed a network of telegraphic stations linking coastal lighthouses from New Jersey to North Carolina, a total of over 500 miles. (Figure 2) Observers at these stations had orders to describe all passing vessels and report the signal flags or lights they displayed. Using wig-wag, the observers sent messages to the ships. Although its primary goal was to warn of approaching storms, thus saving lives and property, the seacoast network also provided entertainment for passengers and vacationers who wished to send messages and keep in touch with the national news and weather as reported from Washington.[32] Not all of these messages were important or welcome. Out of devotion to his patron, Myer sent U.S. Grant a cablegram at 3A.M. when his daughter Nellie and her new husband arrived safely in England. In 1873 he began sending Grant daily weather forecasts for the entire country. He even

[32] See for example H.H.C. Dunwoody to W.G. Atkinson, chairman, meteorological committee, Merchant's Exchange, Baltimore, 27 June 1877, National Archives and Record Administration (NARA), RG27, Office of the Chief Signal Officer, Miscellaneous Letters Sent.

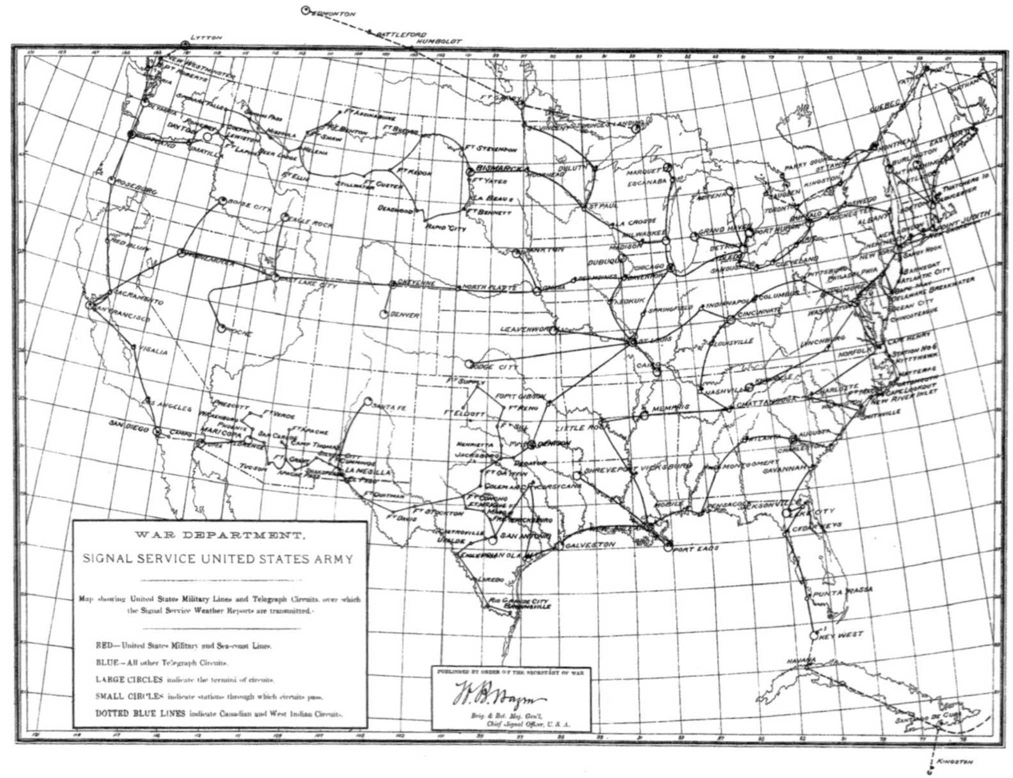

Figure 2. Map published in 1881 depicting U.S. military (dark lines throughout the Midwest, West Coast, and the San Francisco–Chicago line) and sea-coast (light lines in Texas and the southwest, northern Great Plains and Rocky Mountains, and one on the Eastern Seaboard) lines in relationship to commercial telegraph lines. From *Annual Report of the Chief Signal Officer* [1881], between pp. 250–51.

sent reports to Long Branch during Grant's summer vacation. One day Myer received a very brief reply: "Stop them—Grant."[33]

The Signal Service also considered it their mission to help solve the problem of western locusts by cooperating with the U.S. Entomological Commission and other government agencies. They collected specialized data and produced maps, charts and annual reports on how temperature and humidity conditions affected locust hatching, how winds might influence locust migration, and the status of crops and other forage that supported locust swarms. The Signal Service was then able to issue warnings of potential locust outbreaks and supervise the destruction of eggs and young before a swarm developed.[34]

The fact that America now had a well-funded national weather service did not go unnoticed in Europe. In contrast to their usual disdain of "colonial" science, European scientists and administrators now expressed admiration for the huge scale of the American meteorological service, its generous funding, and its utility (if not its theoretical accomplishments). Myer, representing the United States at the International Meteorological Congress in Vienna in 1873, proposed that the weather services of the nations of the world prepare an international series of simultaneous observations and charts to aid the study of world climatology and weather patterns. The result was the *Bulletin of International Simultaneous Observations* published by the Signal Office from 1875 to 1889. From 1873 his office also issued the still-published *Monthly Weather Review*, begun by Cleveland Abbe, the chief civilian scientist for the Signal Office. As Abbe wrote to his colleague Elias Loomis at Yale, this was the "beginning of what I hope to see grow into something worthy of the human race—for you will see that it is the union of the whole world in an attempt to produce a daily map of the atmosphere. And only thus can we study the atmosphere."[35] In 1879 Myer returned to Europe as a delegate to the second International Meteorological Congress in Rome.

Permanent status for the Signal Corps came late in Myer's life. It became a bureau of the War Department in 1875 and received

---

[33] Letter of D. Marean to Henry E. Williams, 7 April 1922, NARA RG27, Reminiscences of Employees & Miscellaneous Historical Meteorological Information, Item 47.

[34] *Report of the Chief Signal Officer* (1877), p. 500.

[35] Letter of Cleveland Abbe to Elias Loomis, 13 Mar. 1878, Yale University, Beinecke Library, Papers of Elias Loomis.

a permanent enlisted force in 1878. An act of 24 February 1880 established the rank of the Chief Signal Officer as Brigadier General. Albert Myer received that rank on 16 June, just two months before his death; reports of his funeral were reported to the nation instantaneously by telegraph.

The Network Under Siege: The Corps Under Hazen and Greely

Myer, the founder of the Signal Corps and the architect of its postwar surveillance functions to establish and enforce order, employed telegraphy as his preferred technology and meteorology as his source of patronage. The chief signal officers that followed him, however, faced different challenges in managing and attempting to improve a huge operation. They also had to justify their activities before a growing number of critics. Meteorological data collected by the Signal Corps dealt with a specific threat: dangerous weather. But to separate meteorological from political and military data gathering is to introduce an artificial distinction. While the first of Myer's successors attempted to isolate and elevate the scientific functions of the corps, the second reasserted a more broad-ranging intelligence-gathering function in the face of bureaucratic competition and high-level criticism.

William Babcock Hazen (figure 3), a veteran of the Civil War and Indian wars, succeeded Myer as Chief Signal Officer. Hazen was born in Vermont in 1830 and received his military education from 1851 to 1855 at West Point. He was assigned to Fort Davis, Texas in 1858 as a scout and was wounded in a skirmish with Comanches the following year. During the Civil War he participated in General William T. Sherman's infamous march to the sea. In 1866 he was appointed Acting Inspector-General of the Department of the Platte and served in the Plains Indian campaigns. In 1870 he was appointed Superintendent of Indian Affairs at Fort Smith, Arkansas. Following several tours of duty in Europe as an attaché and observer of the Prussian military, Hazen was promoted to Brigadier General and Chief Signal Officer in 1880.[36] Hazen's interest in meteorology derived

[36] George W. Cullum, *Biographical Register of the Officers and Graduates of the U.S. Military Academy at West Point, N.Y. from Its Establishment, in 1802 to 1890 with the Early History of the United States Military Academy*, 3rd ed. (Boston, 1891), *s.v.* Hazen, William B., no. 1704, II:632–35. See also Paul Andrew Hutton (ed.), *Soldiers West: biographies from the military frontier* (Lincoln, Neb., 1987).

Figure 3. General William Babcock Hazen (1830–1887). From generalsandbrevets.com.

in part from his experiences countering the exaggerated claims of land speculators and railroad companies about the beneficent climate of the west, especially the rainfall of prairies.[37]

Under Hazen's leadership the Signal Office established a scientific study room, appointed a permanent advisory committee of members of the National Academy of Sciences, improved testing and recruitment of new officers, published a new journal *Signal Service Notes*, issued William Ferrel's text book, *Recent Advances in Meteorology*, compiled a multi-volume international *Bibliography of Meteorology*, and responded to its critics in Congress on the issue of whether the military should be supporting a national weather service.[38] The Signal Office, however, did not fare well in the 1880s. Its funding was reduced and some stations were closed. Its military intelligence missions were largely curtailed.

In 1881 Secretary of War Robert Todd Lincoln ordered a thorough investigation of the Signal Corps and issued a report stating that the weather service had "no natural connection whatever with the military service." He recommended an end to meteorology in the Army. Senator John A. Logan responded by introducing a bill into the Fourty-Seventh Congress to transfer the weather service to the Department of the Interior. Three years later a committee on meteorology of the National Academy of Sciences chaired by Montgomery C. Meigs recommended transfer of the weather service to civilian control. Also in 1884, a joint bipartisan congressional committee chaired by Senator William B. Allison began its investigation of the scientific bureaus of the government, including the weather service of the Signal Office.[39] One focus of enquiry was the extent to which government employees were engaged in abstract scientific research rather than in public service. Weather service employees testified that their work was wholly practical. The majority report of the com-

[37] William Babcock Hazen, *Our Barren Lands: the interior of the United States west of the 100th meridian and east of the Sierra Nevadas* (Cincinnati, Ohio, 1875).

[38] In April 1881 the National Academy of Science appointed a committee on meteorology to confer and cooperate with the chief signal officer. Members included Simon Newcomb (chairman), Elias Loomis, Wolcott Gibbs, H.A. Newton, William Ferrel, Charles A. Schott, Samuel P. Langley, Ogden N. Rood, and Charles A. Young. Fleming and Goodman, *International Bibliography of Meteorology* (note 20).

[39] "Testimony before the [Allison] commission to consider the present organization of the signal service . . ." U.S. Senate, misc. doc. 82, 49th Congress, first session. See also A. Hunter Dupree, *Science in the Federal Government: a history of policies and activities* (Cambridge, Mass., 1957), pp. 190, 215, *et seq.*

mission, accepted by Congress, upheld the *status quo* in federal science, although a minority report advocated the transfer of the weather service to a civilian department.

In 1885, another controversy with Secretary of War Lincoln over the handling of the Lady Franklin Bay polar expedition (see below) resulted in Hazen's court-martial and temporary removal from office, but subsequent acquittal. That same year, in the final military action of his career, he sent three heliograph signal detachments to Arizona Territory for use in the campaign against the Apaches. Hazen's final trip to Europe was as the U.S. representative to the International Commission of Meteorology in Paris. After completing his memoirs, he died in January, 1887 at the age of 56.[40]

Adolphus W. Greely (figure 4), a long-time signal officer and celebrated Arctic explorer, succeeded Hazen. Greely was born in Newburyport, Massachusetts, in 1844 and attended local schools. During the Civil War he enlisted in the 19th Massachusetts Volunteers, seeing action at the battles of Antietam and Fredericksburg and in the Peninsula Campaign. He was promoted to corporal in 1862 and captain of the 81st U.S. Colored Troops in 1865. After the war Greely served in the frontier army in Wyoming, Utah, and as a signal officer in the 1869 campaign against the Cheyenne in Nebraska. A year later he was transferred to Washington, D.C. to assist Albert Myer in the establishment of the weather service. In addition to serving with the River and Flood Service, Greely was placed in charge of completing the Signal Corps telegraph line across Texas in 1875. He also wrote a professional report on isothermal lines of the United States.

In 1881 Greely volunteered for a scientific expedition to the Arctic to establish circumpolar meteorological stations as part of first International Polar Year (1882–83). He led a 25-man party to Lady Franklin Bay on the east coast of Ellesmere Island where he established a base camp and collected meteorological and geophysical observations. He and his party mapped a stretch of Greenland's coast, explored Ellesmere Island, and achieved a northern record of 83°24'. When supply and relief ships failed to arrive in 1882 and 1883 Greely and his men broke camp and made their way south by boat to Cape Sabine where they were forced to winter over. By the time a rescue mission arrived in June 1884 only seven men were alive, one of whom

[40] William Babcock Hazen, *A Narrative of Military Service* (Boston, 1885).

Figure 4. Major General Adolphus Washington Greely (1844–1935). From photolib.noaa.gov.

soon died. While General Hazen was taking the blame for the failure of the relief efforts and rumors of cannibalism among the men, Greely was quickly absolved of any wrong-doing and received a hero's welcome, collecting explorer's medals from the Royal Geographical Society of London and the Société de géographie de Paris.[41] He spent his recovery time preparing popular and official reports of the expedition, the latter documenting observations on meteorology, magnetism, aurora, ice sheets, and ocean currents in high latitudes.[42]

Greely, now a Brigadier General (he had only made Captain a year earlier!) fought to keep the Signal Corps in existence and staffed with qualified scientists. He expanded the service's work concerning agricultural meteorology and focused largely on research. He even wrote a popular book on *American Weather* (1888). Civilian scientists were employed by the Corps to explore the laws of atmospheric motion, the spatial and temporal distribution of temperature and moisture, cloud photography, atmospheric electricity, the diminution of temperature with altitude, analytical mechanics, and solar physics.

Greely believed that military discipline was necessary to obtain accurate observations, yet, in agreement with earlier critics, advocated the transfer of the weather service to a civilian agency and reorganization of the Signal Corps to increase its efficiency and preserve its military mission. Perhaps too he saw the writing on the wall. As early as 1884 the National Academy of Science recommended a non-military meteorological bureau and, from 1882 to 1890, Congress introduced a number of bills proposing the transfer of the weather service out of the Signal Office.[43] One argument for the separation was that enlisted men were first of all soldiers: "Their

---

[41] Leonard F. Cuttridge, *Ghosts of Cape Sabine: the harrowing true story of the Greely expedition* (New York, 2000). Powell, *List of Officers* (note 12), *s.v.* Greely, Adolphus W., p. 339; and *Dictionary of American Biography* (New York, 1928–36), *s.v.* Greely, Adolphus W.

[42] Adolphus W. Greely, *Three Years of Arctic Service: an account of the Lady Franklin Bay expedition of 1881–84, and the attainment of the farthest north* (New York, 1886); and International Polar Expedition, *Report of the Proceedings of the United States Expedition to Lady Franklin Bay, Grinnell Land*, 49th Congress, first session, House misc. doc., no. 393 (Washington, 1888).

[43] [H. Helm Clayton], *The Transfer of the United States Weather Service to a Civil Bureau* (Boston, 1889). See also Thomas J. Brown, *The Necessity for a Civilian Directorship of the New Weather Bureau, as Against a Military Protectorate* (n.p., 1891), copy in the American Philosophical Society Library, Philadelphia; and "Records Relating to the Transfer of Meteorological Functions from the War Department to the Agriculture Department, 1887" [NARA, RG27].

first duty is not to prosecute natural inquiries, but to obey the commands of their superior officers." But the observers themselves were also college graduates who overwhelmingly preferred a transfer to a civil bureau. One additional factor should be mentioned—the rise of agricultural colleges following the passage of the Morrill Land Grant act of 1862, experiment stations connected to universities or to state extension services, and state weather services provided an alternative model of organization, one that included more scientific control and promised to satisfy the desire of many in agriculture, business, and state government for decentralized (and hopefully more accurate) forecasting.[44] The transfer was enacted by Congress in 1890 and completed on 1 July 1891 when the newly established U.S. Weather Bureau began functioning in the Department of Agriculture.

Conclusion

The US Army Signal Service, established as a special military unit during the Civil War, continued its military mission into peacetime as a national weather service and intelligence gathering agency. From 1870 to 1880, under its founder and chief officer Albert Myer, the Corps pursued with a vengeance stormy weather, striking workers, renegade Indians, and other threats to domestic tranquility. Further removed from the shadow of war, Myer's successors Hazen and Greely focused more on expanding and improving the weather service as they battled their critics and prepared to surrender the weather service to a civilian agency, yet each used signaling technology for both civilian and military purposes.

The need for daily weather reports had provided the rationale, the personnel and the funding for a wide-ranging set of military activities involving telegraphy and surveillance. According to Napoleon Bonaparte, "Le secret de la guerre est dans le secret de communications" and for the U.S. military, telegraphic communication was a new growth area. Using meteorology as leverage for the Signal Service budget, however, was a temporary solution, as was pointed out by William A. Glassford, who knew Myer personally: "To the

[44] N.S. Shaler, "The Weather Service Should be Non-military," *Boston Herald* (10 March 1889), quoted in [Clayton], *Transfer*, p. 15.

legitimate duties of military signaling he added the *utterly foreign* concerns of a meteorologist, with a result well known."[45]

Myer's successors, Generals Hazen and Greely, were not nearly as aggressive in using the weather service for surveillance activities and by 1891 the weather service was moved to the Department of Agriculture. Only a year later, however, the Signal Office, now a purely military outfit, advised the National Guard on the use of signaling devices—heliographs, lanterns, and telegraph lines—in the Homestead riots in Pennsylvania at the Carnegie Steel and Iron Company. During the Chicago Railroad riots of 1894 the Signal Office installed a "very complete" system of visual, telephone, and telegraph communications between department commanders and subordinates. The lessons learned by Myer during the Rail Strike of 1877 were not lost on his successors. Although officially "forgotten" today, they have been effectively reconstituted as homeland security.

General Greely served as CSO until 1906. He directed the Corps' activities in the Spanish-American War and was responsible for relief activities after the San Francisco earthquake of 1906. Under his leadership, the Signal Corps introduced new technologies, including wireless telegraphy, the automobile, and the airplane. Military meteorology reentered the picture in World War I when the Signal Corps and Weather Bureau cooperated in a joint operation headed by the noted physicist Robert A. Millikan investigating battlefield climatology, issuing weather forecasts, calculating the effects of upper air winds on the trajectories of artillery shells, and studying the weather conditions affecting gas warfare. But that is another story.

Dedication

To the memory of David Van Keuren, friend and scholar.

Acknowledgments

This research was supported in part by a research grant from the Interdisciplinary Studies Division of Colby College and by a Frederick

[45] W.A. Glassford, "Historical Sketch of the Signal Corps, United States Army." *Journal of the Military Service Institution* (Governor's Island, New York Harbor, n.d.).

W. Beinecke fellowship in western Americana at Yale University. Earlier versions of this paper were presented at the Smithsonian Institution, History of Science Society, American Geophysical Union, Yale University, the University of Arizona, Harvard University, and the Naval Research Laboratory. I thank Nathan Reingold, Marc Rothenberg, Paul Theerman, Howard Lamar, John Heilbron, Everett Mendelsohn, Rena Selya, David Nickles, Bruce Hevly, and the late David Van Keuren for their valuable comments and suggestions.

CHAPTER SIX

REMNANTS OF TESTING AT THE SANDY HOOK PROVING GROUNDS, SANDY HOOK, NEW JERSEY

Gerard P. Scharfenberger

Introduction

Unexploded Ordnance (UXO) sweeps in the early 1990s within Gateway National Park at Sandy Hook, New Jersey uncovered a huge cache of projectiles, ordnance and military hardware spanning three centuries of military activity—from the late eighteenth century to World War One. UXO sweeps are conducted when an area has the potential to contain possibly live ordnance and proposed construction on the North and Gunnison Beaches necessitated the testing of an area covering 10 acres. During the UXO sweeps at Sandy Hook using subsurface surveillance equipment, over 10,000 artifacts were recovered ranging from unidentified metal fragments to whole projectiles—some live, buried just beneath the surface of beaches used by scores of sunworshippers for decades. Of these, 245 were deemed worthy of further study and retained, with 38 selected for conservation. Much of the assemblage comes from the years 1874–1919, when the Sandy Hook Proving Grounds was at its height as a testing area for new weaponry and increasingly effective defensive armaments.[1] The Proving Grounds made myriad contributions to the technological advancement of the United States armed forces, and examining some of those remains provides insight to that development. Further, these investigations demonstrate the lengthy evolution of Sandy Hook as a strategic military outpost, and it is also worth noting the methods used to conserve its associated artifacts for future generations to understand these instruments of destruction.

[1] Gerard P. Scharfenberger and Timothy R. Sara, *Unexploded Ordnance Sweeps North Beach Center, Sandy Hook Gateway National Recreation Area, New Jersey*, 1996 [report on file with the National Park Service, Denver, Colo. service bureau].

The peninsula of Sandy Hook, New Jersey has been a coveted location for shipping and defense since the earliest days of European exploration. In 1524, Giovanni da Verrazano wrote to Francis I, King of France of, "a very pleasant situation among some steep hills"; a reference to the Navesink Highlands overlooking Sandy Hook and the easily navigated estuary draining into Sandy Hook Bay. It was on the beach of Sandy Hook in 1609, that a landing party from Henry Hudson's *Half Moon* first set foot in New Jersey. According to Richard Juet, first mate on the *Half Moon*, it was, "a very good land to fall in with and a pleasant land to see."[2] As early as 1679, New England Colonial Governor Edmund Andros proposed that, "Beacons for Land or Sea-Marks for Shipping, Sailing in and out, and a Fortification be erected at Sandy Hook."[3]

In 1764 a lighthouse was erected on what was then the northern tip of the peninsula.[4] This lighthouse, which is still standing, is located approximately one and a half miles from the present day shoreline, the result of more than two centuries of littoral drift which extended the peninsula north into New York harbor.[5] The newly constructed lighthouse is depicted on the 1766 *Plan of New York City and its Environs* as the only beacon leading into the busy port of New York.[6] This map also shows a northeasterly oriented appendage called the "False Hook," likely due to its ephemeral condition along the everchanging coastline.

During the American Revolution, Sandy Hook was a haven for the Tory Refugees, a band of local Loyalists who enjoyed the protection of the massive royal fleet anchored in the bay.[7] After the loss of New York by the colonists, the British installed a permanent military base on Sandy Hook to monitor shipping and protect supply vessels sailing in and out of New York Harbor. For several years bracketing the American Revolution, Sandy Hook was actually an island; the result of a brutal nor'easter in 1777 that eroded the land bridge connecting the peninsula with the mainland. After the Battle

[2] Edwin Salter, *Salter's History of Monmouth and Ocean Counties* (Bayonne, NJ, 1890), pp. 6–9.

[3] Franklin Ellis, *History of Monmouth County, New Jersey* (Philadelphia, 1885), p. 29.

[4] *Ibid.*, p. 547.

[5] Scharfenberger and Sara, *Unexploded Ordnance Sweeps*, p. 6.

[6] John Montresor, *A Plan of the City of New York and its Environs to Greenwich on the North or Hudsons River* (London, 1766).

[7] Ellis, *History of Monmouth County*, p. 195.

of Monmouth in 1778, British troops were severely hampered in their attempt to escape to New York via Sandy Hook by the lack of an easily accessible landlocked peninsula to traverse. By 1810, currents had re-joined Sandy Hook to the barrier beach, to which it is still connected today. This phenomenon would be repeated on an average cycle of 50 years, as frequent storms and powerful currents would alternately erode and re-deposit the narrow land bridge connecting Sandy Hook to the mainland.[8] Presently, state and federally funded beach replenishment programs keep the coastline intact in spite of fierce storms that severely erode the beach on an annual basis.

The Proving Grounds

Beginning in 1790, the federal government began to acquire land on Sandy Hook for a military reservation; at the time of the War of 1812, a small wooden fort stood on the peninsula. Work was begun on a large granite fortification in 1859, but was never completed.[9] Technological advances in artillery during and after the Civil War had made masonry fortifications obsolete.[10] Since the fourteenth century, round cannonballs fired from smoothbore cannons caused little damage to masonry walls, but the spinning effect of elongated shells fired from rifled guns was extremely effective on masonry, stone, or brick targets.[11] Similar masonry forts had been reduced to rubble as a result of incredibly powerful, high velocity projectiles fired from rifled cannons.[12] And it was these developments in artillery projectiles for which Sandy Hook played an important part as the preeminent testing ground on the east coast before the opening of the Aberdeen Proving Grounds in Maryland during World War One (see Grier, this volume). Nevertheless, after several revisions to the plans for the Sandy Hook fortifications, the army finally suspended

---

[8] George H. Moss, *Another Look at Nauvoo to the Hook* (Sea Bright, NJ, 1990), pp. 139–141.

[9] *Ibid.*

[10] Emmanuel Raymond Lewis, *Seacoast Fortifications of the United States: an introductory history* (Annapolis, Maryland, 1970), p. 66.

[11] *Ibid.*, p. 67.

[12] Edwin C. Bearss, *Historic Resource Study: the Sandy Hook Proving Ground 1874–1919*, Gateway National Recreation Area, Sandy Hook Unit (Denver, 1983), p. 28.

construction on the granite fort in 1868.[13] Once abandoned, the sea wall was eventually covered by sand deposited by ocean currents. As recently as 1962 a storm eroded 200 feet of beach exposing the granite wall, which was then slowly re-covered by natural deposition.

In 1874, Sandy Hook was selected as the U.S. Army Ordnance Department's first official proving ground for the testing of weapons and munitions.[14] The Proving Grounds at Sandy Hook was the result of a post-Civil War military restructuring, necessitated by a military malaise prevalent during the peacetime decade following the conflict. The United States, financially and psychologically drained from the war, had fallen well behind the European nations in updating their defensive and offensive weapons and fortifications.[15] Prototype weapons, along with modified or improved versions of existing weaponry would be tested for nearly fifty years at the site. The first tests conducted at the proving grounds centered largely around the conversion of pre-Civil War smooth-bore guns into the newer, more powerful rifled guns and testing their effectiveness against ironclad targets.[16]

The need for the United States to bring their defensive and offensive weapons and fortifications up to date was only part of the reason for the establishment of the Sandy Hook Proving Grounds. The combination of an increased sense of nationalism around the time of the country's centennial in 1876 and advent of the American manifestation of the Industrial Revolution facilitated a near-boundless creative environment for the development of weaponry more efficient and accurate than ever dreamed of in the past. Patent after patent appeared for a newer form of gun, torpedo, mine, or other military weapon, and the military services of the U.S. began their own testing and research as well. Moreover, the fact that European powers were busily constructing powerful, state-of-the art navies was not lost on the United States Congress.[17] Thus, driven by the need for greater defenses and the apprehension of falling behind other world powers, the United States government began to take the legislative steps to ensure a foothold in the advancing technological revolutions.

---

[13] *Ibid.*, p. 2.
[14] *Ibid.*, p. 13.
[15] *Ibid.*, p. 15.
[16] *Ibid.*, pp. 18–19.
[17] *Ibid.*, p. 1.

Sandy Hook, however, was not the ideal location the government had hoped for, but rather the setting with the minimum requisite attributes for the proposed experiments. A board of ordnance officers formed to locate such a site determined that it must be "comparatively level [and] of easy access, be free of any highways or extensive watercourse, uninhabited and sufficiently removed from any settlements to avoid any possible accident and embrace an extension of land seven to eight miles in length and from one-half to one mile in width."[18] In addition, the need to have reasonable access to a rail network in the vicinity but without a major line on the site was imperative.[19]

Several sites were considered including Babylon, Deer Park, Islip, and Bay Shore on Long Island and Squan and Island Beaches in New Jersey. Of these sites, Squan Beach, approximately eight miles south of Sandy Hook, was the one that came closest to the government's specifications. It was then up to Congress to approve the funds for the purchase and modification of the Squan Beach site. In the event that Congress turned down the request, it was decided to build a temporary proving grounds at Sandy Hook, which was already owned by the government. In the end, Sandy Hook's close proximity to New York City gave it a superior geographic advantage over Squan, as many of the investors, military men, and dignitaries found easy access to the city more logical and convenient. In August 1874, Secretary of War Belknap officially sanctioned the creation of the Sandy Hook Proving Grounds, thus conveying permanent status, before Congress had even voted on the Squan Beach site.[20]

Military Suppliers

The design and production of the myriad weapons tested at the Sandy Hook Proving Grounds were the result of the collective efforts of an array of manufacturers ranging from private individual inventors to military engineers. In addition, expedient modifications to

[18] *Executive Documents. Printed by the Order of the House of Representatives, During the 2nd Session of the 4th Congress, 1874–1875* (Washington, D.C., 1875), p. 9.
[19] Bearss, *Historic Resource Study* (note 12), p. 9.
[20] *Ibid.*, pp. 11–13.

existing ordnance were undertaken frequently by the military personnel charged with testing the new weaponry. The sources of these weapons can be broken down into three general categories: the Army Gun Factory at Watervliet, New York, other outside arsenals, and private contractors,[21] including a number of European firms.[22]

The major catalyst for determining the type of ordnance testing and fortifications constructed at Sandy Hook was the United States Army Corps of Engineers. Mindful of the rapid technological advances made during the Civil War, they did not want to invest large sums on artillery and works that would be outmoded upon completion. The large, easily-targeted sails of previous years disappeared on naval vessels during the second half of the nineteenth century.[23] No longer reliant on wind-powered wooden warships, steam-propelled ironclads offered increased mobility and tactical flexibility.[24] This in turn necessitated the development of longer range offensive weapons in the form of rifled guns. The government oversight of the Sandy Hook operations is exemplified by a congressional appropriations bill signed on June 6, 1872 by President Ulysses S. Grant which apportioned $270,000, "for experiments and tests of heavy rifled ordnance to be designated by a board of officers."[25]

Interestingly, a form of testing at Sandy Hook had actually been undertaken by a private concern several years earlier. This organization, known as the Boiler Testing Commission, was a federation of railroad companies interested in finding the causes for boiler explosions.[26] An engraving appearing in *Harper's Weekly* dated 23 December 1871 depicts several individuals outside of a simple enclosed complex observing the detonation of a boiler near the coastline. The conducting of tests by the privately owned Boiler Testing Commission on government land is a microcosm of the intricate relationship between the private and public spheres that has characterized Sandy Hook since its first European settlement.[27]

---

[21] *Ibid.*, p. 1.

[22] Thomas J. Hoffman, *The Sandy Hook Ordnance Proving Ground: an overview history from 1874–1919* (Highlands, NJ, 1979). Report on file at Gateway National Recreation Area, Sandy Hook Unit.

[23] Lewis, *Seacoast Fortifications* (note 10), p. 166.

[24] Bearss, *Historic Resource Study* (note 12), p. 5.

[25] *Executive Documents. Printed by the Order of the House of Representatives, During the 3rd Session of the 4th Congress, 1872–1873* (Washington, DC, 1873).

[26] Moss, *Another Look* (note 8), p. 42.

[27] *Ibid.*

As a result of President Grant's directive, Secretary of War William K. Belknap convened a seven member board to review proposals for the manufacture of muzzle and breech loading weapons to be tested at Sandy Hook. The board met several times in New York during the summer of 1872. A total of forty proposals were received, with several inventors making personal presentations and a few, even bringing demonstration models.[28] Those chosen include Dr. W.E. Woodbridge and Alonzo Hitchcock for the manufacture of muzzle-loading guns, Friedrich Krupp, E.A. Sutcliffe and Nathan Thompson, along with French and Swedish systems for the manufacture of breech-loading guns and H.F. Mann and William Lyman for miscellaneous weapons.[29] It is, somewhat puzzling as to why the government commissioned manufacturers to produce muzzle-loaders since breech-loaders had clearly eclipsed the muzzle-loaders by the end of the Civil War.[30] Breechloaders were significantly more accurate than the muzzle-loaders and were much more efficiently and safely loaded by the gun crew, but the breech mechanisms on earlier models were somewhat cumbersome and complex which may have prevented the complete abandonment of the muzzle-loading design during the post-Civil War period.[31] While the government funded these projects, the individual inventors maintained complete control and were allowed to dictate how their inventions were handled. However, the government would still oversee the actual production of the guns.[32]

The first tests at Sandy Hook were conducted in October of 1874 when Du Pont hexagonal powder was used to fire a 170-pound projectile from a Rodman gun.[33] Rodman guns were developed by Army

[28] Bearss, *Historic Resource Study* (note 12), p. 7.
[29] *U.S. Executive Documents 1872–1873.*
[30] Harold L. Peterson, *Round Shot and Rammers* (South Bend, Ind., 1969), p. 63.
[31] Jack Coggins, *Arms and Equipment of the Civil War* (Fairfax, Virg., 1983), p. 85. The other impediment was the lack of any large-scale cannon foundry in the U.S. capable or willing to immediately switch to breechloader manufacture after the Civil War. The West Point Foundry and the South Boston Foundry (and to a lesser extent the Fort Pitt Foundry), the main suppliers to the Union during the war, all steadfastly failed to modernize. In 1872 WPF and SBF petitioned the government to establish a national foundry under, not surprisingly, their joint direction, but to no avail. By the 1870s both foundries had to find the majority of their business outside the armaments sector as the technology had largely passed them by. Steven A. Walton, personal communication, June 2004.
[32] Lewis, *Seacoast Fortifications* (note 10).
[33] Bearss, *Historic Resource Study* (note 12), p. 18.

ordnance officer Thomas J. Rodman during the 1850s. These guns were a departure from earlier cannon in that they were devoid of all decoration and designed solely as a functional weapon based on the distribution of gas pressures within the unit while firing.[34] The Rodman's main competitor, rifled Parrott guns, developed by former army ordnance officer Robert P. Parrott at the onset of the Civil War, had been produced at the West Point Foundry since 1861.[35] While most Rodman and Parrott guns were rifled, providing much more accurate shots, the majority of U.S. artillery was still of the older smoothbore variety. The conversion of smooth-bore cannon to rifled was undertaken with wrought-iron tube inserts manufactured by both the West Point and South Boston foundries.[36] Among the first ordnance likely fired from the newly converted cannons were surplus Civil War projectiles that included Parrott, Shenkl, Sawyer, and James shells.[37] In 1874, the Chief of Ordnance, General Benet, initially asked for $250,000 to fund the conversion, but the next year Congress only agreed to an appropriation of $75,000, less than a third of what was originally requested.[38] Nevertheless, this was a remarkable sum considering that between the years 1873–1877, the United States was mired in the worst depression in its history.[39] This underscores the urgency felt by the Grant administration to strengthen the country's antiquated defense system.

During the years 1876–77, the first cannon produced as a result of the act of 6 June 1872 were tested at Sandy Hook. Among these were the Mann, Sutcliffe and Thompson rifles, all of which had been previously unveiled at Philadelphia's Centennial Exposition.[40] Hale war rockets were also tested at Sandy Hook in 1877. A contemporary engraving from *Frank Leslie's Illustrated Newspaper* depicts the firing of the rockets from the shore into the bay while members of the United States Ordnance Board look on. Interestingly, a number of

---

[34] Lewis, *Seacoast Fortifications*, p. 60.

[35] *Ibid.*, p. 67, and Peterson (note 30), p. 92.

[36] Bearss, *Historic Resource Study*, pp. 17, 27.

[37] Hoffman, *Sandy Hook* (note 22), pp. 50–51.

[38] *Executive Documents. Printed by the Order of the House of Representatives, During the 1st Session of the 44th Congress, 1875–1876* (Washington, DC, 1876).

[39] James A. Henretta, W. Elliot Brownlee, David Brody, and Susan Ware, *America's History: Volume 1 to 1877*, 2nd ed. (New York, 1993), p. 507 and Bearss, *Historic Resource Study* (note 12), pp. 50–51.

[40] Bearss, *Historic Resource Study*, p. 34.

civilians appear to be among the observers of the tests situated on an enclosed porch some distance behind the cannon and gun. An individual calculating the weapon's trajectory on a plotting bench is also among the spectators. Another engraving in the 15 November 1879 issue of *Harper's Weekly* depicts U.S. soldiers testing what appears to be either a 12-pounder bronze Confederate field gun or a bronze Confederate 3-inch field rifle.[41] This suggests that surplus Civil War weapons from the Confederacy were also tried at Sandy Hook, perhaps emblematic of the severe shortage of available weapons and military funding in the latter half of the nineteenth century. In fact, Civil War-era smooth-bore guns were used as late as 1882 and possibly after for the testing of cannister shells.[42]

The actual cannon used at Sandy Hook came from a variety of locations, and often from Europe. This may be an indication of the preeminence of European weapon designs of the day, as well as the desire of the United States to avail themselves of the most advanced equipment available in addition to developing our own weaponry. During the forty-five year span of the proving ground's use, guns produced in England, France, Germany, Italy, and Switzerland were routinely tested at Sandy Hook. In fact, visits by foreign military personnel were common over the years of the proving ground's existence. In 1886, the Brazilian navy ship *Almirante Barroso* sailed into New York harbor and during this visit U.S. Secretary of State Thomas G. Bayard arranged for the ship's captain and his officers to tour the proving grounds and observe gunnery practice.[43] Perhaps the most remarkable example of this was the docking of the German submarine UB-88 at Sandy Hook in early 1917 to observe the testing at the proving grounds. Its commanding officer, Kapitanlieutenant Walter Schweiger was in command of the U-20 submarine that had torpedoed and sunk the *Lusitania* in 1915, but the visit to Sandy Hook occurred about two months before the United States entry into World War I, and about seven months before the UB-88 submarine sank in the waters between Ireland and Scotland, killing Schweiger and his entire crew.[44]

---

[41] Warren Ripley, *Artillery and Ammunition of the Civil War* (New York, 1970), pp. 27, 178. Also, see Hoffman, *Sandy Hook* (note 22), p. 2.

[42] Hoffman, *Sandy Hook*, p. 21.

[43] Bearss, *Historic Resource Study*, p. 91.

[44] Moss, *Another Look* (note 8), p. 63.

The main American arsenal providing arms to the Sandy Hook proving grounds was the Watervliet Army Gun Factory in Watervliet, New York.[45] This arsenal itself is the nation's oldest, having been established in 1813 to supply the military during the War of 1812 and in 1888 Congressional legislation established the Gun Factory there.[46] General S.V. Benet, Chief of Ordnance from 1875 to 1891 called the Watervliet Gun factory, "a triumph of American scientific and mechanical skill."[47] The first guns produced by the new factory were delivered to Sandy Hook in late 1889.[48] Around this time, a gun factory was also approved for the Navy Yard in Washington, D.C., but a third gun factory at the Frankford Arsenal, failed to garner the needed support of the House of Representatives and was not approved.[49] Other companies involved in producing various components of the Proving Ground equipment include the Otis Iron and Steel Company, Cleveland, Ohio; the Trenton Iron Company, Trenton, New Jersey; and the Midvale Steel Company, Philadelphia, Pennsylvania (where Frederick W. Taylor would later implement his ideas of "scientific management" in the first decades of the twentieth century).[50]

Arguably one of the most significant innovations to come from the activities at Sandy Hook was the development of the Lyle gun. In April 1875, Secretary of War Belknap received a request from the Secretary of the Treasury Benjamin H. Bristow to actively help in the development of an improved gun for the U.S. Lifesaving Service. Guns used to fire lifelines to stranded ships had been woefully inadequate up to that point. Captain J.H. Merryman was charged with conducting the experiments, but also assigned to the project was Lietenant D.A. Lyle of the Ordnance Department. These experiments were monitored by the Board of Experimental Guns under the direct order of Secretary Belknap.[51] Lyle's goal was to reduce the weight of the mortar without compromising efficiency to extend

[45] Bearss, *Historic Resource Study*, p. 22.
[46] *Ibid.*, p. 83.
[47] *Ibid.*, p. 5.
[48] *Ibid.*, p. 108.
[49] *Ibid.*, p. 83.
[50] *Ibid.*, p. 4.
[51] Benet to Crispin, 16 April 1875. Washington, DC, National Archives and Records Administration, RG 156, Records of the Office of the Chief of Ordnance [hereafter NARA RG156], Letters Sent, 1875.

the reach of a ship-to-shore rescue line.[52] The research resulted in a range of nearly 695 yards, far exceeding any previous distance.[53]

In-house developments such as the Lyle gun were made possible by the Sandy Hook machine shop, which housed both machinists and mechanics. Among those employed were civilian laborers and mechanics that were paid between $1.60 and $2.75 per day.[54] In fact, with the exception of officers and small numbers of enlisted men occasionally assigned to Sandy Hook, the proving grounds were staffed entirely by civilian laborers and mechanics.[55] (In 1887, in addition to supplying weapons and equipment, the Watervliet, Allegheny, Watertown, Frankford and Springfield Arsenals also sent enlisted personnel to replace civilian workers at the proving grounds when Congress failed to enact the annual Fortification Bill.[56] Prior to this, approximately 80% of the proving ground labor force was civilian, even though the proving grounds were a military base.) This machine shop performed a variety of functions such as the modification of guns and carriages to be tested at Sandy Hook, experimental alterations to both guns and ordnance, the manufacture and repair of spare parts, and carriage alterations for installations throughout the Mid-Atlantic region. Some of these locations included Forts Wadsworth, Point, Delaware and Schuyler, Willets Point, and the Benicia Arsenal.[57]

Although the majority of the testing conducted at the Sandy Hook proving grounds involved heavy artillery, small arms were also tested in significant numbers. The relatively recent development of breech-loaded cartridge rifles coupled with reports of exceedingly long distance infantry fire of up to a mile and a half during the 1877–1878 Turko-Russian War piqued the interest of the Chief of Ordnance to test small arms. Among these were the U.S. Army Model 1873 .45-caliber rifle, the British Army .577–450 Martini Henry, and the Sharps-Borchardt Model 1878 rifle.[58] The results of the tests were significant. Firing tests achieved a maximum distance of 3,680 yards (2.1 miles) and also penetrated into sand eight or more inches after

[52] Bearss, *Historic Resource Study* (note 12), p. 30.
[53] *Ibid.*, p. 51.
[54] Crispin to Benet, 3 June 1879, NARA RG156, Letters Received, 1879.
[55] Bearss, *Historic Resource Study*, p. 76.
[56] *Ibid.*, pp. 83–84.
[57] Smith to Benet 25 October 1881, NARA RG156, Letters Sent, 1881.
[58] W. John Farquaharson, ".45.70 at Two Miles," *Rifle Magazine* (Nov/Dec 1977), pp. 7–15.

passing through a wooden plank one to three inches thick. This proved the effectiveness of these weapons on enemy troops even if they were partially shielded.[59] This is all the more remarkable when considering the distance between Civil War infantry combatants was often 250 to 300 yards and the effectiveness of the popular Springfield Model of 1861 was a mere 500–600 yards.[60]

One of the more well-known cartridge weapons tested at Sandy Hook was the Gatling gun. First patented in 1862 by Dr. Richard Gatling, it was modified in 1866 to accommodate a rimfire cartridge.[61] The Gatling gun, while revolutionary in many respects, was flawed in its initial design, causing the gun to frequently jam and misfire and resulting in inconsistent velocity and accuracy.[62] In 1876, the New York Armory sent two Gatling guns to Sandy Hook for test firing. That same year, a new five-barrel model was tested. In 1887, 10-barrel (long barrel) and 6-barrel (short barrel) models were tested. Gatling guns were tested periodically over the next twenty years, before being phased out in 1910.[63] All in all, the Sandy Hooks Proving Grounds saw tests of most of the new weapons considered by the U.S. armed forces in the half century after the Civil War. The remnants of these tests, excavated in the UXO sweeps on the beaches provide a useful lens through which to understand this testing.

The Human Toll of Testing

The quest for advancement in the volatile arena of military weaponry is subject to the usual trial and error setbacks inherent in all scientific experimentation. While the experiments conducted at Sandy Hook resulted in remarkable advances in our defensive capabilities, it often came at a heavy price. The very nature of experimental weaponry would result in erratic accuracy and misplaced targeting. This wreaked havoc with maintaining the complex infrastructure of Sandy Hook's civilian and military personnel. Employees of the machine shop, Signal Service, and Western Union Telegraph along with their fam-

[59] Hoffman, *Sandy Hook* (note 22), p. 3.
[60] Coggins, *Arms and Equipment* (note 31), pp. 24, 32.
[61] Hoffman, *Sandy Hook*, p. 5.
[62] Coggins, *Arms and Equipment*, pp. 29, 45.
[63] Hoffman, *Sandy Hook*, pp. 20–22.

ilies were all dangerously close to the test guns.[64] Exploded gun fragments had already landed near the Officers Quarters and the coastal fog horns. Over the years, numerous accidents and misfires resulted in the killing or wounding of dozens of men. The first accident occurred in 1886 when a 12-inch shell burst, killing first lieutenant William M. Medcalfe and PFC Joseph Knox and wounding sergeant John Abbott and corporals George Clark and Walter Goodino. In 1895, lieutenant Fremont H. Peck was killed while testing a 4.7-inch Hotchkiss rifle. A year later, corporal Robert Doyle and PFC Frank Conway were killed and lieutenant George Montgomery, PFCs William McDonald, and Patrick Ryan and PSC James Coyne were wounded when a metallic cartridge prematurely exploded while testing a Canet rapid fire gun.[65]

Non-military personnel were also periodically involved in testing mishaps. In 1898, an employee of the Postal Telegraph Company was injured when a five-inch rapid fire gun burst.[66] In 1899, Henry Murphy, a civilian recording clerk, was killed and privates James Harrington and Charles Dieman were injured when a 10-inch breech-loading rifle burst.[67] Over the years, numerous "close calls" involving the U.S. Lifesaving Station, located at Sandy Hook were also recorded. The problem stemmed mainly from the fact that the station, which predated the proving grounds by about two years, was situated directly downrange of the line of fire. The custom then became to evacuate the station when testing was being performed. This proved fortuitous, as projectiles would routinely ricochet off of their wooden targets and land within several yards of the station. Efforts to move the station to other parts of the peninsula only provided a temporary respite from the dangers of the proving grounds. As more batteries were built and the range of projectile ranges became ever longer, the viable areas to place the station dwindled almost to none.[68]

One of the worst accidents at the Proving Grounds did not even involve individuals associated with Sandy Hook. On 9 July 1892, a 500-pound projectile was deflected, striking and sinking the schooner

[64] Bearss, *Historic Resource Study* (note 12), p. 93.
[65] *Ibid.*, p. 144.
[66] Hoffman, *Sandy Hook*, p. 25.
[67] *Ibid.*, p. 26.
[68] *Ibid.*, pp. 16–17.

Henry R. Tilton approximately three-quarters of a mile off the coast. The cause of the accident was determined by the test supervisor Capt. Heath to be the result of a large powder charge and an unexpected angle of impact on the beach, which sent the projectile farther out into the ocean than originally planned.[69] Most of the ordnance, however, hit its targets on the beach and despite some attempt by the military to clear the resulting fragments of ordnance and targets, a good deal of artifacts remained buried in the sand until the mid-1990s.

The Artifacts

In advance for construction of public access facilities at North Beach Center, an unexploded ordnance (UXO) sweep was conducted to locate, identify, and remove materials that could have been potentially hazardous to the public. The UXO sweeps were conducted by MTA, Inc., on behalf of the U.S. Army Corps of Engineers (CORPS) and the National Park Service (NPS) between 1994 and 1995. The artifacts retained were selectively recovered by Louis Berger and Associates from an array of metal objects unearthed and stockpiled by MTA during the UXO sweeps at North Beach Center. Metal objects were recovered from pre-established survey grids in the vicinity of the Gunnison Battery and the Proof Battery, both of which were part of the historic Sandy Hook Proving Ground. North Beach Center is located within the National Historic District of Fort Hancock, near the north end of the Sandy Hook peninsula.

As a result of the UXO sweeps at Sandy Hook, a total of 10,000 artifacts was recovered, of which 245 were deemed worthy of further research and catalogued. Out of these, 38 of the most significant and unusual pieces were conserved using either electrolytic reduction or sodium sulfite methods of conservation. The artifacts discussed below are among the most unusual found during the UXO sweeps, and represent rare or prototypical specimens associated with the Sandy Hook proving grounds or the military during the late nineteenth and early twentieth centuries. Several had been expediently modified by proving grounds personnel for specific, on-site test-

[69] Moss, *Another Look* (note 8).

ing strategies. The recovered assemblage ran the technological gamut from whole projectiles to individual elements, such as fuzes and rifling bands. In addition, non-military items, such as lifesaving equipment, and elements of the proving grounds rail infrastructure illustrate the complexities of the activities at Sandy Hook.

Although a variety of military and civilian technological experiments were undertaken at the proving grounds, the most important testing was the testing of ordnance. Numerous shells, both live and test rounds were recovered. One notable example is an iron projectile, almost whole (minus the base), with the body measuring 8 inches in diameter. (figure 1) The head has a threaded fuze hole and no markings are evident. This piece has a Eureka-pattern sabot band and is from a very early period in the proving ground's history. This artifact, initially thought to be a Civil War Parrott round is now believed to be an Arrick model, first introduced ca. 1870.[70] A Eureka-pattern sabot band, misshapen from firing, possibly came from this shell. Brass sabot bands replaced ones made of iron when the latter proved unsatisfactory during the Civil War.[71] Eight-inch Rodman rifles capable of firing this type of shell, were the first weapon tested at the Sandy Hook proving grounds in 1874.[72]

Another projectile is a 12-inch diameter shell that was broken into five sections as a result of shape-charging. (figure 2) During the course of excavations, it was not always readily apparent whether or not a projectile's explosive charge was live or inert, so rather than having the author find out the hard way, those pieces in question were taken to an isolated area and shape-charged, a method of controlled detonation whereby plastic explosives are strategically placed, usually at two or more points along a shell. Each applied point directs an explosive charge to the center of the projectile which in effect, detonates the piece and at times, fractures it resulting in an easily mended condition as seen in figure 2. The tip section, along with the lower body and base sections are pictured after electrolytic reduction treatment, while the upper body section is shown before treatment. This shell had a stamp on the underside of the base, "WATERVLIET ARSENAL", which indicates its manufacture in the arsenal at Watervliet,

[70] See Benet to Crispin (note 52).

[71] Ripley (note 41), pp. 290–291. The sabot band was a band of softer metal attached to the iron shell that would engage with the rifling in the barrel.

[72] Hoffman, *Sandy Hook* (note 22), p. 33.

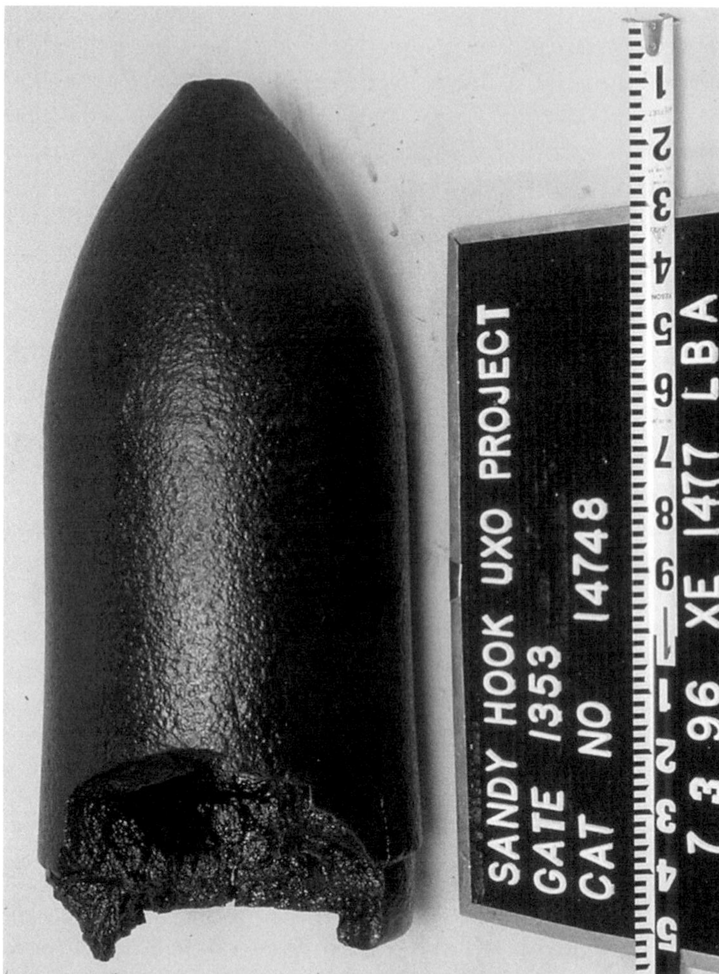

Figure 1. 8-inch iron shell, thought to be an Arrick model of c.1870 fired from an 8-inch Rodman gun at Sandy Hook Proving Grounds (SHPG) in the later 1870s. Excavated at Sandy Hook National Park. Photo Courtesy of Louis Berger Associates, Inc., East Orange, NJ.

New York. This type of stamp suggests the time of manufacture as circa 1880, since later products of this arsenal had only the single letter "W" as a manufacturer's mark.[73] The fifth piece from this projectile is a 2-inch-wide copper rotating band with lands and grooves. Prior to firing, these bands are smooth, but when the gun is fired the band expands into the rifling of the tube, causing the projectile to rotate in the bore and hence take on this characteristic appearance.[74]

A third example is a large ferrous alloy projectile head from a 12-inch round with partial body section. (figure 3) This piece is 31 inches long, measured from nose tip to the lowest fracture point on the body. The body is hollow, with a maximum remaining interior cavity diameter of 5.5 inches. This artifact known as a "needle nose" is a high capacity shell which may date from the earliest years of the proving grounds. The narrow nose shape is designed to have high penetrating power in response to the changing needs of projectiles during the second half of the nineteenth century. In particular, iron clad ships became commonplace after the Civil War, necessitating the need for armor-piercing artillery.[75]

Along with testing new ordnance at the proving grounds, innovative methods of defensive armaments were developed and tested. To achieve this, projectiles with exaggerated weights were employed during testing exercises. This was done to test the strength of armor plating used on ships and on land fortifications. Several of the shells recovered during the UXO sweeps had cement or mortar as an interior packing, or as an exterior coating (figure 4). Different weights would simulate different types of shells. In addition to cement and mortar, other methods were employed to test the various sized ordnance and their armor-piercing capabilities. One test round recovered during the UXO sweeps is a large, ferrous alloy 12-inch projectile. The body is hollow, with a remaining maximum interior cavity of 7 inches. Lead balls in a matrix of mortar or very fine sand were present in the cavity. The sand and lead balls were put into the cavity for balance or to bring it up to the proper weight. This artifact

---

[73] Joe Vann, UXO Specialist, Mare Island Naval Complex, Vallejo, California, personal communication, 1999.

[74] U.S. Navy, *The Bluejackets Manual.* 12th ed. (Annapolis, Maryland, 1944), p. 252.

[75] Lewis, *Seacoast Fortifications* (note 10), p. 66.

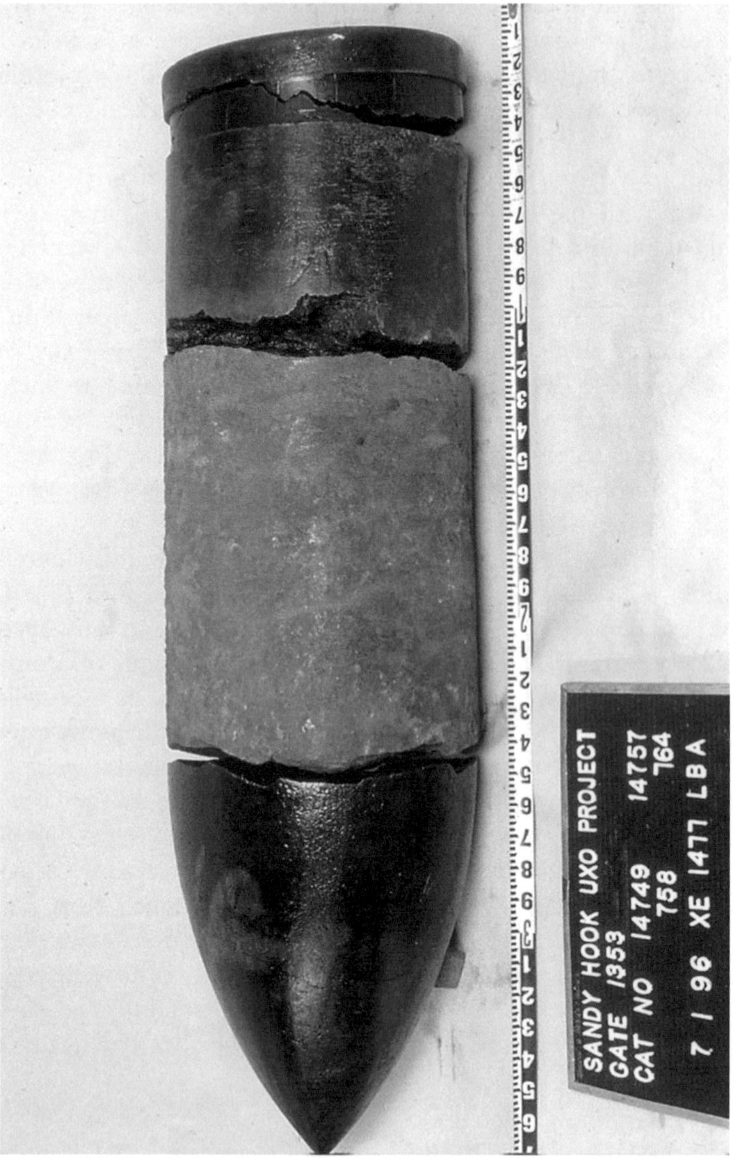

Figure 2. 12-inch shell manufactured at the Watervliet Arsenal, Troy, NY c. 1880. The intact shell was shape-charged in the UXO sweeps. Excavated at Sandy Hook National Park. Photo Courtesy of Louis Berger Associates, Inc., East Orange, NJ.

Figure 3. Fragment of a "needle-nose" 12-inch projectile, early 19th century. Excavated at Sandy Hook National Park. Photo Courtesy of Louis Berger Associates, Inc., East Orange, NJ.

exhibits dimples on the head for securing a windscreen, with no other markings evident.

Smaller projectiles designed for portable cannon were also recovered from the proving grounds. One is nearly complete with a partial rotating-band visible, ending at a fracture point near the base. (figure 5) This piece is a 3-pounder or 47 mm Hotchkiss shell fitted for the Demarest nose fuze, and was manufactured by Hotchkiss Ordnance of Paris, France.[76] Hotchkiss Ordnance was one of the largest manufacturers of arms during the Civil War. They moved their factory to France after the end of the war, but remained a major supplier to the United States up to World War II.[77] Records indicate that Hotchkiss revolving cannons were tested at the proving grounds in 1876,[78] and later cannons of this type were used by the United States Army during the conflict at Wounded Knee, South Dakota in 1890.[79]

A 3.6-inch projectile consisting of a mendable tip and base is an example of a production technique that was devised to increase the effectiveness of projectiles without increasing their size. (figure 6) The piece has a hollow-core and a round slotted base plug specifically designed for use with the new 3.6-inch breech-loading gun. The projectile's unique method of construction involved casting in a chilled iron mold where the tip cooled faster and became harder than the base section, as evidenced by the different rates of rusting on the artifact. This process allowed the hardened tip to penetrate armor and explode inside a ship or fortified wall. The charge was able to withstand the shock of initial impact, and a delayed-action fuze ignited seconds after the projectile landed. Armor-piercing shells such as this were first introduced in 1864, and came into widespread use around 1880.[80] The 1864 patent for this production technique belongs to Capt. Pallisar of the Hungarian Army, which illustrates the range of the international sources of new technology used at the proving grounds.[81]

[76] Scharfenberger and Sara (note 1), p. 76.
[77] Joe Vann, pers. comm.
[78] Hoffman, *Sandy Hook* (note 22), p. 33.
[79] Dee Brown, *Bury My Heart At Wounded Knee* (New York, 1970).
[80] Hoffman, *Sandy Hook*, pp. 46, 64.
[81] *Ibid.*, p. 46.

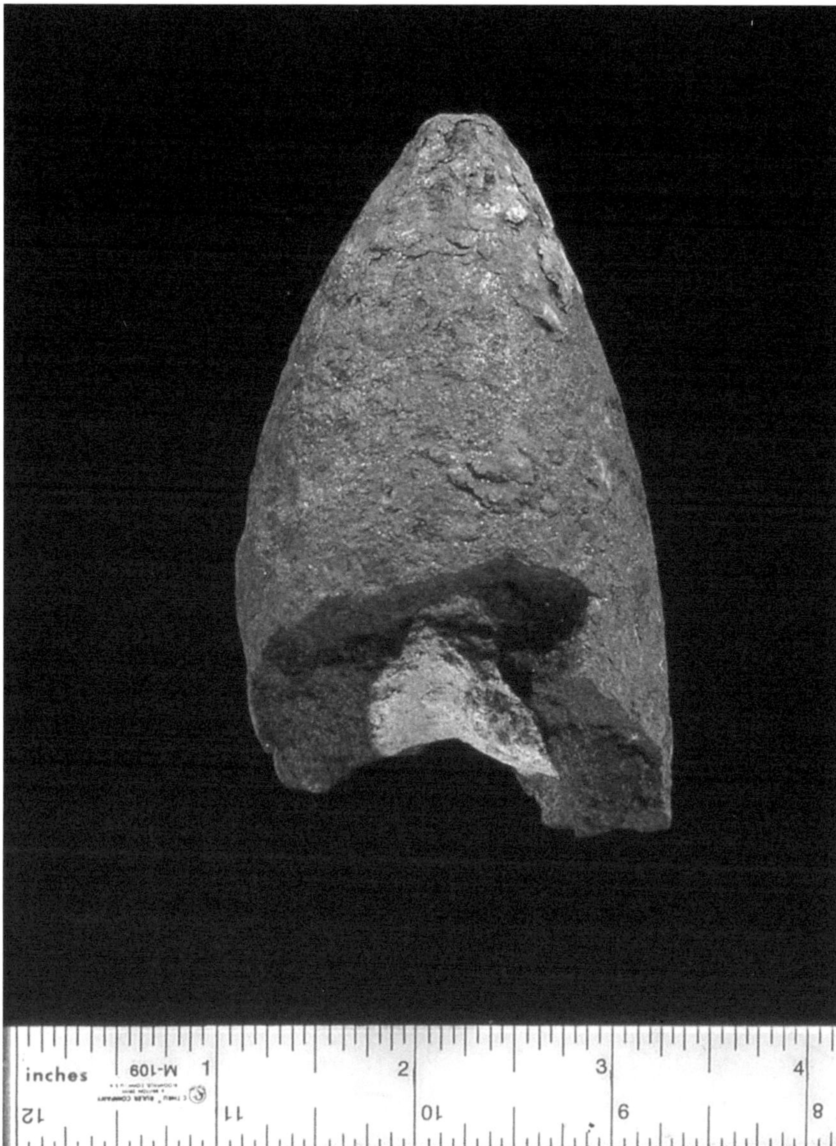

Figure 4. 3.6-inch shell with interior packing of cement, used to simulate the weight of a charge during practice firing. Excavated at Sandy Hook National Park. Photo Courtesy of Louis Berger Associates, Inc., East Orange, NJ.

Figure 5. 47 mm (3 pound) Hotchkiss shell fitted for the Demarest nose fuze, Hotchkiss Ordnance Co., Paris, tested at SHPG in 1876. Excavated at Sandy Hook National Park. Photo Courtesy of Louis Berger Associates, Inc., East Orange, NJ.

One of the recovered artifacts relating solely to the testing process of projectiles at Sandy Hook is a massive circular iron target, 38 inches in diameter and 3 inches thick at the center, tapering toward the outside. (figure 7) In its original condition, this piece would have been solid and flat, but the impact from a projectile has opened a hole in the center and concaved the remaining frame. Targets such as these were used in shore to shore testing exercises at the Sandy Hook proving grounds to test both the accuracy of rifled cannons, and the strength of armored plating against various sized shells.[82] During testing in 1892, the remarkable accuracy of the new rifled guns was exemplified by an 80% strike rate at a distance of one mile on a target area of twenty by twenty-one inches. At a range of 3,000 yards or just under two miles, six shots impacted on a target of one and a half by four feet, similar to the piece shown here as it would have been in its original size.[83]

The need for anti-aircraft artillery had been a reality long before the advent of mechanized flight. During the Civil War, hydrogen-filled spy balloons were used successfully, flying too high to be effectively shot down, or else benefiting from the lack of sufficient fuses to effectively time shots from existing artillery to bring them down.[84] Technological advances associated with the relatively new concept of aerial warfare is represented by is a brass 21-second timing fuse from the first decade of the twentieth century. (figure 8) Several identifying markings are visible on this piece which denote it as a Mark 3 1907 M anti-aircraft fuse manufactured by the Picatinny Arsenal.[85] Timing fuses are designed to function at a predetermined time after the shell leaves the gun. They are used for both anti-aircraft high-explosive shells and for shrapnel intended for ground forces. Projectiles may also contain a percussion element which explodes on impact either by design, or as a back-up in case the timing element fails.

---

[82] Joe Vann, pers. comm.

[83] Bearss, *Historic Resource Study* (note 12), p. 119.

[84] Coggins, *Arms and Equipment* (note 31), p. 110.

[85] Major William C. Foote, C.A. (compiler), *Notes On Ammunition: Bulletins 216-R2 and 287* (Fort Monroe, Virg., 1918); U.S. War Department, *Miscellaneous Ammunition and Fuses, Technical Report No. 1370-A* (Washington, D.C., March 24, 1931); and Scahrfenberger and Sara (note 1).

Figure 6. 3.6-inch armor-piercing shell with mendable tip and base, c. 1864–1880s. note the differential rusting, indicating a hardened tip. Excavated at Sandy Hook National Park. Photo Courtesy of Louis Berger Associates, Inc., East Orange, NJ.

Figure 7. Iron target used for target practice at SHPG, after impact. 38 inches in diameter and 3 inches thick. Excavated at Sandy Hook National Park. Photo Courtesy of Louis Berger Associates, Inc., East Orange, NJ.

Components of larger projectiles were tested and modified in a variety of designs and dimensions. An excellent example of this process is a brass Drigg's Patent fuze recovered in the sweeps. (figure 9) A nearly complete stamp is present that, if complete, would have read, "D.S.G. & Co./ DRIGGS PATENT/ JAN. 7–90/9–99." indicating that the fuze was patented in July 1890 and manufactured by the Driggs-Seabury Gun & Ammunition Company in September 1899. In addition, the Ordnance Department stamp comprised of a circular belt, crossed Rodman guns and a flaming bomb would be present. The version of this fuze prior to 1898 had a lug at the rear for insertion, rather than the spanner holes on this example. However, this example has had a post-production modification made by filing down of the base in two places to allow the use of the old style lug wrench.

Part of the process of deploying projectiles and equipment to the various battery locations prior to testing was through the use of a rail network. This was an outgrowth of the Civil War that saw railroads become an integral part of military mobility for moving both troops and heavy equipment.[86] In fact, the presence of a rail system on Sandy Hook figured into the government's decision to situate a proving ground on the site.[87] One of the more significant artifacts associated with the proving grounds rail system was an intact track switch in near-perfect condition with the embossment, "Ramapo Iron Works Pat. Oct. 19 1897 No. 6." (figure 10) According to Chard, the iron works was organized in 1866 by H.L. Pierson as the Ramapo Wheel and Foundry Company.[88] Beginning in 1893, the War Department began to expand the existing railroad by purchasing six miles of track, sidings and switches from the New Jersey Southern Railroad.[89] This allowed guns and ordnance to be shipped from as far away as the Watervliet Arsenal in New York and the Rock Island Arsenal in Illinois, as well as receiving anthracite coal from Cattaraugas, Pennsylvania at a fraction of the previous cost.[90] In addition, munitions could be transported easily to even the most remote coastal

[86] Coggins, *Arms and Equipment* (note 31), p. 9.
[87] Bearss, *Historic Resource Study*, p. 9.
[88] Jack Chard, "The Ramapo Works," *The North Jersey Highlander* (Fall-Winter 1982).
[89] Bearss, *Historic Resource Study*, p. 148.
[90] *Ibid.*, p. 158.

Figure 8. Mark 3 1907M, 21-second anti-aircraft brass timing fuse, manufactured by the Picatinny Arsenal. Excavated at Sandy Hook National Park. Photo Courtesy of Louis Berger Associates, Inc., East Orange, NJ.

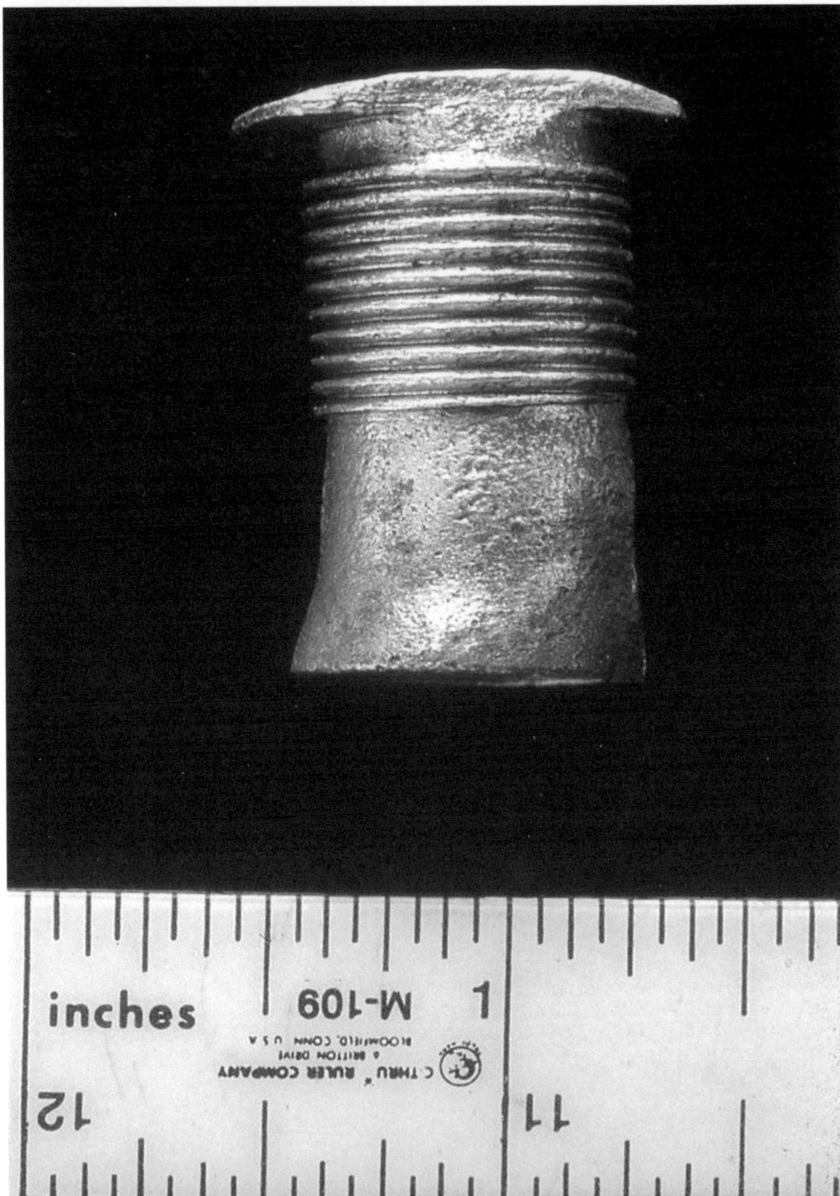

Figure 9. Drigg's Patent fuze (second model), 1899. Excavated at Sandy Hook National Park. Photo Courtesy of Louis Berger Associates, Inc., East Orange, NJ.

batteries. This switch demonstrates the integrated nature of the systems at Sandy Hook and reiterates the idea that instrumental testing is not simply a matter of bench-top experiments, especially in the case of full-scale military hardware.

One of the more unusual items recovered during the UXO sweeps is an iron tip paling measuring 17 inches from the tip to the end of the strap. (figure 11) Possibly dating from the eighteenth century, this piece was originally fastened to the end of a wooden post allowing it to be driven into the ground. The artifact has four flanges, one of which has two handwrought nails affixed while the other three each hold one nail. This piece clearly predates the proving grounds and may represent the military occupation of Sandy Hook by the British during the eighteenth-century. While not representative of a particular technological advance attributable to Sandy Hook, it is characteristic of equipment specifically designed for use at a colonial-period beach location military installation. Further, it shows that the laboratory space of a site like Sandy Hook is not closed as would be a typical laboratory, but had to accommodate all sorts of people and groups for various military purposes over the centuries.

While the proving grounds were a major source of military activity at Sandy Hook during the second half of the nineteenth-century, the testing of life-saving equipment and techniques played a pivotal role since the late eighteenth-century. One of the most novel innovations developed and implemented at Sandy Hook, was the rescue device known as the "Francis Life-Car." Firing a line from the shore to a stranded ship, a pulley system was employed to ferry passengers to shore. The cable was drawn taught and a capsule-shaped container was transported out to the ship, and returned to shore carrying four or five passengers at a time.[91] (figure 12) Ironically, the testing of the life cars involved the use of non-defensive projectiles as a means of delivering the cable. A Hale War Rocket, tested in 1877 for the U.S. Lifesaving Service, was recovered at Sandy Hook. Hale War rockets, a Civil War-era improvement over the Congreve rocket, were still somewhat undependable and even occasionally posed a danger to the firing crews.[92] Two Francis Life-Cars were found in dunes on the North beach; one in near perfect condition, and the other approximately 80% complete. The car measures 11 feet in

[91] Moss, *Another Look* (note 8), pp. 91–92.
[92] Coggins, *Arms and Equipment* (note 31), p. 97.

Figure 10. Ramapo Iron Works rail track switch from SHPG rail system, post 1897. Excavated at Sandy Hook National Park. Photo Courtesy of Louis Berger Associates, Inc., East Orange, NJ.

Figure 11. Iron tipped paling, 17 inches long, possibly dating from the late 18th century. Excavated at Sandy Hook National Park. Photo Courtesy of Louis Berger Associates, Inc., East Orange, NJ.

length and 4 feet wide, with an opening 20 by 21 inches. Eight diamond-shaped sets of pinholes designed to allow air inside, are distributed on the surface of the car. Handles fitted along the interior of the shell allowed passengers to steady themselves during the tumultuous trip to safety. The first years resulted in a maximum reach of 400 yards for throwing a line from the shore to a foundering ship. By 1880, the introduction of the Lyle gun as mentioned above had improved the attainable distance to one-half mile.[93] First developed around 1850, the Francis Life Cars were eventually replaced by the cheaper, faster Breeches Buoy around 1899 although the nature of shipping changed toward the end of the nineteenth-century—from immigrant ships to cargo ships with small crews, which eliminated the need to ferry large numbers of passengers from a shipwreck.[94] The astounding success rate of the Francis Life-Cars is exemplified by the shipwreck of the *Aylshire* off the coast of New Jersey in 1851, where of the 201 passengers on board, 200 were brought safely to shore. The lone fatality occurred when one passenger, afraid of being confined in the cramped iron capsule, chose instead to ride atop where he was quickly swept into the pounding surf and drowned.[95]

The development of the Francis Life Car for the U.S. Life Saving Service, which was the forerunner of the U.S. Coast Guard, typified the unified efforts of the various branches of the military and civilian interests at Sandy Hook. These collaborations utilizing army projectiles, Coast Guard manpower, naval conditions and civilian technology resulted in a monumental advancement that transcended military/civilian boundaries. Historically, this relationship has existed since the eighteenth century. In fact, the motivation behind the construction of the Sandy Hook lighthouse in 1764, nearly one hundred years after a military fort was first proposed, was to protect civilian shipping interests. This lighthouse, built with government funds by a private contractor to help stem the losses from shipwrecked commercial vessels, became a military target during the American Revolution.[96]

[93] John Bailey Lloyd, *Six Miles At Sea: a pictorial history of Long Beach Island, New Jersey* (Harvey Cedars, NJ, 1990), pp. 28–29.

[94] Van R. Field, *Wrecks and Rescues on Long Island: the story of the U.S. Life Saving Service* (East Patchogue, NY, 1997), p. 10 and Paul Giambarbra, *Surfmen and Lifesavers* (Cape Cod, Mass., 1985), p. 61.

[95] *Ibid*., p. 8.

[96] Moss, *Another Look* (note 8), pp. 10–11.

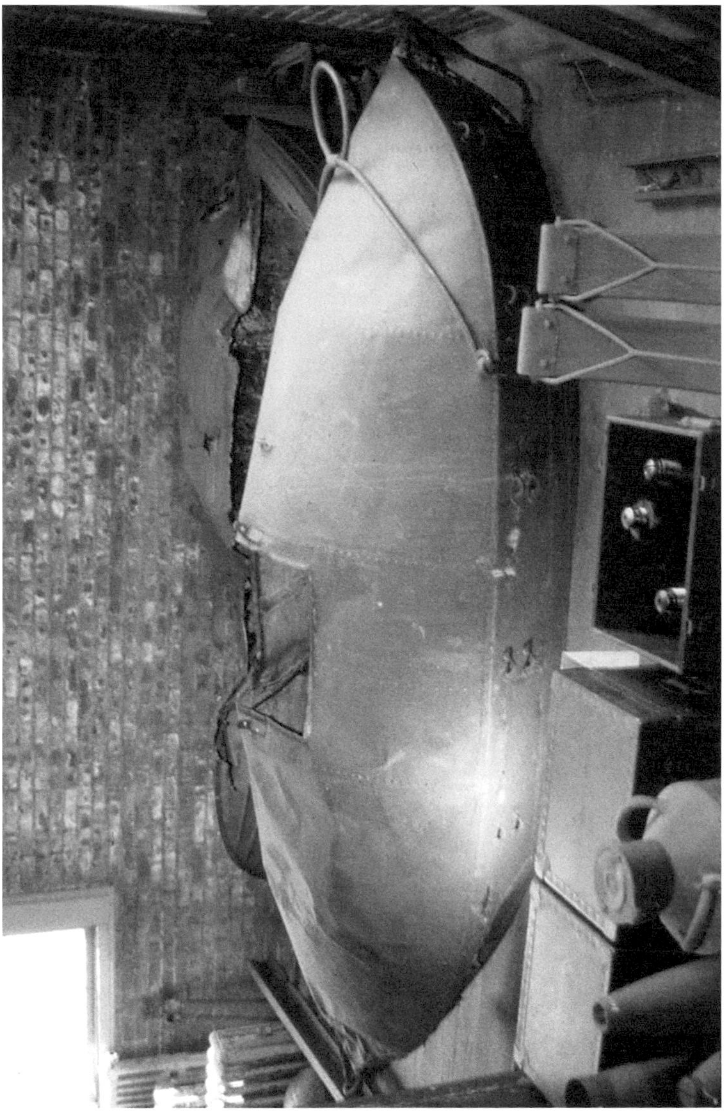

Figure 12. Francis Life-Car maritime evacuation system excavated at Sandy Hook National Recreation Area. The car is 11 feet long and 4 feet wide and was suspended from a cable fired to a grounded ship by a rocket. Photo by the author.

Conclusion

The proving grounds at Sandy Hook played an integral role in the development of America's defenses in the period immediately following the Civil War. Weapons tested at Sandy Hook were essential to the United States military efforts during the Spanish-American War, the Phillipines Insurrection, and World War I.[97] These weapons came from a variety of sources: public and private; individual and corporation; national and international; all of which combined to provide the most effective, economical means of ensuring our military supremacy. Ironically, it was the same advances in technology that led to the establishment of the Sandy Hook Proving Grounds that facilitated its demise. As the range capabilities of the newer, more powerful weapons increased with each passing year, the limited length of Sandy Hook's shore became a liability. As a result, the Proving Grounds in Maryland were opened in 1918 (see Grier, this volume), and the Sandy Hook Proving Grounds were phased out the following year.

This stage in the evolution of Sandy Hook as both a military and maritime fixture on the coastal landscape, was the catalyst for the nation's emergence as a world power at the dawn of the twentieth century. Technology that would benefit all Americans not only in times of war, but in peacetime as well was developed at Sandy Hook, as guns designed to repel our enemies were tested alongside guns that fired lifelines to stranded ships. The research conducted at the proving grounds resulted in vastly improved armaments that helped thrust the nation into global military prominence. The consequences for the United States during the first and second world wars would have been dire had the superiority of European weaponry produced in the latter half of the nineteenth century not been met and challenged by technologies developed and tested at the Sandy Hook proving grounds.

[97] Hoffman, *Sandy Hook* (note 22), p. 16.

Acknowledgements

The author would like to thank National Park Service Archaeologist Dana Linck and Curator Felice Ciccione of the Gateway National Recreation Area for their assistance on the project and their input in the original version of this paper. Special thanks to Richard Veit for the on-site photos and Rob Tucher for his consistently magnificent artifact photos. I would also like to acknowledge UXO specialist Joe Vann for sharing his vast knowledge of the weapons, armaments and arsenals associated with the Sandy Hook Proving Grounds. Steve Walton and two anonymous reviewers for reading and critiquing early drafts of this paper. Any mistakes or omissions, of course, remain my own.

CHAPTER SEVEN

FROM MEASURING PROGRESS TO TECHNOLOGICAL INNOVATION: THE PREWAR ANNAPOLIS ENGINEERING EXPERIMENT STATION

William M. McBride

The establishment of the Naval Research Laboratory (NRL) in 1921 has attracted the attention of historians as an example of the extension of nascent American industrial research and development to the federal government. The focus on NRL was a by-product of its key role in the scientific research that supported the development of long-range naval radio and radar before World War Two.[1] On the other hand, the less-known U.S. Naval Engineering Experiment Station (EES), across the Severn River from the U.S. Naval Academy in Annapolis, predated the Naval Research Laboratory by over a decade and its contributions to the war were arguably just as significant. Yet EES has received little historical attention, perhaps because the engineering side of technological development is often overshadowed by the scientific side.

[1] For the history of NRL and treatments of naval research before World War Two, see David K. Allison, *New Eye for the Navy: the origin of radar at the Naval Research Laboratory* (Washington, 1981); Bruce W. Hevly, "Basic Research within a Military Context: the Naval Research Laboratory and the foundations of extreme ultraviolet and X-ray astronomy, 1923–1960," Ph.D. dissertation, Baltimore, 1987; and William M. McBride, "The 'Greatest Patron of Science'?: the Navy-academia alliance and U.S. naval research, 1896–1923," *Journal of Military History* 56 (1992): 7–33. Also of interest is Susan J. Douglas, "The Navy Adopts the Radio, 1899–1919," in Merritt Roe Smith (ed.), *Military Enterprise and Technological Change: perspectives on the American experience* (Cambridge, Mass., 1985), pp. 117–73. For an introduction to early American industrial research, see Leonard S. Reich, *The Making of American Industrial Research: science and business at GE and Bell, 1876–1926* (Cambridge, 1985). For the differentiation between industrial engineering research and industrial scientific research before World War Two, see Thomas C. Lassman, "Industrial Research Transformed: Edward Condon at the Westinghouse Electric Manufacturing Company, 1935–1942," *Technology & Culture* 44 (2003): 306–39.

Initially, the Engineering Experiment Station's activities were much more in line with contemporary industrial testing laboratories where "scientists and engineers labored to assure consistency and efficiency."[2] These differed slightly from the industrial research facilities established in 1900 by General Electric (GE) and in 1911 by American Telephone & Telegraph (AT&T) that were designed to solve specific technical problems and to "rationalize the process of technological innovation."[3] As EES matured it evolved from a simple testing facility and emulated the industrial research at GE and AT&T. EES paralleled Leonard Reich's criteria for an industrial research facility since it was "staffed by people trained in science and advanced engineering who work toward a deeper understanding of corporate-related science and technology," and was "organized and administered" to keep it "insulated from immediate demands yet responsive to long-term company needs."[4] As a fully developed industrial research facility during the 1920–30s, EES advanced general engineering knowledge, naval engineering, and naval power so that the U.S. Navy entered World War Two with strategically superior ship propulsion systems.

American naval engineers perceived the need for a facility such as EES as they struggled to design and operate machinery to support the American naval renaissance during the 1880s and 1890s. The modern American navy began humbly with the 1883 authorization of its first steel-hulled ships, the cruisers *Atlanta, Boston, Chicago*, and the dispatch ship *Dolphin*. Authorization for the second-class battleships *Texas* and *Maine* soon followed. After decades of rapid scientific and engineering change, and confusion over the optimal technological basis of naval power, the world's major navies were concentrating on heavily armored, steam-propelled, steel battleships by the late 1880s. In 1890 Congress authorized the first three "seagoing" battleships for the U.S. Navy. By 1898 the U.S. Navy had four battleships in commission, and the navy's efficacy against Spain in 1898 stood in stark contrast to the small and technologically inferior American navy of 1889 characterized by Secretary of the Navy Benjamin F. Tracy as subordinate, not only to the major naval powers, but to lesser ones such as Turkey, China, and Sweden.[5] Modern

---

[2] Reich, *Making of American Industrial Research*, p. 2.

[3] Lassman, "Industrial Research Transformed", p. 307.

[4] Reich, *Making of American Industrial Research*, p. 3.

[5] Benjamin F. Cooling, *Benjamin Franklin Tracy: father of the American fighting navy* (Hamden, Conn., 1973), p. 75.

American naval technology enabled a nascent American empire in 1898 and this translated into appropriations for fourteen more battleships by 1905—Theodore Roosevelt's much-desired blue-water navy.

Due to its limited engineering bureaucracy, the late-nineteenth-century navy relied on the private sector for the design and construction of the complex technological components of its modern ships. This left the navy vulnerable to the unsubstantiated performance claims and vagaries of its industrial suppliers and private shipyards. Naval engineers were not comfortable relying on the good will of the private sector to supply quality machinery. The navy had no specifications for machinery or material and no viable means, save for noting failures and excluding vendors, to ensure quality control. Unlike the antiquated force of the 1870s, the technologically complex modern navy needed a testing laboratory such as EES. Congress agreed with naval engineers and authorized it in 1903 and EES was in operation by 1908.

One contemporary measure of the importance of the work conducted at Annapolis was the international attention it received. In 1925 the Italian naval ministry directed its Washington attaché, Commander E. Sommati di Mombello, to request a "complete and detailed description" of the Experiment Station from the U.S. Navy's director of naval intelligence.[6] In his surprisingly candid and extensive reply to the Italians, EES's officer-in-charge described the station's history, organization, and work. The Experiment Station's mission, as conveyed to the Italians, was drawn from Lieutenant Commander Ernest J. King's 1913 observation: "there will always be progress; and it is the work of the Experiment Station to assist in determining what is progress—and what is not."[7]

---

[6] Commander E. Sommati di Mombello, attaché of the Royal Italian Navy to the United States, to Captain W.W. Galbraith, USN, acting director of naval intelligence, letter dated 12 August 1925 in Records of the United States Naval Academy, U.S. Naval Academy Nimitz Library, Annapolis, Md., Record Group 405, Records of the Superintendent, General Correspondence [hereafter RG405]; Support Facilities, Naval Ship Research and Design Laboratory; Organization and Mission; Station Orders and Regulations; Buildings and Grounds, 1928[*sic*]–69, Box 1, Folder 1.

[7] "The U.S. Naval Engineering Experiment Station, Annapolis, Maryland," enclosure (a) to 3rd endorsement (Serial NP16/P11–2–(2173–Q–BK)) by Officer-in-Charge, Engineering Experiment Station to Superintendent, U.S. Naval Academy, dated 29 August 1925, RG405, Support Facilities, Naval Ship Research and Design Laboratory; Organization and Mission; Station Orders and Regulations; Buildings and Grounds, 1928[*sic*]–69, Box 1, Folder 1, 19. The mission statement was drawn from Lieutenant Commander E.J. King, U.S. Navy, "The United States Naval Engineering Experiment

In the years after World War One, EES was more than just a yardstick of progress. As an industrial research laboratory in the tradition of GE and AT&T, EES took the lead in a wide variety of fields related to marine engineering. EES was instrumental to the continual refinement of naval propulsion technology and to the development of strategically important marine propulsion machinery that enabled American long-range naval operations in the Pacific during World War Two. A great deal of EES materiel evaluation and research extended beyond the military and also benefited the fields of electrical power generation, diesel propulsion, metallurgy, lubrication, and the fracture analysis of metals.

The naval engineers who pushed for the creation of EES *circa* 1900 routinely touted it as a place that would undertake "scientific" research. However, as EES matured, its research differed from the more scientific work conducted at the Naval Research Laboratory. EES innovations often accrued from mundane engineering tasks that involved developing instrumentation and measuring and analyzing marine engineering processes. As Thomas Parke Hughes observed almost thirty years ago, this more common, every-day, engineering development work, critical to the process of technological change, "lacks the presumed excitement of invention."[8]

The lineage of Cold War science and technology draws heavily on World War Two, and includes the Office of Naval Research, Vannevar Bush's Office of Scientific Research and Development, and the Naval Research Laboratory. However, the history of the Engineering Experiment Station reflects another important fusion, that of scientific knowledge and engineering practice and also a special type of military-based engineering knowledge. It was the Kuhnian "normal" engineering development and research, but not design work *per se* conducted at EES that periodically led to what Walter Vincenti would have considered "radical" innovations of strategic significance.[9] It is no understatement to attribute successful American naval operations during World War Two to EES innovations that led to efficient

Station, Annapolis, Md.," *Journal of the American Society of Naval Engineers* [hereafter *ASNE Journal*] 25 (1913): 426–46 at p. 446.

[8] Thomas P. Hughes quoted in Allison, *New Eye for the Navy* (note 1), p. 187.

[9] Walter G. Vincenti, *What Engineers Know and How They Know It: analytical studies from aeronautical history* (Baltimore, 1990), p. 8. See Thomas Kuhn, *The Structure of Scientific Revolutions* (Chicago, 1970).

and powerful high-pressure, high-temperature steam turbine propulsion and long-range diesel engines for submarines.

Genesis

The origin of the Engineering Experiment Station can be traced to an 1861 federal law requiring that no "patented article" could be used in any "steam vessel of war" until it had been reviewed by a "competent board of naval engineers" and recommended by them for use.[10] These boards were *ad hoc* until 1876 when the naval leadership appointed Chief Engineer Benjamin F. Isherwood, former chief of the Bureau of Steam Engineering and designer of the innovative but ill-fated steam warship *Wampanoag*, as senior member of a permanent experimental board. The shortage of engineer officers in the wake of the restrictive 1882 Naval Appropriations Act resulted in the end of the permanent examining board and a return to *ad hoc* reviews.[11] Both the permanent and *ad hoc* boards were unfunded, so the cost of the navy's examination was born by the applicant. Since the law stipulated that only patented items could be examined, the navy could not assess promising, non-patented technologies.[12]

The shifting membership of the *ad hoc* review boards meant there were no standardized procedures, no funding for testing instruments, and no corporate memory. As Lieutenant Commander King observed, this meant that amid the building of the modern navy, there was "no continuity of purpose or of effort in regard to engineering practice in general."[13] The machinery the navy acquired often did not work as it should have and the navy had no effective way to develop engineering principles relevant to the design, improvement, and operation of new machinery. This had a deleterious effect on the rise of American naval power during the 1880s and 1890s.

---

[10] King, "U.S. Naval EES" (note 7), pp. 426–27.

[11] Successful congressional lobbying by the Line Officers Association resulted in the statutory limitation of the total engineer officers in the navy to one hundred and combined the Naval Academy's cadet midshipmen and cadet engineers into "naval cadets," with a corresponding homogenous curriculum. See William M. McBride, *Technological Change and the United States Navy, 1865–1945* (Baltimore, 2000), pp. 21–24.

[12] King, "U.S. Naval EES" (note 7), p. 427.

[13] *Ibid.*

In an 1895 issue of the *Journal of the American Society of Naval Engineers*, a junior naval officer, Passed Assistant Engineer F.C. Bieg, claimed that naval engineering officers desired "some place where engineering experiments could be carried on in a thorough and scientific manner." Bieg believed in the "value of scientific research in naval engineering" and proposed the augmentation of the *ad hoc* experimental boards by a "properly-equipped Engineering Experimental Station or School." To Bieg, the ideal model was found in civilian education where "every engineering school is supplied with a laboratory, not only for the instruction of the students, but for the experimental uses of the professors." Emulation of the engineering research conducted at Cornell University, the Massachusetts Institute of Technology, and the Stevens Institute would advance the "interests of the Navy."[14]

Bieg argued for an experiment station near "still and deep water" and at a "convenient distance" from "manufacturing and engineering centers." He wanted his experiment station under the control of the chief of the Bureau of Steam Engineering and commanded by an "engineer officer of rank." Co-locating the station at any navy yard "would load it with useless routine and be disastrous to good results." Therefore, Bieg wanted the "testing outfits" and "apparatus for the standardization of indicators, gauges, and other instruments" moved from the New York Navy Yard to the new station.[15]

Bieg considered the station essential to engineering research and engineering education for the navy and a key element in improving the quality of engineering in the fleet. In addition to the director and two other naval engineers, Bieg proposed temporarily staffing the station with engineer officers whose proposals for research were approved by the chief of the Bureau of Steam Engineering. Instead of sending assistant engineers to navy yards for a short time after their graduation from the Naval Academy to "pick up" what engineering knowledge they could, Bieg wanted them assigned briefly to the experiment station. There they would learn "management, care and repair of machinery, the estimating of work, &c., aided by the

[14] Passed Assistant Engineer F.C. Bieg, USN, "On the Necessity and Value of Scientific Research in Naval Engineering Matters as Related to the U.S. Navy, and the Necessity of Engineer Training for the Younger Members of the Engineer Corps of the U.S. Navy," *ASNE Journal* 7 (1895): 449–53 at pp. 450 and 449.

[15] *Ibid.*, p. 451.

practical work of trial trips and assisting in the experiments." Such an experience would prepare these officers for shipboard engineering duty and "give them the knowledge which they might not pick up by themselves in years." By serving as assistants to the "experimenters" these junior officers would become "scientific and careful experimenters" and be prepared to study engineering "phenomena" throughout their careers.[16]

Bieg was preaching to the choir and naval engineers' response to his proposal was overwhelmingly positive. The most important endorsement came from the famous Isherwood:

> The substitution of exact experimental knowledge for 'rule of thumb,' of the laws of nature for caprice, of mathematical demonstration for guesswork, are what is needed, and these can only be had by some such means as Mr. Bieg proposes. The sporadic work of a few gifted individuals can no more compensate the efforts of a large, compact, trained body acting under a well devised system of instruction and experience, than the brilliant performance of a few heroes can equal the results of a thoroughly drilled, armed and disciplined mass of regular soldiers. Such a body will always supply geniuses enough to lead them.[17]

Some key members of the Executive Branch and Congress also recognized the centrality of science-based engineering to the nation's pursuit of modern naval power. In 1897 Assistant Secretary of the Navy Theodore Roosevelt pushed the idea—resurrected from his Civil War predecessor, Gustavus Fox—that every officer in the navy must be an engineer. This controversial amalgamation of the Engineer Corps into the line came as a result of legislation enacted in 1899.[18] Many older line and engineer officers were chary of this new fusion. Most engineers were pleased that amalgamation resulted in rank and pay parity with the seamen officers of the line but many were worried about maintaining the sufficiency of engineering knowledge and practice. A significant number of sea-going line officers had no desire to become "mechanics" and, thanks to their lobbying, the 1899 amalgamation law had no means to force them. Engineering knowledge

---

[16] *Ibid.*, pp. 452–53.

[17] Benjamin F. Isherwood, untitled commentary on Bieg's article in *ASNE Journal* 7 (1895): 456.

[18] For an overview of the line-engineer amalgamation, see McBride, *Technological Change* (note 11), pp. 32–33. For Assistant Secretary of the Navy Gustavus Fox's support of naval engineers, see *ibid.*, pp. 14 and 247, note 25.

among the new, nominally homogenous, post-1899 officer corps was a thin veneer.

The chief of the Bureau of Steam Engineering, Rear Admiral George W. Melville, was very concerned about the state of American engineering education, especially within the navy. Melville thought the slavish devotion to tradition in officer education was not helpful in moving the naval officer corps into the twentieth century. He considered the navy's 1897 decision to acquire a new, sail-powered training ship for the Naval Academy "incredible" since the cadet-midshipmen would soon "serve on and command vessels entirely propelled by machinery and without any sails." Melville believed the construction of a sail-powered training vessel a flawed policy and the argument "exactly the same as that given by the classicists for the study of the dead languages."[19]

Melville's opinion on the type of naval profession needed for the twentieth century was echoed by retired naval engineer Charles Manning writing in the *Journal of the American Society of Naval Engineers*: "The naval career today is a technical one, and those having a distaste for mechanical affairs had better look elsewhere for a career, for the principal work at least of the junior naval officers will hereafter relate to electrical, pneumatic, hydraulic, and steam appliances. There will be scant time for poetry, and mooning, and vain regrets for the disappearance of the braces and the halliards."[20]

The burgeoning navy faced a significant shortage of qualified engineering officers. Melville saw an experimental station such as Bieg proposed as a means to enhance engineering practices in the navy and to increase engineering knowledge within the officer corps. Melville was unimpressed by the Naval Academy's engineering curriculum and endorsed a rump plan to bring technical school graduates into the navy and commission them after a one-year officer-training course. Melville, with some glee, thought an alternate to the Naval Academy a good thing: "I think it would raise the standard of the curriculum and the standard of the naval school [Academy], too. I think that they would find that these bright fellows, who would be the select of these technical schools and col-

[19] Melville, quoted in *ibid.*, p. 32.

[20] Passed Assistant Engineer Charles H. Manning, USN (Retired), "The Education of Naval Engineers for Future Needs," *ASNE Journal* 15 (1903): 850–58 at p. 855.

leges, if they should come with us, would make them stir their stumps."[21]

Melville's periodic digs at the Naval Academy's curriculum raised hackles. Melville was a forceful, knowledgeable, and well-respected officer who had no tolerance for inefficiency and retrograde thought. However, he was one of the few senior officers who was not a Naval Academy graduate and underestimated the importance his colleagues placed on the acculturation the Naval Academy provided.

The Bureau of Steam Engineering selected a small, waterfront site at the Naval Academy for its proposed experimental station. Melville told the House Naval Affairs Committee that Secretary of the Navy John D. Long favored the Annapolis site, provided the cost for the experiment station did not come out of the $8 million appropriated to build the "new" Academy then being designed. However, the academy superintendent, Captain W.H. Brownson, was lukewarm to Melville's plan and wanted no part of the experimental side of the proposed experimental station: "Any scheme to advance the education of the midshipmen in engineering meets with my approval, but an experiment school, as proposed at the Academy, where they would carry on experiments only, is a different proposition. It does not seem to me that is the place for it. It seems to me the place would be the navy-yard at New York."[22]

Whether at Annapolis or not, Melville saw a clear need for an experimental station to support the development of American naval power. He held up the experimental practices of the predominant British Royal Navy as an ideal to Congress. Even though the U.S. Navy was building modern ships as a by-product of the government's pursuit of great power status, it did not have the funds or facilities to evaluate engineering technology. Melville pointed out that the British had spent "hundreds of thousands of dollars" and outfitted similar ships with different boilers to determine experimentally which

[21] "Statement of Rear-Admiral George W. Melville, Engineer-in-Chief, U.S. Navy," 10 December 1902, *Hearings Before the Committee on Naval Affairs, House of Representatives, on Appropriation Bill Subjects* (Washington, 1903), [hereafter *Naval Affairs Committee Hearings, 1903*], no. 8, p. 9.

[22] "Naval Academy—Hearing of Captain W.H. Brownson," 11 December 1902, *Naval Affairs Committee Hearings, 1903*, no. 9, 7. As Bieg, "On the Necessity" (note 14), mentioned in his proposal, the New York Navy Yard conducted some material testing.

was superior. In contrast, Melville told the committee "We do not spend a cent, except what we filch and steal" and reported that he routinely "begged" fuel for "experimental purposes." Melville argued that the navy needed an experimental station for full-scale machinery and equipment testing: "it is no use to experiment in the laboratory with a little boiler burning a pound or cubic foot of oil; you have got to do it on a grand scale."[23]

Melville envisioned the experiment station as a tool of progressive efficiency that would allow individual inventors and engineers a public venue, outside of the corporate world, in which their ideas could be evaluated:

> A naval engineering laboratory would be greatly appreciated by mechanics and inventors who have not the financial ability to perfect and test patented appliances which may be of substantial benefit to the naval service. While it is always within the power of wealthy corporations to determine the worth of inventions submitted to them, the humble inventor is often deprived of his just recompense because he is compelled to permit his more fortunate neighbors to determine the value of his appliance. The naval engineering laboratory and experiment station would test all appliances relating to naval purposes at slight cost to the inventor. The persons submitting the appliances for tests would be assured that the absolute results of all tests and experiments would be furnished them and that a just and impartial professional report would be submitted.[24]

The solution for the U.S. Navy and for the American engineering "world," according to Melville, was to emulate the German engineering school at Charlottenberg and have the proposed experiment station aid the cause of "scientific engineering training." The supremacy of German scientific engineering, and its effect on maritime affairs, was clear to Melville: "because of the experiments in Germany, today Germany is driving the English merchant ships off the seas. The best ships today are the German ships, like the *Deutschland, Kaiser Wilhelm*, and others, and the reason is that the Germans, who do everything that way, on the scientific basis much more so than the French, have demonstrated just what they can do with their engines and boilers, and have the fastest ships in the world to-day." Melville also wanted to emulate the free dissemination of engineering exper-

---

[23] "Statement of Rear-Admiral George W. Melville," (note 21), p. 10.
[24] *Ibid.*

imentation in Germany, based on the testing at Charlottenberg. He promised Congress that if it authorized an experiment station at Annapolis "every private party would have the benefit of it."[25]

The House Naval Affairs Committee accepted Melville's linkage of research and experimentation at Charlottenberg to the burgeoning naval power of rival Germany. The committee also accepted the utility of an experiment station at the Naval Academy where "every man has to graduate as a midshipman, as a sailor man, and an engineer."[26] In its report to accompany H.R. 17288, the Naval Appropriation Bill of 1903, the House of Representatives increased Melville's request and appropriated $400,000 for "an experiment station and testing laboratory in the department of marine engineering and naval construction" of the Bureau of Steam Engineering.[27] The legislation specified $250,000 for the building and $150,000 for "equipment" and the "necessary apparatus."[28]

The location of the new experiment station at the Naval Academy generated praise among old-school naval engineers. Retired Chief Engineer Charles Manning was pleased that the engineering experiment station would serve two educational missions: to show the "cadets the scope and character of engineering work" and to allow the Academy's "Academic Board to carry on post-graduate work of the most advanced nature." In addition to its educational benefits, the laboratory would "make for future naval engineering efficiency."[29]

According to the official history, Melville, on the eve of his retirement in 1903, came to consider the site at the Naval Academy "cramped" and with "no opportunity for expansion."[30] Melville also thought the mechanical tests conducted at EES would be too disruptive to the Naval Academy. When the head of the Academy's

[25] *Ibid.*, p. 11.

[26] *Ibid.*, p. 13.

[27] *The Statutes at Large of the United States of America from December, 1901 to March, 1903, Concurrent Resolutions of the Two Houses of Congress and Recent Treaties, Conventions, and Executive Proclamations, Vol. XXXII—Part 1* (Washington, 1903), p. 1194.

[28] "House of Representatives, Report No. 3554 on the Naval Appropriation Bill," *Naval Affairs Committee Hearings, 1903* (note 21), p. 11.

[29] Manning, "Education of Naval Engineers" (note 20), pp. 856–57. The only engineering graduate program for naval officers was the naval architecture course at the Massachusetts Institute of Technology, established in 1901. See McBride, "'Greatest Patron of Science'?" (note 1).

[30] Rear Admiral George W. Melville to Captain W.R. Brownson, superintendent, U.S. Naval Academy, quoted in Rodney P. Carlisle, *Where the Fleet Begins: a history of the David Taylor Research Center, 1898–1998* (Washington, 1998), p. 37.

Steam Engineering Department agreed, a board under Rear Admiral W.C. Field selected a ten-acre site across the Severn River from the Academy. This new site removed the experimental station from easy access by midshipmen, limiting its pedagogical uses, but did fulfill Bieg's desire for a station accessible by deep-draft ships since only slight dredging was necessary to allow large navy ships to moor at the station.[31]

After a delay to obtain clear title to the land across from the Naval Academy and two years of construction, the experiment station's main building was ready in early 1908. On 28 March 1908, EES director Commander Walter F. Worthington informed the Bureau of Steam Engineering that all the new equipment, along with a Parsons turbine and other apparatus stored at the Naval Academy, had been installed.[32] The new building housed a chemical laboratory for the analysis of "waters, metals, etc." and the examination of lubricating oils. The machinery laboratory contained three, 500HP Parsons turbines each connected to a different style of condenser; different types of navy freshwater distillers and evaporators; and various types of turbine-powered forced-draft blowers to pressurize the firerooms containing the various boilers.

By August 1908 EES had begun operations in earnest.[33] Worthington asked the Bureau of Steam Engineering for permission to hire three mechanical engineers and one chemist. Worthington required the engineers to have a degree in mechanical engineering from a "university with a fully equipped engineering laboratory" and ten years of mechanical engineering work after their graduation. Preference was given to applicants "engaged in testing or experimental work or original research." The successful chemist applicant had to possess a university degree in chemistry and six years experience as an "analytical chemist."[34]

Worthington delineated the initial work for his new civilian staff: tests on the new Parsons turbines, surface condensers, lubricants, metals for bearings, and fire-room blower engines. EES purchased

[31] *Ibid.*, p. 38.

[32] Commander Walter F. Worthington, USN to chief, Bureau of Steam Engineering, letter dated 28 March 1908, RG405, Entry 33, Box 28, Folder 37.

[33] King, "U.S. Naval EES" (note 7), 430.

[34] Commander Walter F. Worthington, USN to chief, Bureau of Steam Engineering, letter dated 23 April 1908, RG405, Entry 33, Box 28, Folder 37.

a CO_2 recording apparatus and coal calorimeters to measure the efficiency of boilers. The staff tested water gauges (a critical safety instrument to maintain proper feedwater levels in shipboard boilers) and also sheet and roll packing, used in valves and pumps to prevent leaks, for acceptance for naval use. These early tests marked the beginning of EES's lengthy evaluations of private-sector machinery, fittings, instruments, gauges, and material for naval use and the process of developing and refining navy specifications to insure suppliers met minimum standards. The electrical laboratory began testing a Curtis turbogenerator.[35] In keeping with the pedagogical ideals touted during the push for the experimental station, and in spite of EES's less-convenient location across the Severn River, two hundred senior midshipmen were allowed to operate the Parsons turbine and gain some familiarity with this latest form of marine propulsion.[36]

In 1909 Worthington submitted his first annual report. The Bureau of Steam Engineering had authorized fifty-six tests at EES of which twenty-three were completed, twenty-two were ongoing, and two were cancelled. Some of these tests were made for "corporations or individuals," such as the evaluation of valve packing for the Clement Restein Company of Philadelphia, who paid "all expenses." Worthington underscored the practical nature of these tests that allowed the navy to "frame better specifications" for the acquisition of materiel for shipboard use.[37] The reciprocating steam engines and steam turbines used in navy warships both received their steam from main propulsion boilers and a great deal of the work at EES centered on refining boiler components and testing and evaluating various boiler-related pumps, gaskets, valves, gauges, and instruments.

EES also served as the primary engineering laboratory for the navy's two-year postgraduate course in marine engineering established at the Naval Academy in 1909. Although the ten officers initially assigned to the course were "hampered" by their "inadequacy" in "mathematics, mechanics and thermodynamics," the secretary of

[35] "Report of Chief of Bureau of Steam Engineering," in *Annual Reports of the Navy Department for the Fiscal Year 1909* (Washington, 1909), pp. 693–95 [hereafter "*Annual Reports . . .*" with year].

[36] "Report of Chief of Bureau of Steam Engineering," in *Annual Reports . . . 1908*, pp. 683–84.

[37] "Report of Chief of Bureau of Steam Engineering," in *Annual Reports . . . 1910*, pp. 489–90. Also see Bernard O. Scott, Clement Restein Company, to Captain J.M. Bowyer, USN letter dated 23 June 1910, RG405, Entry 33, Box 28, Folder 37.

the navy thought the excellent laboratory facility at EES where "experiments are regularly conducted on a large scale to determine questions of great interest to naval engineers" could partially compensate.[38]

Addressing the lingering engineering shortfall within the amalgamated navy, the secretary of the navy opined that the marine engineering postgraduate school would "continue to stimulate the interest of the service in engineering matters." In addition, its graduates would "disseminate engineering information throughout the fleet." To expose students to current engineering knowledge, the curriculum included lectures by eminent American engineers who were invited to "criticize most freely naval practices" and "to develop as far as possible any weakness of naval engineering." These lecturers included George Melville, now retired and a past president of the American Society for Mechanical Engineers, William Le Roy Emmet, General Electric's point man for the Niagara Falls hydroelectric project; electrical engineer and Harvard professor A.E. Kennely, and Frederick W. Taylor who discussed shop management.[39]

The Engineering Experiment Station also affected the undergraduate engineering education at the Naval Academy, just as Melville had hoped. The 1910 annual report by the secretary of the navy touted the new engineering focus of the curriculum since "almost every part of the line officer's duty of to-day has to do with machinery in some form." The engineering experiences at EES contributed to the secretary's assessment that the "regular course of instruction of midshipmen at the Naval Academy has been developed along engineering lines until it is to-day superior to that formerly given cadet engineers alone and as a practical course is probably the best given by any engineering school in the country."[40] The perceived quality of the Naval Academy engineering curriculum had come full circle. It had been eliminated for almost a decade, thanks to anti-engineer lobbying by line officers on the 1882 Naval Appropriations

---

[38] The assessment of the academic shortfalls of the first students is in Lieutenant Commander John Halligan, Jr., USN, "Post Graduate Education in Naval Engineering," *ASNE Journal* 28 (1916): 215–29, at p. 219. The secretary's comments are from "Report of the Secretary of the Navy" in *Annual Reports . . . 1910*, [hereafter "Report of the Secretary of the Navy, 1910"], p. 27.

[39] *Ibid.*, pp. 29 and 27–28.

[40] *Ibid.*, p. 26.

Act. This cancellation occurred just four years after the curriculum had received the Diplôme de Medaille d'Or at the Paris Universal Exposition in 1878.[41]

In 1912 the "School of Marine Engineering" became the "Post Graduate Department, U.S. Naval Academy," and its curriculum expanded to include graduate work in naval construction (naval architecture), ordnance, and training "electrical, radio and civil engineers." By 1916 engineering-oriented naval officers were calling for a split, two-year postgraduate curriculum with the first year spent at Annapolis and the final year of study at a civilian engineering school to broaden the officers' intellectual and cultural horizons. One perceived drawback of this desired civilian option was the loss of the excellent pedagogical framework provided by the "well-equipped laboratories of the Naval Academy, and the proximity of the Naval Engineering Experiment Station."[42]

Senior navy engineers considered the postgraduate engineering training essential to alleviate the shortages of fleet naval engineers, especially in providing support at navy yards and repair facilities. EES was the perfect instructional laboratory in which officer students could learn a wide range of "practical" engineering work and each student spent three to four months working at EES. There, the postgraduate engineering students undertook work in engineering chemistry; photomicrography and tests in the strength of materials; work on boiler water chemistry and the prevention of boiler corrosion; testing of gauges, packing, and boiler accessories; the design, set-up, and carrying out of machinery tests; the calibration of testing apparatus and instruments; and the generation of conclusions based upon tests.[43]

In addition to its educational service, EES's testing work during its first eighteen months of operation fulfilled predictions and the Bureau of Steam Engineering leadership believed that EES contributed to the bureau's efficiency and to the material readiness of the navy. Enumerations of EES work began to appear in the *Journal of the American Society of Naval Engineers*.[44] In his annual report, the

[41] McBride, *Technological Change* (note 11), p. 22.
[42] Halligan, "Post Graduate Education," 220.
[43] King, "U.S. Naval EES" (note 7), 438.
[44] "Experimental-Station Reports," *ASNE Journal* 22 (1910): 1254–65.

navy's engineer-in-chief, Hutch Cone, listed the accomplishments of the Annapolis experiment station for the secretary of the navy:

> By the aid of the work performed at this station [EES] the bureau has been successful in eliminating from competition makers of inferior packing, has reduced the number and types of [valve and pump] packing for use in the service, and has materially reduced the cost per pound. Through the operation of this station the bureau has also been able to determine which are the best brands of oil for gasoline engines, and light is also being obtained on the subject of oils for forced lubrication. A lengthy study has been made of boiler corrosion and of boiler compounds for use in inhibiting corrosion and preventing the formation of boiler scale. As a result the bureau has issued standard rules for the care of boilers and has adopted as a standard a boiler compound that can be purchased much more cheaply than commercial proprietary compounds.[45]

Cone publicly echoed his praise of EES in his address on "Naval Engineering Progress" to the Naval War College in August 1910. There, he told senior naval officers about the engineering innovations being introduced into the navy and the role of EES.[46]

Testing and Research

By 1913, EES was well established and its core staff of technical experts was in place. The experimental station's director supervised two officers whose duties included "management of the plant" and testing. The civilian staff included two mechanical engineers who supervised "general machinery tests." A third mechanical engineer focused solely on packing for valves and pumps, gasket materials, boiler fittings, and gauge glasses. A chemist and chemist's assistant undertook the analysis of "metals, compounds, oils, fuels, etc." Approximately sixty additional personnel worked as machinists, pattern makers, sheet-metal fabricators, and general laborers.[47]

---

[45] "Report of Chief of Bureau of Steam Engineering," in *Annual Reports . . . 1911*, p. 266.

[46] Engineer-in-Chief Hutch I. Cone, USN, "Naval Engineering Progress: A Lecture by Engineer-in-Chief H.I. Cone, USN, before the Naval War College, August 9, 1910," *ASNE Journal* 22 (1910) 1013–37 at p. 1036.

[47] King, "U.S. Naval EES", p. 436.

EES's executive officer, Lieutenant Commander Ernest J. King, provided a clear elucidation of the nature of work and research at EES. King considered the title "Experiment Station" a misnomer and classified the work of EES under five areas: testing, research, instruction, the increase and improvement of equipment, and repair and maintenance. The demarcation between testing and research was often unclear and this was complicated administratively by the fact that each investigation was assigned a "test number." EES work considered to be "testing" included "(a) Examination for conformity with specifications (b) Examination for the framing of specifications (c) Performance and endurance of machinery (d) Performance and endurance of apparatus and accessories [and] (e) Performance and composition of materials."[48]

Paralleling civilian industrial research, the management of EES defined "research" to be work in which "(a) Existing data is not systematic or sufficient (b) A new combination of mechanical principles is desired or proposed for a specific purpose (c) A new combination of physical principles is desired or proposed for a special policy (d) A new combination of mechanical or physical principles is desired or proposed for a special purpose (e) A new material or compound is to be developed for a special purpose."[49]

A central issue for a navy built upon a metal material culture was the issue of homogeneity of metals, their resistance to fracture, and analysis of metal fatigue and failure. EES work in this area mostly depended upon photomicrography—the analysis of crystalline structure. The EES staff routinely provided instruction in photomicrographic principles and practices to the staff of the inspectors of materials who served at navy yards and in the navy's Central Tool Plant located at the Philadelphia Navy Yard.

A significant amount of staff effort at EES involved evaluating existing, private sector testing instruments and in the "evolution, development and improvement of apparatus to meet the special needs of the Experiment Station."[50] These "special needs" of EES referred to the necessity of conducting evaluations in environments that duplicated conditions found in navy ships.

[48] *Ibid.*, p. 437.
[49] *Ibid.*, pp. 437–38.
[50] *Ibid.*, pp. 438–39.

EES's "repair and maintenance" work, in addition to the normal upkeep required in any industrial plant, included maintenance of the "hydroaeroplanes" [seaplanes] assigned to the Navy Aviation Camp during the years prior to naval aviation's move to Pensacola, Florida in 1913. EES also maintained the Annapolis Reserve Torpedo Division, whose torpedo boats served as test platforms for radio tests and other experiments.[51]

In keeping with Melville's progressive ideal, EES performed work for a wide variety of non-naval clients during its first five years. EES conducted safety tests on valves for the Isthmian Canal Commission, worked on auxiliary radios for the Bureau of Commerce, tested and evaluated boiler safety valves used on locomotives for the Interstate Commerce Commission, and even tested airplane motors for acceptance by the U.S. Army Signal Corps.[52]

In something reminiscent of a Baconian taxonomy, the naval staff divided EES engineering work into twelve categories: machinery; heat transmission apparatus; boiler fittings and accessories; pump and engine fittings and accessories; valves and pipe fittings and accessories; fuels; lubricants; packing and gasket materials; strength and composition of metals; chemical analyses; miscellaneous tests (propeller stuffing boxes, whistles, flexible couplings, thermometers, etc.); and miscellaneous investigations. The miscellaneous investigations included a study of water circulation in boilers, development of machines to test oils, radio tests, heat transmission through metals, parametric analysis of steam condensers, and the variation of the specific heat of oils with temperature.[53]

Five years of EES work translated into development of an improved, stronger boiler gauge glass with mica (to prevent boiler casualties); delineation of six classifications of packing material and a reduction in its price; development of oil testing machines to evaluate the performance of oil in service; development of a boiler water test outfit for shipboard use to prevent corrosion in boiler water tubes; development of an instrument to test the heat transfer between steam-air and steam-water interfaces; and photomicrographic methods to examine the grain structure of metals and to determine the probable cause of failures as part of EES's metallographic analyses. Photomicrography

[51] *Ibid.*, p. 439.
[52] *Ibid.*
[53] *Ibid.*, pp. 441–42.

would serve the navy well over the years and, early on, the chief of the Bureau of Steam Engineering touted the usefulness of the EES photomicrographic laboratory that had "contributed much to our knowledge of castings and forgings, and has demonstrated the practicability of making photomicrographic examination of a shaft before it is installed in a ship."[54]

In 1914 the *Journal of the American Society of Naval Engineers* published an article by two EES staff members, Lieutenant Harold Bowen and civilian mechanical engineer Leo Loeb, who touted EES's success to a larger audience. In recounting EES's work, Bowen and Loeb made a forceful, public case for the utility of EES. EES had reduced the cost of standard navy engineering material by substituting less-expensive alternates, revealed minor defects in machinery designs through testing, eliminated faulty material through endurance testing and metallurgical analyses, provided design specifications for the Bureau of Steam Engineering, and protected the Bureau from "freak designs" and "cranks."[55]

The chief of the Bureau of Steam Engineering echoed Bowen and Loeb's praise of EES work and the budget for EES for the 1915 fiscal year included $60,000 for "experimental and research work" and $20,000 for new equipment. While the work of EES had been of "great value to the service" and had "paved the way for marked improvement in design," the chief complained that EES's "limited personnel gives small opportunity for much work beyond that involved in conducting tests of material and of apparatus of various kinds to determine their suitability for use in the Navy."[56] External requests, such as one from the Department of Commerce for work on the corrosion of metals, exacerbated the personnel shortfall at EES and resulting limitation of its work.[57]

---

[54] "Report of Chief of Bureau of Steam Engineering," in *Annual Reports . . . 1913*, p. 229.

[55] H.G. Bowen, Lieutenant, U.S. Navy, and Leo Loeb, Mechanical Engineer, "The Engineering Experiment Station: Some Results," in *ASNE Journal* 26 (1914): 706–23, at p. 723.

[56] "Report of Chief of Bureau of Steam Engineering," in *Annual Reports . . . 1915*, pp. 343 and 347.

[57] See William A. Redfield, secretary of commerce, to Josephus Daniels, secretary of the navy, letter dated 20 February 1915, and associated endorsements by the superintendent, U.S. Naval Academy and chief of the Bureau of Steam Engineering, and also the letter from Josephus Daniels to the secretary of commerce dated 10 March 1915, all in RG405, Entry 36, Box 19, Folder 5A.

EES contributed to the increased material readiness of the navy but the bulk of engineering innovation still came from the private sector. EES scanned this environment for useful technologies, instruments, and processes. This paralleled contemporary industrial work where corporate research organizations did not invent new technologies but used the work of individual inventors, often in a creative synthesis.[58] Lieutenant Commander H.C. Dinger underscored the navy's reliance on the private sector in 1915:

> The Navy relies on the commercial engineering field for the excellence of the products from which the material matters of our naval forces are constructed. It relies on it for the development of tools, methods of work and the training of artisans by which our fighting weapons are produced in superior form and efficiency. Without a high state of engineering ability and progress in the country at large, the highest character of excellence in navy material can not be realized. The capacity of our commercial engineering plants is the principal asset of our naval engineering reserve of material.[59]

With such a deep reliance on acquisition from the private sector—the navy bought more than any other government agency—it required a "thoroughly organized and widely distributed inspection service for inspecting engineering supplies and material than any other Government department."[60] EES filled this role.

The First World War

Despite the engineering station's accomplishments, the war in Europe presented a new challenge EES was ill-suited to solve. The stunning success of German submarines in 1914–15 caused understandable concern within the U.S. Navy. In December 1914 the commander of the Atlantic Fleet, Rear Admiral Frank F. Fletcher, admitted to the House Naval Affairs Committee that no non-visual means existed to detect submerged submarines or to keep them away from battleships. He reported that proposed methods of fighting submarines—

[58] See John Kenly Smith, Jr., "The Scientific Tradition in American Industrial Research," *Technology & Culture* 31 (1990): 121–31, at p. 126.

[59] Lieutenant Commander H.C. Dinger, USN, "The Reserve Forces of Naval Material: Cooperation Between the Navy and the Producers of Naval Material," *ASNE Journal* 27 (1915): 853–72, at p. 856.

[60] *Ibid.*, p. 864.

torpedoes, rapid-fire guns, and grappling hooks—had not proven "absolutely positive as yet."[61]

As useful as EES had proven, developing a means to detect submerged submarines was a major research project rooted in physics. EES's small staff and budget, testing work load, lack of acoustic physicists, and record of industrial synthesis in developing technology left EES ill-equipped to take the lead in antisubmarine efforts. An additional complication in 1915 was Secretary of the Navy's Josephus Daniels's infatuation with Thomas Edison and Daniels's use of Edison's Naval Consulting Board as a public relations ploy to blunt public fears of the submarine and to provide the perception of enhanced naval preparedness in the face of the European war.[62]

In 1916 the Bureau of Steam Engineering sidestepped the Naval Consulting Board and resorted to scientific entrepreneurs to solve the submarine problem. University of Chicago physicist Robert Millikan convinced the Bureau of Steam Engineering to fund AT&T engineers to re-open the bureau's evaluation of a previously rejected detection device. Millikan was acting as a protégé of George Ellery Hale who linked the war in Europe with expanded scientific research and had convinced President Woodrow Wilson to establish the National Research Council. The chief of the Bureau of Steam Engineering established an antisubmarine experimental station, under Millikan's leadership, at Nahant, Massachusetts. This Nahant station included representatives from AT&T, Westinghouse, General Electric, and the Submarine Signal Company.[63]

After American entry into the war in April 1917, the Bureau established a special scientific research group at New London, Connecticut under Naval Academy alumnus and University of Chicago Nobel physicist-laureate Albert Michelson. The New London facility harnessed academic scientists to support the industrial efforts at Nahant.

---

[61] Testimony of Rear Admiral Frank F. Fletcher, USN, 14 December 1914, *Hearings before the Committee on Naval Affairs of the House of Representatives on Estimates Submitted by the Secretary of the Navy, 1915* (Washington, 1915), pp. 514–15.

[62] For histories of the Consulting Board see Daniel J. Kevles, *The Physicists: the history of a scientific community in modern America* (New York, 1979), chaps. 8 and 9; Thomas P. Hughes, *Elmer Sperry: inventor and engineer* (Baltimore, 1971), chap. 9; Allison, *New Eye for the Navy* (note 1); and Allison, "The Origins of the Naval Research Laboratory," *U.S. Naval Institute Proceedings* 104 (1979): 63–69; and McBride, "'Greatest Patron of Science'?" (note 1).

[63] McBride, "'Greatest Patron of Science'?" p. 24.

By the end of the war, Michelson was presiding over thirty-two professors drawn from academia to develop submarine acoustic detection devices.[64] The Bureau of Steam Engineering spent $2 million on its industrial-academia research work to develop a means to detect submarines.[65]

EES's location on the shallow Chesapeake Bay—the opposite of the relevant open-ocean, antisubmarine environment—worked against a central role for EES in the antisubmarine effort. Yet after the armistice, the navy shifted its antisubmarine work from New London to Annapolis. At EES, navy researchers completed development of the MV (multiple variable) hydrophone, a purely passive detection device of limited utility, and developed a sonic depth finder. The predominantly scientific, rather than engineering, basis of antisubmarine acoustic research made for a poor fit at the overworked Annapolis station.[66] The navy moved the antisubmarine research to NRL in 1923 where its scientific staff pursued submarine detection and localization using acoustic echoes from its hull, and by 1924 NRL had work on radio and underwater sound well underway.[67]

While the navy pursued submarine countermeasures, the government continued more traditional means to ensure American maritime security in the face of the European war and its aftermath. In the summer of 1916, Congress passed, and President Wilson signed, a massive $588 million naval appropriation. The 1916 building program authorized ten superdreadnought battleships, six battlecruisers, and numerous smaller warships and auxiliary vessels. The navy selected General Electric's revolutionary turboelectric propulsion machinery for the superdreadnoughts and battlecruisers.[68] Battlecruisers were extremely large ships equipped with battleship guns but had

[64] Kevles, *The Physicists*, p. 126.

[65] *History of the Bureau of Engineering during the World War, Navy Department Publication #5, Office of Navy Records and Library* (Washington, 1922), p. 73.

[66] Harold G. Bowen, *Ships Machinery and Mossbacks: the autobiography of a naval engineer* (Princeton, 1954), pp. 194–95. Also see "Report of the Chief of the Bureau of Engineering," in *Annual Reports . . . 1922* [hereafter "Report of the Chief of the Bureau of Engineering, 1922"], pp. 235–37 and "Report of the Chief of the Bureau of Engineering," in *Annual Reports . . . 1923*, p. 324.

[67] "Report of the Chief of the Bureau of Engineering," in *Annual Reports . . . 1924*, p. 302.

[68] For the turboelectric propulsion in the 1916 program capital ships, see William M. McBride, "Strategic Determinism in Technology Selection: The Electric Battleship and U.S. Naval-Industrial Relations," *Technology & Culture* 33 (1992): 248–77.

to sacrifice battleship armor in order to carry the propulsion machinery to achieve high speed (approximately 33 knots compared to 21 knots for the superdreadnoughts). The battlecruisers required propeller shafts capable of transmitting approximately four times as much horsepower as the superdreadnoughts. These large amounts of horsepower pushed the envelope for power transmission through a propeller shaft.

Testing the machinery and propulsion shafting of the 1916 program ships played to EES's cutting-edge work in metallographic analysis. (figure 1) EES's leading metallographist, Dr. D.J. McAdam, considered it a "branch of physical chemistry" while also being, in its "practical application" a "branch of metallurgy." In testing metals, the EES staff employed "chemical analysis, physical tests, and metallographic examination." EES also tested metal components that failed in service use. These included machinery shafts, gun parts, rivets, stays, bolts, boiler plate, pistons, valve stems, chain links, turbine blades, and miscellaneous articles made of brass, bronze, and other alloys.[69]

Drawing on early EES work in metallography, the Bureau of Steam Engineering required metallographic analyses of propeller shafting starting in 1914. To preclude failure in machinery shafts and couplings, Navy Department Specification 49S2b required the metallographic examination of these parts, that is, an examination of the microscopic structure of metals in relation to the laws of crystal origin, growth, and transformation.[70] If any material exhibited a non-uniform crystalline structure, EES required the manufacturers to apply an additional heat treatment to the material in question. Some suppliers ignored EES instructions and simply annealed the rejected piece.[71] In the past, annealing may have been sufficient to cover any obvious defect. However, metallographic reexamination caught such cosmetic repairs and underscored that annealing was insufficient to deliver the requisite fine-grained, uniform crystalline structure required in cutting-edge propulsion machinery.

---

[69] D.J. McAdam, Jr., "Testing of Metals at the Engineering Experiment Station," *ASNE Journal* 28 (1916): 361–68 at pp. 362 and 366.

[70] *Ibid.*, pp. 361–62.

[71] Annealing involves heating and slow cooling in order to toughen metal and reduce its brittleness.

Figure 1. The U.S. Navy's Engineering Experiment Station staff use instruments developed at the Station to determine the homogeneity of propeller shaft castings as part of a metallographic analysis. From the *Journal of the American Society of Naval Engineers* 28 (1916).

As a result of EES's inspection standard, a "great improvement" occurred in the shafting furnished to the navy. McAdam conceded the relative inexactness of metallographic specifications versus tensile specifications. Metallography, however, was often much easier to undertake than tensile tests, especially on large samples. McAdam reported that manufacturers had "come to an understanding of the grade of metallographic structure that is considered acceptable." By mid-1916, remanufacturing (additional heat treatment) was only necessary in a "small percentage of cases" thereby streamlining the manufacturing and assembly process of ships' machinery while ensuring its quality.[72] An additional advantage of metallography was EES's development of a method to conduct metallographic analyses of large objects, such as a battleship propeller shaft, *in situ*. By 1916, EES was averaging about twenty tests per month on new shafting.

The expansion of the navy after the U.S. entered the war in April 1917, and the need to analyze the effect of high-tempo wartime operations on machinery, translated into an increased workload for EES and an increased staff. At its peak, the EES staff included six officers, twenty-eight sailors (including nine female yeomen), seventy-eight civilian employees of the "Technical, Drafting, and Clerical Force," and 145 "Mechanics, Helpers, and Laborers." Wartime research included new work on electric welding, analysis of the failure of airplane parts, research on aviation oils, the investigation of fabric for use in aerial balloons, methods to produce hydrogen gas for balloon work, the evaluation of lubricants for use in depth charges (a wartime invention to destroy German submarines), and a method to fire crushed coal using compressed air.[73]

By July 1918 the EES staff had been reduced slightly without a reduction in the work load. The EES staff numbered 168 and the navy component was somewhat of an oddity consisting of one rear admiral (Thomas Kinkaid), one naval reserve ensign, and a chief petty officer. The civilian staff was built around five mechanical engineers, four chemists, three metallographists (one temporary), one

[72] McAdam, "Testing", pp. 365–66.

[73] Rear Admiral T.W. Kinkaid, USN to Karl Sigewald, secretary, Historical Division, Maryland Council of Defense, letter serial No. 121–ES–20.HS dated 12 January 1920, RG405, Entry 36, Box 169, Folder 65–A. The navy enlisted its first female sailors in early 1917 as a result of the 1916 Naval Act. They served as administrative workers (yeomen) and all were enlisted in the Naval Reserve.

draftsman, and assorted support staff of laboratory assistants and laborers.[74] Rear Admiral Kinkaid reported that EES was "fulfilling its mission by prosecuting research work on and making tests of appliances, materials, and methods which relate to the motive plants of ships of war." As before the war, a "considerable part of the Station's activity" involved the "metallography of steel and other metals." In spite of the amount of work and small staff, Kinkaid reported proudly that the "results of the Station's steel samples are usually dispatched to the district inspectors within 24 hours of the receipt of the specimens."[75]

Although the EES workload remained high after the armistice, Rear Admiral Robert Griffin, chief of the Bureau of Steam Engineering, ordered EES to prepare a list of employees for severance as part of a reduction in force.[76] In 1920 the staff included three naval officers, three mechanical engineers, three metallographists, two chemists, two physicists, three "technicists," eighteen "laboratorians," fourteen laboratory helpers, and 95 mechanics and laborers.[77] Griffin was not amused when EES offered to sacrifice its hydrophone development staff, shifted from the wartime antisubmarine work in New England, in order to protect EES's core workforce and focus.[78]

Postwar Work

In 1920 EES was conducting the majority of the "marine engineering research and experimental work" for the Bureau of Steam Engineering and its services were of the "greatest value, not only in the designing work of the Bureau, but in the maintenance of the machinery

---

[74] *Ibid.*

[75] Rear Admiral T.W. Kinkaid, USN, "Annual Report of Head of Naval Engineering Experiment Station" contained in letter to the Bureau of Steam Engineering, serial 2475–ES–19.HS dated 1 July 1918, RG405, Entry 36, Box 76, Folder 2B.

[76] Rear Admiral R.S. Griffin, chief, Bureau of Steam Engineering to head of Engineering Experiment Station, letter serial 493509–604–21–K dated 27 May 1920, RG405, Entry 36, Box 169, Folder 65–B.

[77] Rear Admiral A.H. Scales, USN, Superintendent, U.S. Naval Academy to Alfred D. Flinn, secretary, The Engineering Foundation, letter serial No. 65–48–W dated 7 October 1920, RG405, Entry 36, Box 169, Folder 65–B.

[78] Rear Admiral R.S. Griffin, chief, Bureau of Steam Engineering to head of Engineering Experiment Station, letter serial 498439–635–23–Kc dated 11 June 1920, RG405, Entry 36, Box 169, Folder 65–B.

of the Fleet."[79] EES was working on methods to balance high-speed machinery, refine gasoline engines, develop substitutes for metals used in bearings, develop refrigerating machinery, methods of burning coal, the means to calibrate aeroplane propellers, and centrifugal and reciprocating pumps.[80] As the postwar staff reduction loomed, EES maintained its traditional, and cutting-edge, metallurgical work by undertaking comparative tests of contractor-supplied turbine blades for the electric-drive capital ships of the 1916 building program: battleships 49–52 and battlecruisers 2 and 4.[81]

After the war, EES continued to research machinery problems in the fleet. The navy had acquired a number of destroyers propelled by 27,000HP impulse-type steam turbines powering propellers through reduction gearing. Most of these destroyers were delivered to the navy during 1918, passed their acceptance trials, and steamed thousands of miles under war conditions without a single turbine-blade failure. However, a series of turbine blade failures occurred in these ships during late summer in 1919. The low-pressure turbines in which the failures occurred had five stages. The blades of the first two stages were made of Monel (a nickel-copper alloy resistant to salt-water corrosion) and the last three stages had steel blades. Some destroyers had blades made of chrome-vanadium steel while others had chrome-nickel steel blades.[82] The crew of the repair ship USS *Vestal* investigated the blade failures during replacement of the damaged turbines. *Vestal*'s repair force could not determine why the blades had failed and it fell to EES to solve the mystery.

The bureau sent EES damaged chrome-vanadium steel blades from three destroyers and directed EES to determine the blades' condition and the cause of the "repeated failures." The staff of the Metallurgical Laboratory at EES went to work under McAdam's supervision. They performed a chemical analysis on each blade and employed metallography to examine the grain structure, looking for non-metallic inclusions that would have contributed to blade failure.

[79] Lieutenant Commander William L. Cathcart, USNRF, "The Achievements of Naval Engineering in this War," *ASNE Journal* 31 (1919): 27–28.

[80] Kinkaid to Sigewald, 12 January 1920 (note 73).

[81] Bureau of Engineering (S.M. Robinson) to head, Engineering Experiment Station, letter serial 583855–49–B–9–Db dated 21 October 1921, RG405, Entry 36, Box 223, Folder 3.

[82] Lieutenant Commander D.F. Ducey, USN, "Low Pressure Turbine Blade Failures in Destroyers," *ASNE Journal* 33 (1921): 512–40, at pp. 512–13.

The EES staff tested the blades for brittleness, tension, impact shear, and then heated them to test the effect of temperature on brittleness. These extensive tests revealed only slight, insignificant, variations from navy specifications regarding the alloy composition of the steel. The various "heat-treatment, impact-shear, and bend tests" demonstrated that the material was properly ductile but, in some cases, there was a slight tendency toward brittleness. EES's judgment was that "under the worst conditions" the metal in the failed blades was not "excessively brittle."[83]

The special bending experiments proved the key to solving the turbine-blade mystery. The commanding officer of USS *Vestal* reported difficulty in fitting replacement blades to the damaged turbines. Many of the replacement blades were "cracked completely" as they were bent to fit into the turbine shrouding. EES concluded that other blades suffered minor, unnoticed cracks during installation that increased to the point of failure under "service fatigue stresses." EES's careful analysis of the navy's turbine blade failures resulted in specific, public recommendations regarding turbine manufacture and adherence to metallurgical standards.[84] EES had solved a navy problem and provided a valuable service for turbine users in both the maritime and electric-power industries.

In addition to its fleet support, EES continued to test private-sector engineering devices and materials ranging from steam turbines to spark plugs. In 1922, for example, the station staff conducted twenty-eight major investigations including an analysis of Krupp steel, continued its long-standing work on boiler tube corrosion, and analyzed hub bolts for airplane propellers. The station developed eight products, including a muffler system for internal-combustion engines, a sonic range finder, a navigation sound source, and manufactured microphone receiving lines for twelve navy ships.[85]

Despite EES's engineering achievements, the newly created Naval Research Laboratory eclipsed EES at the 1925 annual banquet of the American Society for Naval Engineers. The physicist Michael Pupin, professor of electromechanics at Columbia University and

[83] *Ibid.*, pp. 522–23 and 538.

[84] *Ibid.*, pp. 538–40.

[85] "Report of the Chief of the Bureau of Engineering, 1922," (note 66), pp. 235–37.

president of the American Association of the Advancement of Science, presented the keynote speech. Pupin addressed the historical bifurcation between science and engineering observing that "a generation ago no one would have ventured to discuss in an assembly like this a double-headed subject—Science and Engineering. In those days a scientist's opinion of engineering did not go for very much and the engineer's opinion of science went for even less."[86]

The new cooperation between engineering and science, according to Pupin, was rooted in the scientific research movement that started at Johns Hopkins University after its formation in 1876. The original seven members of the Hopkins faculty provided the "moving power" to the American "scientific movement." Pupin praised the "flame of scientific idealism" that allowed "American universities and the American industries" to form "one harmonious unity, supplementing each other's activity."[87] This was exactly the model of aggressive, cooperative research found at Nahant and New London during the navy-academia-industrial research into submarine detection devices during the First World War. According to Pupin,

> The industries discovered that the best investment they could make was the investment in scientific research and scientific engineers. The type of men employed to-day by the universities in their most advanced work and the type of men employed by the industries are the same. They are simply highly trained men, trained in science and engineering. The result is that to-day American science and American engineering are welded and they form one organic unit of living force. They helped us to advance to the front rank of scientific achievement, and before many years have passed we will be the leaders of the world in science and engineering.[88]

In its first two decades, EES was a Pupinesque "organic unit of living force" in which "highly trained" staff performed the essence of "normal" engineering work that sometimes led to technological innovation. More significant technological innovations lay ahead.

[86] Dr. M.I. Pupin, "Science and Engineering," *ASNE Journal* 37 (1926): 270–74, at p. 270.
[87] *Ibid.*, p. 272.
[88] *Ibid.*, pp. 272–73.

Innovation and Arms Control

EES tended to be reactive to industrial development and often merely measured progress, rather than initiating it. However, a more creative and innovative research dynamic would arise during the 1930s stimulated by international naval arms limitations. The Five-Power Treaty that resulted from the 1921 Washington Naval Conference measured naval power in terms of ship displacement (weight). The treaty restricted capital ships (battleships and battlecruisers displacing more than 10,000 long tons with guns whose bores exceeded eight inches) and aircraft carriers but placed no limits on other types of warships. Renewed naval competition was possible, especially in the unrestricted cruiser category displacing less than 10,000 long tons. The Harding and Coolidge administrations were chary of renewed naval competition in cruisers; however, it was not until the 1927 Geneva Conference failed that the United States pursued new cruiser construction in earnest.[89]

Designing effective warships displacing under 10,000 tons was difficult and weight savings, always important in ship designs, became even more critical. EES played a key role in developing more efficient, lighter propulsion machinery both for new construction and replacement of older machinery in existing ships. In 1925, the chief of the Bureau of Engineering reported that

> Higher pressures and higher temperatures are constantly being advocated, which means that the metals must be subjected to stresses never before attempted. Before such new materials can be incorporated in the design of machinery a vast amount of research and experimental work must be performed. The bureau is constantly taking advantage of the facilities offered at the experiment station at Annapolis in the way of testing and developing engineering materials . . . There is no substitute for the activities of the . . . [EES] for investigation test, development, and research work in naval engineering.[90]

The pursuit of lighter propulsion machinery with increased power density, in response to the naval arms treaty, yielded new technical problems. As the chief of the Bureau of Engineering wrote: "Due

---

[89] For the cruiser competition, see McBride, *Technological Change* (note 11), p. 160.

[90] "Annual Report of the Report of the Chief of the Bureau of Engineering," in *Annual Reports . . . 1925* [hereafter "Annual Report of the Chief of the Bureau of Engineering, 1925"], p. 272.

to the increase in speed and power of main propelling turbines and the necessity for minimum weight, difficulties have been encountered in turbine construction, particularly in connection with the elimination of dangerous wheel and bucket vibrations, and the development of suitable blade materials." The navy relied on "the most recently developed methods of testing" at EES to eliminate the "possibility of danger in present and future reconstruction from these causes."[91]

Amid this push to develop more efficient propulsion plants, EES suffered under the same fiscal restraints imposed on the rest of the navy by the Harding and Coolidge administrations. The EES budget fell from $200,000 in 1922 to $175,000 per year from 1923 through 1929.[92] In 1924 the three EES laboratories—Metallurgical, Chemical, and Mechanical—began work on 244 new tests and generated 418 reports on "tests of engineering materials, appliances, and apparatus for use in the Navy."[93] By 1928 EES work was "an increase of 31.6 per cent over the productive test work during 1927 and of 59.4 per cent over 1926."[94] Tests were conducted on "forgings for the machinery of the new light cruisers, turbine blading, shrouding, boiler and condenser tubes, including the determination of the safe pressure for use in shock testing of condenser tubes, oils, packing, heat insulation, gasket materials, grinding compounds, pump governors, feed pumps, and ball bearings."[95] Yet this increase in work brought no larger budget for EES.

In carrying on its progressive goal of diffusing engineering knowledge, the work of the overstretched EES staff appeared in a series of articles in the *Transactions of the American Society of Steel Treating* and in the *Proceedings* of the American Society for Testing Materials (ASTM). In June 1927 ASTM awarded EES's Dr. D.J. McAdam's its Charles B. Dudley Medal for his paper on "Stress-strain-cycle Relationship and Corrosion Fatigue in Metals."[96]

---

[91] *Ibid.*, p. 278.

[92] Carlisle, *Where the Fleet Begins* (note 30), p. 95.

[93] "Report of the Chief of the Bureau of Engineering," in *Annual Reports . . . 1924*, p. 308.

[94] "Report of the Chief of the Bureau of Engineering," in *Annual Reports . . . 1928*, p. 297.

[95] *Ibid.*

[96] "Report of the Chief of the Bureau of Engineering," in *Annual Reports . . . 1927*, pp. 252–53. ASTM established its Charles B. Dudley Medal in 1925 "to stimulate research leading to standardization, extend knowledge in a specific field of interest to the Society, and recognize meritorious contributions to the publications of the

In 1929, the head of EES, Captain H.C. Dinger, argued that EES should coordinate all navy testing, not just that done for the Bureau of Engineering, and, on an *ad hoc* basis, for the other navy bureaus as well. Dinger based his argument on a critical theme of the New Era—fiscal efficiency—and claimed that an additional $50,000 in annual funding would increase the "useful output" of EES by 50 percent.[97] Dinger's proposal to be more efficient by spending more money fell on deaf ears. In 1930 the EES staff completed 604 tests and 1037 reports. While engineers in and out of the navy might be impressed by EES's accomplishments, EES, like the remainder of the navy, was now feeling the economic pinch of the Great Depression and would have to wait for New Deal funds for any budgetary increase.[98]

During the 1930s, EES became a truly innovative industrial research facility and generated significant engineering developments in five areas: quantification of the relationship among stress and the corrosion fatigue of metals; welding of machinery components; the development of a "work factor" for lubricating oil; design improvements leading to long-range, efficient diesel engines; and high-pressure, high-temperature steam turbine propulsion machinery. The last two would prove to be critical strategic technologies for the U.S. Navy during World War Two. Years of EES research on metallic stress and corrosion led to an equation that reflected a "definite relationship between stress range and the rate of damage due to combined stress and corrosion." The chief of the Bureau of Engineering characterized this knowledge as being of "great importance in preventing failures in structures."[99] In 1931 the Bureau of Engineering reported EES's development of a new welding technique that "guarantees the production of a weld superior in physical characteristics to the parent metal." These welds were used on the boiler drums in four new-construction *New Orleans*-class cruisers and provided these ships "ade-

Society." American Society for Testing and Materials; see Membership/Society Awards online at <http://www.astm.org/membership/socawards.htm>.

[97] Carlisle, *Where the Fleet Begins* (note 30), p. 96.

[98] "Report of the Chief of the Bureau of Engineering," in *Annual Reports . . . 1929*, p. 281. On New Deal funding for the navy, see William M. McBride, "The Unstable Dynamics of a Strategic Technology: Disarmament, Unemployment, and the Interwar Battleship," *Technology & Culture* 38 (1997): 386–423.

[99] "Report of the Chief of the Bureau of Engineering," in *Annual Reports . . . 1930*, p. 283.

quate strength and considerable savings in weight."[100] While weight savings were always important in ship design, the savings derived from an increased use of welds was underscored by the extensive displacement limits established by the London Naval Treaty of 1930.

One of the more far-reaching developments at EES was a measure of the effectiveness of lubricating oils. The head of EES's Chemistry Laboratory, James G. O'Neill, developed a comparative test—using similar machines—that measured the loss of lubricating ability. This "Navy Standard Work Factor Test," indicated that clean "old" oil still retained most of its lubricating abilities. This contradicted the oil industry's recommendations to change oil often and reduced navy expenditures on lubricating oil. Agencies ranging from the Coast Guard to the Department of Agriculture employed O'Neill's test. By 1932, the American Society of Mechanical Engineers (ASME) accepted EES's Work Factor Test as a valid quantification of the "service performance quality" of oil.[101]

The EES Internal Combustion Engine Laboratory, established in 1933, evaluated private-sector diesel engines and diesel fuels. The laboratory's work on diesel engines marked a shift from reactive work generating test-based specifications to the creation of experimentally-defined design specifications. This was the essence of engineering design, development, and evolution and led to technological innovation. As the official history observed, "private firms were building bureau-designed engines. The nature of the work had moved, gradually and unobtrusively, away from passive testing to active design development."[102] The diesel engines designed as a result of EES evaluations propelled a new generation of navy "fleet" submarines that bore the brunt of the long-range submarine war against Japan, sank 51 percent of all Japanese tonnage, and isolated Japan from the raw materials it had seized in Indonesia and Southeast Asia.

The vastness of the Pacific Ocean drove U.S. warship design since the rise of the modern, steam-powered navy during the 1880s. Propulsion machinery efficiency, coupled with fuel capacity, determined a ship's radius of operations. American commercial interests

---

[100] "Report of the Chief of the Bureau of Engineering," in *Annual Reports . . . 1931*, p. 317.

[101] Carlisle, *Where the Fleet Begins* (note 30), pp. 99–100.

[102] *Ibid.*, p. 106.

in distant China and southeastern Asia, coupled with a lack of overseas bases, made propulsion efficiency a much greater design imperative for the U.S. Navy than other maritime powers.[103] During the early 1930s, the navy began serious development work on high-pressure, high-temperature steam propulsion machinery with much of the work done at EES. By the late 1930s, a standard, high-pressure, high-temperature steam turbine propulsion suite had been designed and the navy installed it in most of its warships used during World War Two. No other maritime power pursued this path. The British Royal Navy, whose technical inquisitiveness and experimental expenditures had been praised by Rear Admiral Melville in 1902, had become a service that lacked the financial means and institutional energy to pursue such an innovative path. Interestingly, the engineer-in-chief of the Royal Navy took a very conservative position regarding propulsive innovation during the 1930s: "let the Americans and Germans do it, and if it succeeds we will copy them."[104]

The new high-pressure, high-temperature propulsion plant formed the "background of propulsion engineering during World War II" and its excellent fuel economy translated into a significant increase in the radius of fleet operations. To ascribe the U.S. Navy's successful, long-range naval war against Japan to this innovative propulsion plant would not be hyperbole. According to Vice Admiral Earle W. Mills, later chief of the Bureau of Ships, U.S. naval operations in the Pacific "would not have been possible without it."[105]

The superiority of American propulsion machinery during World War Two was quite significant. For example, the *North Carolina*-class battleship *Washington* (1937 design), equipped with high-pressure, high-temperature steam propulsion machinery, burned 39 percent less fuel at low speeds and was 30 percent more fuel efficient overall than the British *King George V* class designed in 1936.[106] In fact, the cruising radius of the *North Carolina*-class battleships was nearly double

[103] Captain C.W. Dyson, USN, "The Development of Machinery in the U.S. Navy During the Past Ten Years," *ASNE Journal* 29 (1917): 217.

[104] The Royal Navy's engineer-in-chief quoted in Bowen, *Ships, Machinery, and Mossbacks* (note 66), p. 100.

[105] *Ibid.*, p. 111.

[106] The *King George V–Washington* comparison is based on comments by Admiral Sir James Somerville, RN, commander of Force H which pursued the German battleship *Bismarck*, and Admiral Sir John Tovey, RN, commander-in-chief, Home Fleet in *ibid.*, p. 113.

that of the *King George V* class at lower speeds and 20 percent greater at higher speeds. The effect of high-pressure, high-temperature steam propulsion on other classes of warships was even greater. In April 1942, Commander Paul F. Lee, an assistant American naval attaché to the United Kingdom, reported that the limited cruising radius of their ships was

> a very serious question with the British and is having a marked effect on their naval operations. Even in their newest ships the fuel consumption is at least 50% higher and in some cases almost 100% higher than we have in our modern designs. Due to war conditions the normal peacetime cruising radius of their ships has been reduced by as much as 50% in some classes. This latter, combined with the poor fuel economy, has given their ships a comparatively short cruising radius. They are now fully alive to the mistakes they made in their pre-war designs.[107]

The legacy of British maritime technological supremacy had ended. The words of the prewar British engineer-in-chief proved prophetic. The Royal Navy copied the U.S. Navy's high-pressure, high-temperature approach in the *Daring*-class destroyers designed for the British 1944 program. Not only was the United States Navy emancipated from Britain in its engineering but the historical flow of technological innovation had been reversed. By the end of World War Two, the British were using pressurized-casing boilers produced by the American firms of Foster Wheeler and Babcock & Wilcox.[108]

The disparities in propulsion efficiency were not limited to the British Royal Navy. The *North Carolina*-class battleships used only 0.64 tons of fuel per nautical mile at 20 knots compared to Germany's *Bismarck* which burned 0.83 tons of fuel per nautical mile at the same speed—approximately a 30 percent difference. Japan's *Yamato* (1937), built along British low-pressure lines, used 0.88 tons of fuel per nautical mile at 18 knots.[109] Although exact figures are unavailable,

---

[107] Commander Paul F. Lee, USN to Vice Admiral Harold G. Bowen, USN, letter reprinted in *ibid.*, p. 114.

[108] For the shift in propulsion design in the *Daring* class, see Edgar G. March, *British Destroyers: a history of development, 1892–1953; drawn by Admiralty permission from official records & returns, ships' covers & building plans* (London, 1966), pp. 465–66. High-pressure, high-temperature steam turbine propulsion as a symbol of American engineering emancipation from Britain is addressed in Bowen, *Ships Machinery, and Mossbacks*, p. 51.

[109] Fuel consumption figures calculated from information in William H. Garzke,

Yamato would have used even more fuel at 20 knots—probably 40 percent more—increasing its disadvantage *vis-à-vis* the U.S. design. High-pressure, high-temperature steam propulsion systems did not have the broad postwar spillover of radar, but it was very significant to the successful outcome of the war.

Conclusions

The Annapolis Engineering Experiment Station was a Progressive Era creation designed to contribute to the advancement of American naval power and American engineering knowledge. EES was successful in both areas although its contribution to naval engineering education fell short of its proponents' desires and EES soon was replaced by formal graduate education at civilian universities. EES was based on the German scientific engineering model at Charlottenberg and, like the German experience, produced useful maritime engineering innovations. Technological innovation at EES—with the high-pressure, high-temperature steam propulsion plant and fuel-efficient diesel engines the most significant—certainly fulfilled the expectations of visionaries such as Bieg and Melville.

EES started out with a humble testing agenda that included the generation of material specifications and the evaluation and improvement of simple instruments such as pressure gauges and boiler gauge glasses. By the 1930s EES had transformed, from a largely derivative testing facility that measured engineering progress in the private sector and filtered it for navy use, to a truly dynamic industrial research institution that advanced engineering design of propulsion machinery, and engineering in general, on a broad front. As its official history observed, "American ships, aircraft, and submarines of World War II would go into battle carrying the ideas, the ingenuity, and the hard work of the [David Taylor Model] basin and [Engineering Experiment] station staff."[110] The United States entered

---

Jr. and Robert O. Dulin, Jr., *Battleships: Axis and neutral battleships in World War II* (Annapolis, Md., 1985), pp. 124 and 292. While the Germans employed high-temperature, high-pressure steam in the *Bismarck* design, they used single-reduction rather than double-reduction gearing and that failed to take advantage of the increased thermal energy of the steam.

[110] Carlisle, *Where the Fleet Begins* (note 30), p. 95.

World War Two in December 1941 with the world's technologically preeminent navy thanks, in no small measure, to the engineering research at the Annapolis Engineering Experiment Station. Ironically, EES's accomplishments have received little historical attention reinforcing the perception that normal engineering, like normal science, performs most often outside the spotlight of history.

CHAPTER EIGHT

DR. VEBLEN AT ABERDEEN: MATHEMATICS, MILITARY APPLICATIONS AND MASS PRODUCTION

David Alan Grier

Mathematics has been a weapon of war as long as it has been an identifiable field of intellectual pursuit. Ancient armies used mathematics to number troops, to array their armies on the field and to construct fortifications. Galileo Galilei did mathematical analyses for the arsenal at Venice and Isaac Newton analyzed ballistic trajectories in the *Principia*. Ken Alder, in his book, *Arming the Revolution*, recounts how in *ancien régime* France mathematicians and regular army officers (often the same people) trained artillery officers. During the eighteenth and nineteenth centuries, the armies of England, continental Europe and America employed a host of mathematicians, whose names are not usually associated with military applications: Taylor, Simpson, Laplace, Gauss, Peirce. In the American Civil War, the Union Army employed several self taught mathematicians, most notably John Dalhgren. In the First World War, Princeton mathematician Oswald Veblen assembled a staff of academically trained mathematicians to work for the Army Ballistics Office. This story is fairly well known[1] but rather than pointing to the military use of a fairly abstract technology, it best illustrates how the methods of mathematics and computation are deeply tied to the techniques of mass production.

In the spring of 1917 Oswald Veblen, then a young Princeton professor, went in search of a commission in the army's ordnance office. The nation's scientists had attempted to establish a civilian scientific research office under the auspices of the National Academy of Sciences, but the army had rejected this idea and had seconded the new committee, called the National Research Council, into its

[1] Herman Goldstine, *The Computer from Pascal to von Neumann* (Princeton, 1972), pp. 72–84; See also David Alan Grier, *When Computers Were Human* (Princeton, 2005).

Signal Corps division.[2] For the two years of American involvement in the war, most military research would be conducted by civilian scientists serving as reserve officers. For example, Harvard chemist James Conant working as a reserve Lieutenant did poisonous gas research at the Army's Washington Chemical Arsenal (now the site of American University).[3] Veblen was successful in his suit and received both a commission and an appointment to the Office of Ballistics Research at the new Aberdeen Proving Ground on the shore of Chesapeake Bay, midway between Baltimore and Philadelphia. He went through basic training in the summer of 1917 and arrived at the proving ground train station on 4 January 1918.[4]

The Aberdeen Proving Ground was the most expensive military project of the war, the Manhattan project of its time. It cost $73 million to construct and occupied thirty-five thousand acres of Maryland's Chesapeake Bay shoreline.[5] The proving ground was the last stop on a massive military production line, a production line designed to manufacture the weapons for the army. "When America accepted the challenge of Germany in 1917", wrote a senior ordnance officer, "part of the range of ordnance had already been produced in moderate quantities in the United States." Moderate quantities were not sufficient to supply the American Expeditionary force, so the Ordnance Office had to borrow the weapons of the French and British while they built production facilities, "on a grand scale and in a minimum of time."[6] When it began operations in January 1918, the Aberdeen Proving Ground served as a test facility. Aberdeen staff would proof newly manufactured weapons before sending them to arsenals and transportation ships.

Veblen's office, the division of experimental ballistics, was the mathematical research facility at the proving ground. Its members

[2] Albert Christman, *Sailors, Scientists and Rockets* (Washington, D.C., 1971), p. 18.

[3] James Herschberg, *James B. Conant, Harvard to Hiroshima and the Making of the Nuclear Age* (Stanford, Cal., 1993), p. 45. American University was in the news in 1999 when an unremediated area of this chemical weapons facility was discovered under a football field.

[4] Oswald Veblen Diaries, 23 March 1917, Oswald Veblen Papers, Library of Congress, Washington, D.C. (herafter "Veblen Diaries").

[5] Benedict Crowell, *America's Munitions: 1917–1918* (Washington, D.C., 1919), pp. 548, 550.

[6] Benedict Crowell and Robert Wilson, *The Armies of Industry* (New Haven, Conn., 1921), p. 21; also see H.A. De Weerd, "American Adoption of French Artillery, 1917–1918," *The Journal of the American Military History Institute* 3.2 (1939): 104–116.

studied the dynamics of weapons, especially the flight of artillery shells and the propagation of explosions. Much of its work could be traced back to the seventeenth-century work of Galileo, who produced a new example of a key product of mathematical ballistics, the calculated range table (range tables had been made before, but they were products of practical experimentation, not mathematical calculation). Range tables predict the power or range of a cannon; they give the distance that a cannon ball would fly from a weapon set with its barrel at a specific angle. On most range tables for field artillery, the angles begin at "point blank" or parallel to the horizon and increase to an elevation angle of 45 degrees. In Galileo's table, which was based on mathematical analysis rather than experiments, the cannon balls flew the farthest when the barrels were set at this angle.[7]

Galileo's theory was the first step in a detailed analysis of projectile flight. His work was far from complete, it did not adequately account for the atmospheric drag on a shell, yet it matched the technology of his time. The cannon of the seventeenth century were smoothbore, muzzle-loading guns. They were not especially accurate, as the cannonballs would careen down the barrels and fly with no stabilizing spin (or worse yet, have a spin at angles to the axis of flight and hook and slice like a bad golf shot). Weapon design improved steadily during the next two and half centuries and the improved cannons spurred mathematicians to develop more detailed theories of ballistics. A substantial change in technology occurred in the middle of the nineteenth century, with conical shells replacing spherical cannonballs and the invention of the accurate breech-loading, rifled bore gun. These guns could place repeated shots within few yards of a fixed target. With these guns came a great effort to refine the mathematical analysis atmospheric drag and produce a workable mathematical model of trajectories. New theories of ballistics were created in France, England, Russia and the United States. The most popular theory was developed by Francesco Siacci, a mathematician at the Turin Academy.[8]

---

[7] Edward McShane, John L. Kelly, and Franklin Reno, *Exterior Ballistics* (Denver, 1953), p. 754. See also Stillman Drake and James MacLachlan, "Galileo's Discovery of the Parabolic Trajectory," *Scientific American* 232.3 (March 1975): 102–10 and R.H. Naylor, "Galileo: the search for the parabolic trajectory," *Annals of Science* 33 (1976): 153–72.

[8] McShane, *et al.*, *Exterior Ballistics*, p. 769; Bernhard Menne, *Krupp or the Lords of*

Siacci's theory was elegant but simple, it reduced the equations of flight to four simple statements and provided tables that could be used to compute the basic aspects of a trajectory. Siacci addressed the most frustrating problem of ballistics, the difficulty in measuring the atmospheric drag on shells, by borrowing data from the French Proving Ground at Gâvre. His method was quickly adopted by artillery officers in Britain, France and Germany.[9] It reached the United States the 1880s and was modified by an American officer, Col. James M. Ingalls. The Ingalls version of the theory became the basis for analyzing the weapons of the U.S. Army and army officers used Ingalls' theory to develop several new gunnery strategies, including indirect fire, hidden fire and rapid battery movement.[10] Ingalls' ballistics texts and tables were still widely used at the time of American entry into the First World War, but his ideas would be substantially revised during Veblen's service at Aberdeen. Though Veblen oversaw a substantial revision of this theory of mathematical ballistics at Aberdeen, his major contribution was the integration of mathematical ballistics into the process of testing new cannon and shells, the combination of mathematics and mass production.

A Primitive Laboratory

The world of mathematical ballistics was in many ways as distant to Veblen's prior mathematical accomplishments as the unfinished camp at Aberdeen was from the well groomed campus of Princeton. "Veblen was a firm believer in the abstract approach to mathematics." wrote his biographer Montgomery Dean. Throughout his career, he had generally shunned applied mathematical problems. His earliest professional work had been in the discipline of geometry, where he had developed an axiomatic foundation for the field. This kind of work, which was part of a major mathematical activity at the

Essen (London, 1937), p. 57; David Zabecki, *Steel Wind, Colonel Georg Bruchmüller and the Birth of Modern Artillery* (Westerport, Conn., 1994), pp. 9–10.

[9] McShane, *et al., Exterior Ballistics*, p. 781; See F. Siacci, "Rational and Practical Ballistics," *Report of Chief of Ordnance, United States Army* (Washington, D.C., 1891), pp. 218–242.

[10] Philip Alger, *The Groundwork of Practical Naval Gunnery* (Annapolis, Maryland, 1915), p. 23; The methods were first explored by European officers, see Zabecki, *Steel Wind*, pp. 13–14.

start of the twentieth century, pushed geometry away from its traditional concern with concrete lines and figures and moved it towards the abstract world of logic. During his first two decades as a professor, Veblen showed no interest in mathematical astronomy or mathematical physics, the traditional applied fields of the age, or in new applied fields of mathematics such as mathematical economics. In the months immediately prior to the war, he had been researching algebraic topology, the study of theoretical surfaces and infinitely elastic deformable shapes.[11]

The mathematical ballistics of 1918 were neither as refined as axiomatic geometry nor as theoretical as algebraic topology. Many mathematicians considered it messy and ill-formed. The English mathematician G.H. Hardy called it "repulsively ugly and intolerable dull," though he conceded that it required "quite elaborate technique."[12] Veblen's army colleague, the mathematical astronomer Forest Ray Moulton, was more willing to consider the subject but had little good to say about its state in 1918. "Upon entering the army," Moulton wrote, "a hasty examination of the classical ballistic methods showed not only that they were wholly inadequate for current demands, but also that they were not well suited to the solution of the problem, even under earlier conditions."[13]

Modern mathematical ballistics is an application of the theory of partial differential equations. By the start of the First World War, mathematicians had divided it into two sub-disciplines: internal ballistics and external ballistics. Internal ballistics modeled the forces inside of a gun barrel produced by the explosion of the propellant charge while external ballistics attempted to capture the flight of a projectile. Internal ballistics was tool of weapons engineers, who used it to develop new propellants, set standard charges, verify weapons designs, and create new shells. Exterior ballistics, while also used by engineers, was also used to plan combat. Artillery officers would use range tables when choosing placements for their batteries and planning artillery campaigns. Some gunnery officers carried range tables into battle but they were usually of little use in the heat of conflict. Few officers had the time to consult a table and do the necessary

[11] Dean Montgomery, "Oswald Veblen," *Bulletin of the American Mathematical Association* 69 (January, 1963): 26–36.
[12] G.H. Hardy, *A Mathematician's Apology* (Cambridge, 1967), p. 138.
[13] Forest Ray Moulton, *New Methods in Exterior Ballistics* (Chicago, 1926), p. ii.

adjustments on the figures before aiming their guns. To aim their guns, crews used small mechanical tools, called fire control devices, that were based upon range tables. The most common of these devices for first world war gunners was a compound slide rule, known popularly as "the Coffin Box."[14]

The range tables of the era usually had four dimensions and then were augmented by other information. The first dimension was the range or distance that a gun could throw a shell. The table then showed how the range was affected by three factors: the muzzle velocity (equivalent to amount of propellant or charge), angle of the barrel, and the type of shell. Many tables then included quantities that could be computed from the shell trajectory, including the time of flight, the maximum altitude, the terminal velocity of the shell and the angle at which the shell struck the ground. All of these quantities were presented for conditions that the army identified as standard: level ground, firm placement, a fixed temperature, no wind and a specific air density, which was a function of both pressure and humidity. In the field, officers would have to adjust their estimates to account for local conditions.[15]

Oswald Veblen arrived at Aberdeen in January 1918, shortly after the facility opened. The base was little more than a few rough buildings, a railroad spur and a tent ground. The perimeter of the facility was far from secure and the Army had only recently finished evacuating the 35,000 residents of the area and moving them to new homes. The new ranges were incomplete, the machine shops were still under construction and base had only a skeleton staff of officers and a small assignment of soldiers.[16] As quickly as he could, Veblen set out to learn the intricacies of range work and the construction of range tables. He had learned the basic procedures of this work at Fort Hancock, New Jersey, where he had received gunnery training.[17]

[14] Geddes Smith, 2nd Lieutenant, "Coast Artillery Training in the War," *Journal of the United States Artillery* 50.1, no. 151 (1919): 1–84. On internal ballistic testing, see Mauskopf, this volume, and on ranging instruments, see Walton, this volume.

[15] Roger Sherman Hoar, "The New Ballistics," *Journal of the United States Artillery* 51 (July 1919): 285–295.

[16] National Archives and Records Administration, Washington, DC, RG156, Records of the Department of Ordnance, Annual Report of the Aberdeen Proving Ground for 1919.

[17] Service Record of Oswald Veblen, RG156, Records of Ordnance Officers, 1915–1919.

He had never done much experimental science and soon discovered that the work was both detailed and demanding. He later confessed that his first test shots were "more picturesque than satisfactory."[18]

At the end of January, Veblen started work on a range table for a 2.95 inch mountain gun, a lightweight cannon that could be carried on horseback. As the new firing ranges were not yet ready, Veblen established an improvised range in an empty field. He mounted the gun in a temporary placement and began firing. After series of rounds, he would stop the guns and "would go down the range on horseback while the firing was suspended for meals, identify the shell holes and place distinctive marks in them to enable them to be identified by the surveyors."[19] Once the markers were in place, Veblen would survey the distances with a common transit. From a baseline that ran through his gun placement, he took multiple angle measurements and recorded these numbers so that his staff could later calculate the distance and complete a range table. The winter weather, one of the coldest on record, hampered his work. Snow storms swept the region and rivers froze solid. The only benefit that Veblen found in such harsh conditions was the fact that the regular deposits of fresh snow covered the evidence of earlier firings and made it easier to identify the craters made by new shots.[20]

The purpose of these range firings was to estimate a certain value needed by Siacci's ballistic theory, a value known as the ballistics coefficient, which was unique for each type of projectile. Once he had a good estimate for the ballistics coefficient, Veblen could then insert the value into Siacci's formulae and produce an entire range table. In this way, he could avoid the task of having to make firings for every entry on the range table. A detailed table might have several hundred entries but these entries could be computed from just a couple dozen test firings.[21]

Veblen gave the task of estimating the ballistics coefficient to three instructors at Fort Hancock, New Jersey, situated at the former Sandy Hook proving grounds (which had been closed after Aberdeen opened),

---

[18] *History of the Range Firing Section of the Proof Department of the Aberdeen Proving Ground*, Army Ballistics Research Laboratory, Technical Report 84 (Aberdeen, Maryland, 1919) [RG156, Technical Reports of Aberdeen Proving Grounds, 1918–1920].

[19] *Ibid.*

[20] *Ibid.*

[21] Siacci, "Rational and Practical Ballistics" (note 9) and Hoar, "The New Ballistics" (note 15).

about one hundred and fifty miles to the north. These three were army lieutenants, who spent most of their day training new volunteers and had to fit the computational work into their spare hours. The computations required these officers to calculate distances from the angle measurements, adjust the distances to account for the cold weather, snow and wind, combine the distances in a formula that produced the ballistics coefficient and then produce the final table. The work had many unforeseen complications. Some of the data were anomalous, the gun was not always placed in the same spot, the muzzle often jumped in firing, and some of the results were contradictory. The three lieutenants badgered Veblen for more information and help with the calculations. In an era when long distance telephone calls were uncommon, they had to send telegrams to Aberdeen with their questions. If Veblen was at Washington or at some other army base, they lost valuable time. Veblen, anxious to complete the work, complained that the "difficulties in long range correspondence about the many technical details were very great."[22] Three times Veblen had to brave the winter weather and travel north to Sandy Hook in order to help the lieutenants with the computations. The last visit occurred on 25 March 1918, when the four of them finally felt that they had fixed all the problems with the data and had a usable value for the ballistics coefficient. They began computing range values late in the day and finished their work well after midnight. Veblen's diary for the day contains the single line "3:00 am. Finished Table."[23]

Integrating Mathematics and Production

As he worked on his first ballistics project, Veblen began to ponder how he might best convert the raw range data into a finished table. He had not anticipated the problem of transferring adequate data from the range to the computers and the difficulties in handling the

[22] "The range firing section of the proof department, Aberdeen proving ground. Its objects, its development and its accomplishments," Army Ballistics Research Laboratory, Report 12 (Aberdeen, Maryland, 1918), p. 2 [RG156, Technical Reports of Aberdeen Proving Grounds, 1918–1920]. For Sandy Hook and Fort Hancock, see Scharfenberger, this volume.

[23] Veblen Diaries, 25 March 1918 (note 4).

complex calculations. As he worked through the issues that he had encountered, he borrowed ideas that others had developed and modified them for the purposes of the Aberdeen Proving Ground. The first of these ideas was to use pre-printed computing sheets to standardized calculations. These forms had blank spaces for numbers, formula to guide the computers and blank space that might be used for scratch work. (figure 1) Such sheets had been developed in the late eighteenth century by Nevil Maskelyne at the Royal Nautical Almanac Office in London. They had been used by various computing offices throughout the nineteenth century. In Veblen's time, they were common fare at the United States' three largest computing laboratories, those of the U.S. Naval Observatory, the Nautical Almanac Office, and the Coast and Geodetic Survey (all, it should be noted, largely staffed by civilians run by the Navy, although by 1918 the CGS was an independent goverment branch). They were generally prepared by a mathematician, typed on a mimeograph stencil and reproduced on a rough, tan paper.[24]

In the early spring of 1918, Veblen acquired mimeograph machines from the A.B. Dick Company and was producing his own computing sheets.[25] As the army prepared its first large range, Veblen recognized that these sheets need not be under the control of a single computer but could pass from individual to individual, where each computer added some small value to the final result. They would allow Veblen to place his computing staff directly on the ranges and put them "under the direct observation of the firing officers."[26] In such a role, the computers would no longer have to query the range staff about anomalies in the data or make uninformed guesses about the nature of the ranges tests but could rely on the judgment of the range officers and quickly adjust the data for anomalies in the shot, for a failure in the gun, for a gust in the wind, for a sudden change in temperature, for the extra warming of a gun barrel by the sun or repeated firings.

Veblen tested these ideas on a range for long guns, which became operational in April 1918. This range, nicknamed the "water range,"

[24] See David Alan Grier, *When Computers Were Human* (Princeton, 2004), chs. 1, 2, 3, and 5.

[25] Order of 1 March 1918, RG156, Circular Orders of the Aberdeen Proving Ground.

[26] *History of the Range Firing Section* (note 18).

A 1 - Form 4. O. B. P. No.------------

Form for Range Wind Computation.

Gun------------- Date of firing. ------------

Az. of wind = Az. from which wind is blowing.

Projectile------------ M. V. = ------------

Note: Values of t are taken as near as possible to $1/5y_0$, $2/5y_0$, etc.

Date						
No. of Rds.						
Elev.						
ϕ,						
α, = Az. of line of Fire						
W_{x_1} {o to t} = $\frac{d_x \text{ in m.}}{t \text{ in sec.}}$						
Wx_2 {t to t_2}						
Wx_3 {t_2 to t_3}						
Wx_4 {t_3 to t_4}						
Wx_5 {t_4 to t_5}						
Weighting factors						
a {o to t}						
b {t to t_2}						
c {t_2 to t_3}						
d {t_3 to t_4}						
e {t_4 to t_5}						
a Wx_1						
b Wx_2						
c Wx_3						
d Wx_4						
e Wx_5						
Sum= Wx						

Computed by-

Checked by-

Date -

Figure 1. Range Wind Sheet, Aberdeen Proving Grounds, c. 1918. From National Archives and Record Administration, Washington, DC, RG 156, Records of Ordnance Proving Grounds 1889–1941, part 11, Aberdeen Proving Ground, MD.

was located on the shore of the Chesapeake Bay. It was within the confines of the Aberdeen compound, but it posed almost as many communications problems as if it had been located at Sandy Hook. The gun placement was far from the main camp, at the end of a long and muddy road. Veblen noted that "it was necessary to haul ammunition and guns over roads which were often two feet deep in mud. The only conveyance which was able to get through was a six-mule team. Even the Ford was powerless."[27] Observers on this new range were even more isolated from the central camp, as they were stationed on towers, on the far shore of the bay. They were transported to their positions by boat and were expected to gather data by noting the splashes made by the shells as they struck the water.

To help him organize his computing staff, Veblen hired a graduate student from Columbia University, Joseph Ritt, and gave him the title of "master computer." Ritt was Veblen's link to the large computing offices of the day. While a young man, Ritt had spent a year in the computing division of the Coast and Geodetic Survey before moving first to the Naval Observatory and then later to the Nautical Almanac computing office. After his computing career, Ritt had returned to college and was finishing a Ph.D. when the United States declared war on Germany.[28]

Ritt brought a thorough knowledge of conventional computing procedures but Veblen was moving beyond conventional ideas. The only part of his plan that resembled a conventional computing office was "a small shack" which "was ordered for this section [the computing group] on Friday morning and the section moved in on Saturday afternoon." The shack had tables, adding machines and a library of mathematical reference books. It was originally occupied by Ritt plus the three lieutenants from Fort Hancock, who transferred to Aberdeen that spring.[29] The shack was more a social center for the staff than active office, as most of the computing staff spent their days on the range. They returned to the building only at the end of the day to finish their work. There, they would gather to play cards and to tell the stories that men tell when they are far

---

[27] *Ibid.*

[28] *American Men of Science*, 5th ed. (New York, 1933), p. 937.

[29] "The range firing section" (note 22).

from home and living a vigorous and active life. One computer remembered only the games of cards and the way that they would tabulate their winnings on the adding machines.[30]

For most of any day, the computers were on the firing ranges doing preliminary computations and managing the flow of data. They were assigned to the observation towers, the gun placement and the range command post. The computers in the towers were connected by a telephone line to the gun placement. They would wait for the notice of a shot from the range officer and would then look for the splash from the shell impact. The computers at two or three stations would combine their data by using a nomogram, a simple graphical computing device that had been designed for this purpose by Veblen. A nomogram (sometimes called a nomograph) is a set of overlapping lines marked off to scale and arranged in such a way that by using a straightedge to connect known values on two lines, an unknown value can be read off at the point of intersection with another line. With the nomogram, they could quickly compute the range of the shot and its deflection to the right or to the left. Once they had completed their calculation, they would communicate their results to a (human) computer at the gun placement, a person usually identified as "the range computer."

While waiting for the numbers from the observation towers, the range computer would record the factors that influenced the shot, including the time of day, including the type of gun, the shell, the size of the charge, the time of flight, the direction of the wind, the temperature, the humidity, the jump of the gun. After the range values arrived, this computer would adjust the raw figures to standard conditions. It did not take long for the computers to refine this process so that they could produce data faster than the gunners could fire the guns. Veblen wrote that "it has been possible to compute the range and deflection of each shot in less than a minute when the observers were at hand, and in general it is possible to know the results of any firing for range within a few minutes after the last shot was fired." By the start of May, this system allowed the gunners of the water range to fire forty rounds per day, the same number of rounds that Veblen had fired during the entire month of February.[31]

---

[30] Norbert Wiener, *Ex-Prodigy* (Cambridge, Mass., 1953), p. 259.
[31] "The range firing section" (note 22).

In early May 1918 Veblen augmented his mathematical staff by recruiting three Princeton graduate students. As the water range began its regular program of gun testing, he found that he needed to expand the staff far beyond his original intent. Veblen remembered that "twice in the course of a few months, quarters for this section were doubled."[32] He recruited new computers from the mathematics departments of Ivy League schools, a select group of young regular officers, and a few organizations outside of academia. At the offices of the *Encyclopedia Americana* he found a young philosophy student who was anxious to get into the war, a Harvard graduate named Norbert Weiner.[33] In the course of the summer, Veblen found that he was often attempting to persuade some young mathematician to join him in Aberdeen rather than take a commission in the army and command a unit on the front lines of France. The young college men of the era were strong supporters of the war and yearned for the elusive glory of battlefield service. Though he regularly lost members of his staff to the army, Veblen never considered bringing female mathematicians to Aberdeen. At the time, there was a substantial female enrollment in college mathematics classes, as mathematics was often the only science that actively recruited women. At several institutions, such as the University of Michigan and the Harvard Astronomical Observatory, there were substantial bastions of women with sufficient training in mathematics. Further, at the time it was not at all unusual to have many female computers engaged in data reduction at all manner of businesses, universities, and government departments. However, Aberdeen was a male domain. The only female employees were a few clerks and stenographers, who were required to leave the base at the end of each working day.

The Aberdeen water range illustrates the tight connections between mathematical analysis, organized computation and mass production. The Aberdeen proving ground was a central stop on a mass production process that began with the iron mines in Minnesota and the coal pits in Pennsylvania, passed through the gun factories at Bethlehem steel and the other military contractors and ended in the Army arsenals after a visit to Aberdeen. That line not only produced

[32] *Ibid.*

[33] Weiner, *Ex Prodigy*, p. 257. Weiner would later become an important MIT professor and founder of the term and field of cybernetics.

weapons, it produced predictions of how those weapons would behave in new circumstances and with new strategies. The flow of data up the production line not only guaranteed that the weapons would function properly, it also suggested new ways that the weapons could be used. By the early summer most of the Aberdeen ranges employed some version of the means of processing data that was employed at the water range. The most innovative was perhaps the anti-aircraft range. Lacking enough range finders to determine the altitude of shots, the army tested its weapons at night and photographed the aerial bursts against the night sky. Using a Nautical Almanac, the computers would measure the distance between the stars and then calculate altitudes.

Veblen's innovations did encourage the computers to consider a major revision of Siacci's theories, though this work is less related to the process of mass production than it is to the increasing sophistication of the weapons and the fact that many of the computers spent long hours at the proving ground with little to do, except take hikes through the woods, play cards, and talk mathematics. One computer reported that the experience "furnished a certain equivalent to that cloistered but enthusiastic intellectual life which I had previously experienced at the English Cambridge."[34]

The revision of the ballistics theory was directed from an office in Washington by Forest Ray Moulton, an astronomy professor at the University of Chicago. This office oversaw the Veblen's work at Aberdeen. It also edited range tables, set standards and began thorough revision of ballistics theory. Moulton directed the revision and developed the central parts of the new theory. He identified a host of technical problems in the theory then used by the army, including incorrect assumptions, improper simplifications, and discontinuous derivatives for the solutions. He reserved some of his harshest words for the ballistics coefficient. "For a quantity that was treated as constant in the mathematical theory," he wrote, "the ballistic coefficient was made to carry a very heavy burden."[35]

Because of his position in the army's department of ordnance, Moulton could assemble a large team to help with the mathematical

---

[34] *Ibid.* pp. 68, 258.

[35] Forrest Ray Moulton, *History of the Ballistics Branch of the Artillery Ammunition Section, Engineering Division of the Ordnance Department for the period April 6, 1918 to April 2, 1919*, U.S. Army Military History Institute (Carlisle, Penn., 1919), p. 10.

research. He and Veblen recruited a number of university mathematicians to help develop the theory. Moulton prepared the overall framework and delegated specific problems to others. A.A. Bennett, who arrived from the University of Texas, prepared numerical methods to solve the ballistics equations. Gilbert Bliss, also a Chicago professor, helped model the effects of high level winds and the rotation of the earth. Veblen worked on both theoretical and practical problems. He adjusted the theory to account for the shape of the shell itself. Once he had developed a mathematical model, he spent hours collecting data to test his ideas. Even though the demands of his command regularly sent him traveling up and down the east coast, Veblen would try to finish his business early and catch an afternoon train so that he could join the artillery crews on the range. He would fire the guns until the dimming light made further observations impossible.[36]

Aberdeen prepared many guns for battle but only a few of its weapons made it to the fields of France. The most notable were the rail cannon, large guns mounted on rail cars. These weapons arrived in Europe in mid-September 1918 and saw action in the final campaigns of the war. (figure 2) The calculations and mathematical work came to a quick end after the armistice on 11 November 1918. Most of the proving ground had little warning that the Germans were close to surrender. Nonetheless, the army moved quickly stop production. In the days that followed the armistice, they telegraphed stop work orders to munitions factories. They made plans to stockpile weapons at Aberdeen. They terminated research projects, reduced experimental firings and brought the work to a close. By June 1919, an officer could report to Veblen that "no ballistics computations are being done here nowadays."

The Nature of Military Mathematics

The subsequent stories of the Aberdeen Proving Ground and of Oswald Veblen suggest a tight connection between the mathematics of external ballistics and the mass production of weaponry. After the war, mathematical ballistics would not command much attention

[36] Veblen Diaries, 23 August 1918 (note 4).

Figure 2. Railroad Naval Gun being tested at Aberdeen Proving Grounds, 1918, Aberdeen, Maryland. From Naval Historical Center, Washington Navy Yard, Washington, DC.

from academic mathematicians. Few articles on ballistics appeared in the mathematical journals. Only rarely does it appear as an example in mathematical paper. A few of the Aberdeen mathematicians attempted to teach courses in mathematical ballistics, but only one, at the University of Chicago, seems to have attracted students in any number.[37] Veblen quickly distanced himself from the subject and by 1924 wrote to a friend, "I am not any longer well enough in touch with the situation [of research in mathematical ballistics] to feel confident of my judgment."[38] He made one effort to link his war service to civilian mathematics in a fund-raising effort he called "Our Debt to Mathematics", an effort that largely failed to make an impression on either mathematicians or potential sponsors.[39]

When the study of mathematical ballistics revived in mid-1930s, the work became ever more connected to the mass production of weapons and, ultimately, to the mass production of artillery shots on the battlefield. Beginning where Veblen had left off, the mathematicians of the Second World War developed mathematical analyses and computing technologies that followed every aspect of production beginning with the raw materials and the initial design, and ending with the finished weapon and the placement of the final shot on a target. Indeed, if we consider the goal of the mass production process the final artillery shot rather than the gun, we see the integration of mathematics and computation advancing through the second world and the years that followed. Oswald Veblen, as one of the nation's senior mathematical scientists, guided this development through the mid-1950s. He watched as his computing sheets were incorporated into computing devices, such as the Bell M-9 Fire Control System and the ENIAC computer. In subsequent years, onboard computing devices carried mathematical analyses with artillery warheads, including the inertial guidance system for the reentry vehicle of the Minuteman missile, the control system for the Tomahawk cruise missile, and the controls for bombs targeted with the Global Positioning System. In studying such instruments of war, we can easily forget the role that mathematical analysis plays in their design

[37] Goldstein, *The Computer from Pascal to Von Neuman* (note 1), p. 132.

[38] Oswald Veblen to Philip Schwartz, 24 March 1924, Oswald Veblen Papers, Library of Congress, Washington, D.C.

[39] Loren Butler Feffer, "Oswald Veblen and the Capitalization of American Mathematics: Raising Money for Research, 1923–1928," *Isis* 89 (1998): 474–497.

manufacture and deployment. As was noted above, mathematicians have been engaged in the conduct of war since the time of Galileo, but we see mathematics integrated into weapons systems with the development of mass production. In the United States, the start of such integration is most clearly seen in the work of Oswald Veblen at the Aberdeen Proving Ground during the First World War.

CHAPTER NINE

NATIONAL NAVAL LABORATORIES AND THE DEVELOPMENT OF FIRE CONTROL GYROCOMPASSES IN INTERWAR BRITAIN AND FRANCE[1]

Sébastien Soubiran

My study deals with the French and British navies' innovation processes from the First World War to the eve of the Second World War, especially with the administration of scientific research within military departments. More precisely, by studying the development of a gyrocompass for fire control systems by both navies, I tried to understand what kind of research was promoted by the military. How was it practised and by whom? And finally, what was at stake in the use of scientific research? The French and British states promoted and framed a science-based development for their military armament in the interwar period.[2] Many links existed in these two countries between industrialists, scientists, and the military which were not mere exceptions. Moreover, Great Britain was certainly leading the world in matters concerned with armament technologies and the development of science-based technical systems. Those preliminary arguments will be discussed in this paper.

In order to understand the development process of a particular technique within a military environment, I certainly had to question current ideas usually found in the historiography:[3]

[1] This paper was based on my Ph.D. work: Sebastien Soubiran, *De l'utilisation contingente des scientifiques dans les systèmes d'innovation des Marines française et britannique entre les deux guerres ; deux exemples : la conduite du tir des navires et la télémécanique*, Ph.D. thesis, University of Paris, 2002. I would like to thank David Edgerton and Dominique Pestre for their advice. My researches would have not been possible without them.

[2] Consequently, one should be more cautious about C.P. Snow's so-called "two cultures"; C.P. Snow, *The Two Cultures* (Cambridge, 1993).

[3] My understanding of Great Britain's science policy during the twentieth century owes much to the work done by David E. Edgerton who offers a dramatic change in our understanding of British Defence policy and who was certainly one of the first to criticize the theory of the Decline so rooted in historical analyses so far. He has synthesized his work on this subject in a book: *Warfare State: militarism, technocracy and expertise in twentieth century Britain* (forthcoming).

First, not to consider the innovation process as linear and not to think scientific knowledge as underlying all technical development.

Second, not to limit my understanding of "science" as only that sought in a university (usually labelled "pure science").

Third, not to rely on categories like "scientists", "engineers", or "military officers" as obvious, in term of practice, knowledge, and know-how, in order to open the field of understanding of my actors.

Last, but not least, not to use the Second World War as a screen to analyse the First World War and the interwar period, particularly concerning the links between scientists and the military.[4]

My work owes much to the work of Donald MacKenzie on inertial missile guidance[5], but differs in many points because I focused mainly on research departments and laboratories that belong to the military, and my period of interest was between the two World Wars. Moreover, I was interested in two European countries, France and Great Britain, allegedly considered peculiarly decadent by historians in their science policy and armed forces, at least compared to the United States or Germany. Thus, I decided to talk about science from the military point of view, and consequently my work relied mainly on military archives. The navy was particularly interesting. On the one hand because of its long history it was possible to understand the specificity of the period I chose, on the other hand, unlike the Air Department, civil applications were less obvious. The use of scientific research was then mainly for military purposes. Last but not least, research work on the relation of navies and science in the interwar period are very rare for both countries.[6] My paper will consist in a brief presentation of the main policies defined by both countries, in terms of military innovation systems and their evolution in

[4] These points are usually connected in studies that conclude in the failure of the military or the governments to develop science-based technical systems before the Second World War. For instance, David Henry, "British Submarine Policy, 1918–1939", in Bryan Ranft (ed.), *Technical Change and British Naval Policy 1860–1939* (London, 1977), pp. 80–107 and Corelli Barnett, *The Audit of War: the illusion and reality of Britain as a great nation* (London, 1986). For a further analyse of the historiography see the introduction of Soubiran, *De l'utilisation* (note 1), pp. 16–41.

[5] Donald MacKenzie, *Inventing Accuracy: a historical sociology of nuclear missile guidance* (Cambridge, Mass., 1990).

[6] There is one for France and one for Great Britain: Benoît Lelong, "L'innovation industrielle et militaire," *Epistémologiques* 1.3–4 (2001): 205–232; Willem Hackmann, *Seek & Strike: sonar, anti-submarine warfare, and the Royal Navy, 1914–54* (London, 1984). Hackmann focuses mainly on research establishments that developed submarine detection devices.

connection with historical contingencies, starting from the First World War and ending at the eve of the Second World War. I will focus mainly on the complexity of the actors involved, and show how the differences of their practical and tacit knowledge led to the creation of different testing arenas and thus ultimately different devices.

The case I will study is the gyro compass. The navigational device known generically as a gyrocompass indicates true north using the mechanical properties of a gyroscope that points an invariant reference position due to the rotation of a heavy mass (the gyro) around an axis with two degrees of freedom in relation to its case. By fixing a weight to the horizontal axes of the gyroscope, it becomes pendular and thus sensitive to gravity. This force keeps its axes parallel to the earth's axis of rotation and the gyrocompass is kept at a constant speed by alternating current with a specific constant frequency. Thus, a gyrocompass provides a good fixed benchmark to calculate a ship's course. If one wants to use it for fire control purposes, one has to make sure that the different movements of a ship do not interfere with gyrocompass data.[7] Donald MacKenzie explained how the gyrocompass, as part of what he calls the "gyro culture", became a symbol of the technical development of the western countries during and after the First World War. Indeed, the dramatic changes introduced by the use of this device, as well as other gyroscopic devices, by the British and the French navies after the First World War make this tool particularly interesting to understand the innovation processes of these two countries' military departments. The use of the gyrocompass for fire control purposes, the numerous problems that both navies faced, and the different answers they developed will provide the frame of my study.

[7] For more technical precision see Jobst Broelmann, "Compass, Gyro-", in Robert Bud and Deborah Jean Warner (eds.), *Instruments of Science* (London, 1998), pp. 132–134. For information about navies' fire control systems see John Brooks, *Fire Control for British Dreadnoughts: choices in technology and supply*, Ph.D. thesis, University of London, 2001; David A. Mindell, "Anti-Aircraft Fire Control and the Development of Integrated Systems at Sperry, 1925–1940," *IEEE Control System* 15 (Apr. 1995): 108–113; Jon Testuro Sumida, *In Defence of Naval Supremacy: finance, technology, and British naval policy, 1889–1914* (Boston, 1989).

Developing Complicated yet Simple High Tech Gyrocompasses

British Gyrocompasses and the Great War

Before the war two industrialists developed gyrocompasses that they thought would in time take the place of a magnetic compass in ships. They were the American Elmer Sperry and the German Dr. Herman Anschütz-Kaempfe.[8] The development of the gyrocompass arose on the one hand with the development of gyroscopic devices at the end of the nineteenth century and on the other hand with the development of iron ships in which use of magnetic compasses became problematic.[9] The collaboration of Sperry with the American navy as early as 1908 certainly helped to convince the other navies to have a look on that new device. Thus, before the First World War, the development of the gyrocompass stayed mainly in industrialists' hands, with Sperry and Anschütz leading all notions in its development. The First World War made Sperry the main supplier of gyrocompass in the world.[10] However, the British navy had her own department and laboratory to ensure the development of the "gyro culture" on her ships; it was certainly the only one of its kind in any navy of the world at that time.

As far as the British navy was concerned, the development of the gyrocompass led to the creation of a specific department. The compass department, directed by a naval officer, was then mainly in charge of simple testing and of the supply of the fleet with new compasses, magnetic as well as gyroscopic.[11] Before the war the British Admiralty also benefited from the skill of a professor, James Henderson (later Sir James), who taught mechanical physics at the Naval College

[8] Thomas Hughes provides us with a good story on the development of the gyrocompass by Elmer Sperry in, *Elmer Sperry: inventor and engineer* (Baltimore, 1971); Hugues' story completely ignores the British navy's involvement in Sperry's gyrocompass. Peter Galison in *How Experiments End* (Chicago, 1987), pp. 34–47, discusses the links between the mechanics of the gyrocompass and the theoretical work of Einstein on gyromagnetism. He also provides useful information on the German gyrocompass and the conflicted relationships between Anschütz and Sperry companies.

[9] MacKenzie, *Inventing Accuracy* (note 4), pp. 33–38 and see below, note 13.

[10] Hugues defined the Sperry Company after the First World War as "a giant American industrial corporation."

[11] For a more complete view on the pre-war organisation of compass development within the British navy, see Antony E. Fanning, *Steady as She Goes: a history of the Compass Department of the Admiralty* (London, 1986), pp. 1–192.

at Greenwich and who happened to be an expert on gyroscopic devices. Between 1907 and 1915 Henderson applied for more than fifteen British patents, of course all secret. One improvement of the Sperry gyrocompass concerning the correction of errors introduced by pitch and roll of the ship was in fact due to Henderson—given to the Sperry Company by the British navy just before the beginning of the war.[12] The compass department and Henderson received their orders and submited their devices to the naval ordnance department. It is important to notice that this department was also in charge of a ship, the HMS *Excellent*, used solely for experimental testing of new devices. Although mainly devoted to the training of gunnery officers, this "stone frigate", as naval officers used to call it, was used also for practical experiments on new types of weapons.[13]

When the European nations went to war, Henderson still worked for the navy. Sperry built workshops in Great Britain to supply the Admiralty's biggest ships with gyrocompasses. Three particular events that occurred during the war in terms of gyrocompass development but also concerning the evolution of fire control techniques deserve close attention. In 1916 the British industrialists John Perry and Sidney George Brown applied for a new patent for a gyrocompass, providing Great Britain with a national gyrocompass builder. The same year, the British Grand Fleet confronted the German Hoch See Flotte off Jutland in the North Sea on the first of June. The Battle of Jutland was the largest naval battle of the War and because of the difficulties encountered by the British navy during that battle and the numerous losses, British headquarters introduced profound changes in their fire control methods.[14] First, it confirmed that fire control commands should be centralised in a single place. The Admiralty adopted the Fire Director system, invented by Admiral Percy Scott, and entrusted the industrial firm Vickers with the production of the new system. Second, it was decided that ships should

[12] Henderson asked for special rewards from the British Navy in 1915. This controversial story obliged the Board of Admiralty to consider the question of whether scientists and other civil servants who worked for the Navy (but not trained by the Navy) had financial rights to their patents arising from naval research. Cf. Soubiran, *De l'utilisation* (note 1): pp. 61–62.

[13] M.M. Postan, *British War Production* (London, 1952), p. 434.

[14] On that particular point see Jon T. Sumida, "A Matter of Timing: the Royal Navy and the tactics of decisive battle, 1912–1916," *Journal of Military History* 67.1 (2003): 85–136.

be able to provide rapid fire not only while sailing a fixed course, but also while the ship accelerated, decelerated, and changed direction.

This last decision had a dramatic impact on the use of the gyrocompass for fire control. The entire wartime Sperry system that relayed the gyrocompass data to a central panel used to point the guns to the target became obsolete.[15] Although perfectly accurate when the ship was steady, the gyrocompass became useless when the ship accelerated or changed its course too fast. Since neither Sperry nor Brown were able to meet these new needs, the British Admiralty decided to create a new establishment, the Admiralty Compass Observatory (ACO), in 1917. This new establishment provided the British Admiralty not only with a workshop where industrial compasses could be tested and assembled, but it also permitted them to pursue practical experiments on gyrocompasses. The mathematician G.T. Bennett was appointed as a scientific adviser concerning the ACO to the Director of the Compass Department. Moreover, the department asked an electrical engineer working for Sperry, Dr. A.L. Rawlings, to be put in charge of the experimentation on gyrocompasses. Rawlings worked with eight military electrical engineers and the technical officers under the commander Geoffrey B. Harrison.

In a word, although during the war the Admiralty equipped its ships with Sperry gyrocompasses, it is important to underline that at the same time many steps had been taken that permitted it to specify its needs and develop its own expertise in terms of gyrocompasses. Rawlings and Harrison's collaboration resulted in the invention in 1919 of a new device using liquid circulation system to temporarily stop the data transmission of the gyrocompass during fast course changes. This device resulted in a huge improvement in the utility of gyrocompasses in ships, but even moreso in aeroplanes.

Though I am aware that I have been very brief in my description of the development of the gyrocompass in the British navy before and during the First World War, it is important to notice that the Admiralty did provide means to control the development of the gyrocompass. Even at this stage, it is possible to see that the British navy contributed considerably to the framing of the gyrocompass as it existed in 1919. Also, notice that so far there has been no reference

---

[15] For a description of Sperry's fire control system, see Hugues, *Elmer Sperry* (note 5), pp. 230–233.

to the Board of Invention and Research (BIR), which had been created in 1915 to coordinate the collaboration of academic scientists with naval officers on technical problems. The BIR is otherwise unavoidable when one considers scientific and technical development in the British navy during the First World War. My research, however, permitted me to argue against those stories which tend to emphasize the role of the BIR in the analysis of scientific collaboration within the military during First World War. By focusing on the story of the BIR, previous studies committed at least two errors: on the one hand they only take into account the discourses of a well defined type of scientists, members of the Royal Society, with high ranks in the university, and they fail to argue about how partial and specific those discourses are; on the other hand these historians are eager to emphasise the ostracism of the naval officers in the case of new technology, based upon the officers' inability to understand technical and scientific developments. Thus, no attention is paid to other forms of collaborations between the British Admiralty and the scientists; only the Royal Society model of collaboration seems to be considered valid. Further, those studies gave no value to the technical and scientific knowledge of naval officers or to research and experiments taking place inside the Navy without the help of the BIR or eminent scientists. Although I do not want to deny the role of the BIR (or minimize the problems in the Admiralty caused by its creation), I would show in this paper the necessity of considering all kinds of knowledge if one wishes to understand the technical development of the gyrocompass in the British and French navies, and also urge us not to consider the model praised by academic scientists as the only obvious one.[16]

[16] Most of the literature available on that topic privileges the point of view of the scientists even if not written by scientists themselves: Jack K. Gusewelle, "Science and the Admiralty during World War I: the case of the B.I.R.," in Gerald Jordan (ed.), *Naval Warfare in the Twentieth Century: essays in honour of Arthur Marder* (London, 1977), pp. 105–17; Roy MacLeod and E. Kay Andrews, "Scientific Advice in the War at Sea, 1915–1917: the Board of Invention and Research" *Journal of Contemporary History* 6 (1971): 3–40; Michael Pattison, "Scientists, Inventors and the Military in Britain, 1915–19: The Munition Inventions Department", *Social Studies of Science* 13 (1983): 521–568; Guy Hartcup, *The War of Invention, Scientific Developments, 1914–18* (London, 1988); Donald S. Cardwell, "Science and World War I," *Proceedings of the Royal Society of London*, series A, 342 (1975): 447–456; J.J. Thomson, *Recollections and Reflexions* (London, 1936); A.S. Eve, *Rutherford* (Cambridge, 1939); G.M. Carve, *William Henry Bragg, 1862–1942: man and scientist* (Cambridge, 1978).

Organizing a science-based technological system

Several committees were created after the war by the British navy which had profound impact on the interwar Admiralty policies. For instance, the Fire Control Requirements Committee, together with the Post-War Questions Committee defined Admiralty policy in matters concerning fire control for battleships, cruisers, and destroyers. Among the various naval officers in charge of technical development on ships, it is remarkable that we also find two civilians, the professor James Henderson and the engineer Sir Keith Elphinstone working for the Elliot Brothers company. This firm specialised in electrical and mechanical devices, and among others, it sold Anschütz gyrocompasses in Great Britain before World War One. Thus the choice of these two individuals can be understood as an expression of the Admiralty interest in gyroscopic devices. The two committees wrote a report that they submitted to the Board of the Admiralty on 20 January 1921. Briefly, it was decided that fire control should be automated as much as possible. It was explicitly said that on the one hand a solution should be based on complicated and high technological devices but on the other hand that their use should be as simple as possible. The gyrocompass appeared to be a singular instrument that would help to achieve these seemingly contradictory goals. By connecting the gyrocompass directly to the central panel of the fire director system it should have been able, it was reasoned, to point the guns directly at the target, regardless of the ship's motion.[17] The Board of the Admiralty approved the directives given by the committees, although it is important to underscore the emphasis on high technical systems in the naval officers' discourses. They were definitely aware of the dramatic changes introduced by the war and appeared eager to develop them further. War clearly appeared as an "ideal" experimental scene which generated dramatic technical change and facilitated the decision process.

In 1919 the British government created yet another committee in charge of the organisation of scientific research within military departments, the Department of Scientific Research (DSR). Two years later this committee decided to build a research laboratory for the navy

[17] Committees' report to the board of Admiralty, 20 January 1921 [Public Record Office, Kew (herafter PRO), ADM 116/2068].

in which mainly physicists would pursue fundamental physics research. While the details of the setting of the organisation of scientific research within the British navy are too complicated to delve into here, it is important to keep in mind that this general scheme had been thought up by a civilian, the consulting engineer Charles Merz, an important figure in the organisation of the electrical distribution systems in Great Britain at the beginning of the twentieth century.[18] Moreover, eminent scientists like the physicists Sir J.J. Thomson, Ernest Rutherford and the chemist Sir Richard Threlfall were involved in the decision. The Treasury unhesitatingly accepted the creation of the Admiralty Research Laboratory (ARL) and authorized the employment of more than twenty civil scientists and the construction of a new building. They also decided to create a research department within the headquarters in charge of all matters that dealt with scientific research and inventions within the navy. The direction of this department was given to a physicist, Sir Frank Smith, who had worked at the National Physical Laboratory (NPL) since its creation in 1900. In 1917 he was director of the electrical department and worked in collaboration with the Admiralty and is said to be the inventor of the first mechanism for magnetic mines used by the British navy. The director of the ARL, Charles Drysdale, was also a physicist who specialized in the invention and development of electrical instruments. He worked for the firm Nalder, Bross, & Co. from 1896 to 1910 as scientific assistant and then was nominated for the teaching of applied physics and electrical engineering at the Northampton Institute, London. Thus, the Admiralty was particularly interested in scientists with knowledge in applied physics and electrical instrumentation and who worked previously in the industry.

The system in place at the NPL building was duplicated for the ARL just few yards away, and all the scientists working for the navy became civil servants. In 1921 the scientific research pursued at the ARL was divided into three sections: acoustic, electrical, and optical. The electrical section led by Drysdale studied all kinds of electrical devices such as oscillographs, alternative current relays, salinimeters, and submarine cables. Among all these studies they were also testing a gyrocompass damping device, although the Admiralty Compass Observatory stayed in charge of the testing and development

[18] Fore more information see Soubiran, *De l'utilisation* (note 1).

of gyrocompass. Rawlings and his team had already assembled their own model of modified Sperry gyrocompass. Nevertheless, it was Henderson who was the most significant expert at that time. The Admiralty agreed to employ Henderson with a new contract as a gyro expert for five years with a salary of £3000, a notable sum, as Sir Frank Smith as the DSR was only paid £1200 and the Third Sea Lord £5000. Such a high salary has many explanations: in 1921 Henderson was certainly the best expert on this topic in Great Britain if not in the world; it was then important for the Admiralty to keep Henderson's knowledge and the patents he applied for to themselves; the Admiralty thought that such a salary would avoid a new conflict with Henderson concerning its patents on gyroscopic devices; and last but not least, the Admiralty was eager to improve its fire control system and the gyrocompass had a crucial part to play. Well aware of his value, Henderson asked a law firm to write his contract with the Admiralty.[19]

As an expert, Henderson was in charge of the organisation of sea trials, conceiving of different shipboard testing methods to judge the value of the gyrocompass experiments. With the director of naval ordnance and the director of the compass department he determined the level of the gyrocompass precision needed to correlate the course of the ship and the fire control. Moreover, in 1923 Henderson was put in charge of the design of a Master Gyro Unit, a device that would permit the transmission of the gyrocompass data directly to the central panel of the fire director. Sea trials were organized every year, usually twice a year. Four models were tested: one from Sperry, one from the British industrial firm Brown, one from the group working at the ACO, and one from Henderson. An Anschütz gyrocompass taken from German ship was judged less efficient than the Sperry and had been abandoned in 1920. Over the course of four years each of the actors designed various models of gyrocompass, tested them in sea trials, and then modified them based upon the results. Although Henderson and the ACO proved very innovative in designing various devices to prevent the gyrocompass from damping and also improved its accuracy, when Henderson's contract expired in 1925, no gyrocompass was yet accurate enough to be used for fire control purposes.[20] The British navy kept the American

[19] PRO, ADM 1/96/4.
[20] Sumida, "A Matter of Timing" (note 14).

model to supply its fleet as it was then considered the best gyrocompass for navigation. The Brown firm failed to have one of its models accepted by the Admiralty but they were nevertheless encouraged to pursue their efforts.

In 1924 the Board of Admiralty made a decision which introduced dramatic changes in the development of a gyrocompass for fire control purpose. Throughout the interwar period, the British government supported scientific research within the Admiralty. The first committee appointed by the government in 1923 in charge of analyzing the administration of the defence departments to reduce spending, clearly set science-based technology as a priority in terms of reorganization of the various departments and for which money should be spend rather than reduced. Nonetheless, the committee clearly expressed the necessity of centralizing all efforts. The cost of Henderson's knowledge was also criticized and the committee suggested that gyrocompass development should be put in the hand of the ARL. Thus, it was decided that Henderson's laboratory should be closed and the professor's contract not renewed. In 1925, gyroscopic researches were then handed over to the Admiralty Research Laboratory, and directed by Rawlings who moved from the Admiralty Compass Observatory.[21]

When the ARL gyrogroup was created Rawlings immediately appeared as the newly recognized gyrocompass expert in the Admiralty. The director of naval ordnance, the director of the compass department, and the Third Sea Lord in charge of the technical development of the British navy all agreed with his advice. Thus when Rawlings argued for the relaxation of requirements on gyrocompass accuracy in order to promote its development, they listened. The gyrogroup was also entrusted to organise the sea trials. One of the first steps taken by Rawlings was to change the methods of sea trial testing that had been established by Henderson. Instead of men, he used cameras fixed above each gyrocompass to record their data, as well as accurate timing devices that permitted comparison of sea data to land observations (which gave the exact position of the ship and judged the precision of the gyrocompass). Moreover, Rawlings applied for a secret patent for a device called a slave gyro based on

[21] Cabinet: Reduction of national expenditure [PRO, T 163/16/14].

the association of a free gyro to the gyrocompass.[22] With this new device, Rawlings planned to obtain the precision required for fire control use more quickly. Although Rawlings and his team were entrusted with helping the industrialists improve their gyrocompass, they in fact started to build their own device, as at that time neither Sperry nor Brown had succeeded in providing the Admiralty with a gyrocompass accurate enough to be used for fire control. The gyrogroup built special testing devices in the laboratory to artificially recreate ship movements. They analysed the results and assessed the quality of the various devices. They also went to the industrialists' workshops to witness preliminary testing and decide if the gyrocompass was ready for laboratory testing and then for sea trials or not. Tensions arose between the gyrogroup and industrialists over the method of testing and on the interpretation of the results. Moreover, Brown twice accused the Admiralty of patent infringement. This crisis led to the departure of Rawlings in 1928. In fact, he was asked to leave the Admiralty, and was replaced by his assistant, W. Burnside, an electrical engineer.[23] Returning to work with Sperry, Rawlings still remained in touch with the Admiralty and although the Director of Naval Ordnance (DNO) asked the gyrogroup to organise sea trials with industrial gyrocompass, from 1928 onwards the team working at the ARL began to develop its own gyrocompass model, assuming that the industrialists could not find a solution on their own.

Ideal Gyrocompass Development by ARL Civil Scientists

In 1931 after new sea trials with both a Sperry and an Anschütz gyrocompass gave no usable results for fire control, the DNO ordered

[22] Rawlings wrote, "By a 'free' gyro in this connection is meant one which has no meridian seeking tendency but the term may include one in which some provision is made to keep the axle substantially horizontal. Suppose a differential arrangement to be provided which indicates the difference between the axles of the compass gyro and the free gyro. Let it do more than indicate—let it control a means of applying a torque to the free gyro in a direction to bring it slowly into line with the compass. Then we may have the free gyro kept closely to the mean of the compass oscillations." Dr. A.L. Rawlings, note on slave gyro [PRO, ADM 212/108].

[23] Unfortunately we did not find more information about this conflict in the archives we examined, and even A.E. Fanning does not really analyze it; see Fanning, *Steady as She Goes* (note 8), pp. 224–226.

that sea trials should be pursued in the future by "scientific methods". In 1933 the gyrogroup defined new scientific criteria for sea trials to judge the accuracy of gyrocompass and the results confirmed the superiority of their gyrocompass to the one provided by industrialists. Thus, the gyrogroup at the ARL convinced the DNO that their gyrocompass was the best, because only theirs could reach the mechanical perfection needed to have an accurate device—a standard they had set for themselves that they knew to be beyond the industrialists' capability at the time.[24] After 1933 the gyrogroup only advanced to sea trials its own model of gyrocompass. Burnside was convinced that errors occurred only because of the mechanical imperfection of the gyrocompass, so he insisted that mechanical perfection was necessary if one was to use a gyrocompass for fire control. In response, the gyrogroup conceived of new method for testing their device involving three models of gyrocompass built using different kinds of complementary internal devices to assure their precision. In 1935 one of these models nearly reached the precision requested for fire control.

This decision to favour the development of an in-house gyrocompass occurred for many reasons. First, between 1928 and 1934 the government organised the administration of civil servant scientists working in various state departments. In 1934 this new administration led to the creation of numerous posts of research assistants, particularly within the military departments. This was in part due to the economic crisis that struck Great Britain in 1931, resulting in a request to the Admiralty to reduce its spending. The First Lord of Admiralty clearly expressed his belief that the best way to reduce the spending was to emphasize the in-house design and to give the chiefs of research and technical department the responsibility for developing various materials.[25] Moreover, he particularly insisted on the necessity to encourage research activities with only long term applications. The development of the gyrocompass at the ARL shows

[24] Letter from the director of the ARL, 6 August 1931 [PRO, ADM 212/108].

[25] For instance, the Admiralty had chosen to design its own model of torpedoes, fire control tables, and wireless sets. The in-house design within the Admiralty and others British armament services are covered in M.M. Postan, *British War Production* (note 13) and M.M. Postan, D. Hay, and J.D. Scott, *Design and Development of Weapons* (London, 1964). For a study of the development of fire-control tables during the interwar period see also John Brooks, *Fire Control for British Dreadnoughts: choices in technology and supply*, Ph.D. thesis, University of London, 2001.

that the First Lord expressed his opinion with more than hollow words. Beginning officially in 1936, rearmament policy gave another opportunity to the gyrogroup to prove their skill in terms of fire control devices. Indeed, fear of airplane raids drove the Admiralty to develop new devices to protect the ships against aerial bombardment, and at the same time, the military services reinforced their control of technical development and armament production.[26] The development of anti-aircraft automatic fire control devices was strongly encouraged within the three military services by the government.[27] This question of parallel development remains quite controversial and has been criticized by historians of the British navy. Without entering into that debate here, it can be said that the committee in charge of organising the anti-aircraft defence within both the navy and the army eventually decided to entrust the Admiralty Research Laboratory with the task of developing *both* anti-aircraft and shipboard fire control systems.[28]

In 1938 a fire control section was created within the ARL and the gyrogroup was attached to it. The gyrogroup's gyrocompass was supposed to be for a new anti-aircraft device, the tachymeter, which automatically pointed the anti-aircraft guns at their target. The development of the tachymetric system was classified as high priority. Fourteen scientists officers working at the ARL—almost half the total numbers there—were involved in the project, along with twelve research assistants, twelve designers, five laboratory assistants, and twenty-five workers. Vickers engineers were also involved in building the framework of the device. The plan to extend the laboratory's workshop to develop more and more fire control devices instead of buying them from industry indicates the quality of the devices conceived by the ARL scientists, in particular their research on electrical transmissions devices, gyroscopic instruments, and stabilisation systems.[29]

---

[26] This question is well illustrated in D. Scott and R. Hughes, *The Administration of War Production* (London, 1955) and also in M.M. Postan, *British War Production* (note 13).

[27] David Edgerton, *England and the Aeroplane: an essay on a militant and technological nation,* (London, 1991).

[28] For more information on that particular point see Soubiran, *De l'utilisation* (note 1), chapter 5.

[29] PRO, ADM 212/73.

Despite all this, one piece was still missing when Great Britain entered World War Two: the ideal gyrocompass. In fact, this ideal device had been built, but it was more a scientific instrument than a technical device useful aboard a ship. Concerning the gyrocompass device, only one goal had been reached from the 1921 directives: the Admiralty built a highly technical device but it was not simple to use, and certainly too complicated for service use by a ship's crew. Nevertheless, because of the electrical transmission systems and the stabilised platform systems they had developed, the ARL was able to mount a Sperry gyrocompass on a ship and then sufficiently improve its precision for fire control purposes.

Although the ARL gyrocompass was ultimately not ready until 1943, one cannot talk of real failure considering the knowledge and know-how developed by the scientific officers during the interwar period. The rising place of the department of research and experiment and the ARL within the Admiralty is striking and should be underscored. However, it is important to point out that the ARL was not the only place where research and experiments were pursued. Many technical departments had their own research facilities, as for example the signal school, which developed a great deal of wireless transmission devices and worked on radar during the 1930s.[30] There was also a research group dedicated to ASDIC (anti-submarine detection) researches, and the mining school pursued important research on magnetic mines as well as gyroscopic devices for torpedoes.[31] The uniqueness of the ARL was that it was directed only by scientific officers and it is interesting to remember this in light of the fact that the gyrocompass became a scientific device useless on board a ship. As a closing comment on the British case, I would like to remind the reader that the British navy entered the Second World War with more than two hundred scientists working within its departments, even before making connections to university academics, whose collaboration is usually supposed by historians to have revolutionized technical systems during the war.

[30] Postan, Hay and Scott, *Design and Development* (note 25), p. 452.
[31] Willem Hackmann, *Seek & Strike* (note 6), pp. 97–138.

Developing a French Gyrocompass at any Cost

British Navy Model—American Sperry Supplier

Before starting the story of the development of the gyrocompass by the French navy it is important to underline various points. On the one hand, the French navy was not as large or important as the British one. Indeed, the Washington conference in 1921 ranked the French navy fourth in world navies, tied with the Italian navy, and behind those of Britain, America and Japan. In France the army was the main interwar military service with the larger budget. Consequently, one should not be surprised by the difference of ambitions between the British and French navy in terms of financial investment or priorities. Nonetheless, historically the French navy should be counted as an innovative navy. Its contribution in the development of a modern navy—steam engines, iron hulls, and even torpedoes—was important, in part because of the creation of a special military engineer corps specializing in naval construction at the beginning of the nineteenth century. Those engineers from the *génie maritime* were trained at the Ecole Polytechnique, one of the most important *Grande Ecole* which provided the country with its elite engineers.[32] Despite these differences, by focussing on the actors involved in the technical development of gyrocompasses, the differences of policy decisions by both navies may be better understood. In addition, the comparison will therefore be more pertinent between these two countries as well as more interesting.

The French navy bought its first gyrocompass from the Sperry Company in 1913 through the committee of naval ordnance, which had charge of all naval experimentation. The technical development of artillery within the French navy was split between the directorate of naval construction, the directorate of naval artillery, and the commission of naval ordnance. These three departments differed deeply in terms of responsibility and staff. On one hand, technical decisions were directed mainly by military engineers. They were in charge of

[32] On the Ecole Polytechnique see Terry Shinn, *Savoir scientifique et pouvoir social, 1794–1914 L'Ecole polytechnique* (Paris, 1980); Bruno Belhoste, Amy Dahan Dalmedico, and Antoine Picon (eds.), *La formation polytechnicienne 1794–1994* (Paris, 1994); and Bruno Belhoste, Amy Dahan Dalmedico, Dominique Pestre, and Antoine Picon (eds.), *La France des X: deux siècles d'histoir* (Paris, 1995).

the collaboration with industry but also had their own laboratories, design offices, workshops, and docks. On the other hand, the committee of naval ordnance was composed mainly of naval officers, trained at the Naval School. These same officers invented the methods of fire control in use in the Navy and taught them to junior officers. Thus, the development of fire control devices was shared between naval officers, military engineers, and industrialists. For instance, one of the main devices for fire control computing data had been invented by a French naval officer, lieutenant Le Prieur, but was built by the French industrialist Echassoux. The industrialists were allowed few initiatives whereas Le Prieur supervised the construction and worked within the Echassoux workshop. Moreover, military engineers were in charge of the general control of the industrial operations. In a word, Echassoux was not asked to be innovative but only to offer a production capacity for the navy.[33]

When France entered the First World War, the committee of naval ordnance was closed because ships and officers could no longer be dedicated to experimental purposes, leaving the direction of technical developments in naval artillery and naval constructions to the military engineers alone. Of course the ordnance department at naval headquarters still made the final decisions but the control of technical directions in experiments, technical development for production, and the production itself increased dramatically throughout the war, and with it, the influence of the military engineers.

The French navy played only a secondary role during the war. Unlike the British navy it was not involved in any major battle. The technical legitimacy of British naval choices in terms of fire control was strong because they benefited from the experience of the Battle of Jutland, and by the end of the war the British had more than ever the best navy in the world. The French, therefore, looked to the British navy as a model for world navies—quite surprising given the famous rivalry between the two nations. However, as the French did not fight any battles on their own, they could not benefit from "true" sea trials like the British. In 1917, both the French and the American navies adopted the fire director system of the British navy for their dreadnoughts. But unlike the Americans, the French navy

[33] Note from military direction of naval constructions, 22 July 1914 [Service historique de la Marine, Château de Vincennes, Paris (hereafter SHM), SS Ed 12].

was nevertheless confronted by a major problem: a severe lack of production capacity. Neither the industrialists nor the technical departments were able to manufacture the devices needed to equip the French dreadnoughts with fire directors. Moreover, Vickers in Britain was not able to build materials for the French navy until 1919.[34] Consequently, at the end of 1917 the French navy eventually decided to equip its dreadnought fleet with a French device, invented by a French officer, which could be built within few months, although the war ended before any dreadnought was equipped with new fire control material. The technical department that failed to provide the navy with innovative armament material and in sufficient quantity blamed their failure on a lack of industrial capacity.[35]

As far as gyrocompasses were concerned, no special department was created during the war, nor was any French industrialist interested in their development. Sperry remained the only official provider. In terms of submarine countermeasures, however, the French navy was in the lead. As British navy submarine detection research led to the creation of a strong collaboration between the navy, academics, and industry, the French navy created a research laboratory at Toulon in the south of France in which naval officers, civil and military engineers, and academic physicists worked together. This collaboration helped the famous French physicist Paul Langevin to invent SONAR. This form of collaboration, contrary to what historians might have said, was pursued after the end of World War One.[36]

Getting Rid of Military Engineers' Dominance

The French government, academics, and the military decided to retain a research laboratory in physics within the Navy Department where civil scientists were employed. Thus, like Great Britain, academics were involved with the committees in charge of organising technical development within the navy after the war. The laboratory together with six new technical commissions constituted the

[34] Note, Ordnance Department, 11 August 1918 [SHM, SS Ed 12].

[35] For further information on this point see my Soubiran, *De l'utilisation* (note 1), part 4, chap. 1.

[36] For a story of Langevin's collaboration with the French navy see Benoit Lelong, "Paul Langevin et la détection sous-marine, 1914–1929. Un physicien acteur de l'innovation industrielle et militaire," *Epistémologiques* 1.3–4 (2001): 205–232.

newly-created *Centre d'Etudes de Toulon*. Many differences appeared compared to the British organisation. First, civilian scientists who worked for the French navy were few compared to the United Kingdom; no more than ten civilian scientists worked within the laboratory during the interwar period. Moreover, a department of scientific research was also created with a naval officer at its head, but no civil scientists at all. In fact, mainly naval officers worked in the centre. Even the direction of the research within the laboratory was shared between a naval officer and a civil scientist, the physicist François Canac. This is a striking difference with Great Britain, because French naval scientists did not have a laboratory of their own. In my opinion, the promotion of scientific research by naval officers was self-serving in that it returned responsibility of technical development to military engineers rather than ceding it to civilian scientists. This caused a problem, however, in that after the war military officers within the navy and the army strongly criticized the control of armament development by the military engineers and accused them of inefficiency during wartime. The military engineers in turn criticized the employment of civil scientists, arguing that only their scientific knowledge was useful for the technical development of navy armament. On the other hand, some naval officers claimed that the place of a naval officer should be on board a ship and not on land. Consequently, from its inception, the *Centre d'Etudes de Toulon* remained very controversial. Nevertheless, the French naval headquarters strongly defended maintaining a strong collaboration with academics, particularly physicists.

With respect to fire control systems, the French navy also defined strong political choices directly inspired from the war. The ministry of the navy stated that its departments should promote the French precision instrument industry. Thus, the first step was to get rid of the British devices and the technical department controlled by the military engineers was asked to help industry.[37] As for the development of new technical devices, the focus was put on the technical commissions created or recreated after the war. Nine commissions existed in 1921 and six of them were attached to the *Centre d'Etudes de Toulon*; three others, like the commission of naval ordnance recreated in 1919, were attached directly to the regional headquarters in

[37] CEPAN report, 1921 [SHM, 1BB[8] 85].

Toulon. Those commissions were mainly in charge of the organisation of sea trials, and then got the final decision to judge the value of a device. These commissions together with the *Centre d'Etudes de Toulon* should have allowed the naval officers to keep control of the technical innovation, rather than the military engineers.

Thus, the commission of naval ordnance together with naval officers at headquarters was trusted to define the technical choices for fire control, and French nationalist preference clearly oriented the choices. For instance, a French fire director system was preferred to the British one for the new dreadnoughts and the computing data device of Le Prieur was chosen to equip new battleships.[38] The navy also asked a mechanical industrialist, Carpentier, to develop what would be a strictly French gyrocompass. Nevertheless, Great Britain still served as a model for technical development in terms of organization. The compass laboratory created in 1920 at Toulon was directly inspired by the British Admiralty Compass Laboratory.[39] Moreover, the commission of naval ordnance used the same methods and materials for testing the accuracy of the Sperry models. However both navies made a strikingly different choice concerning gyrocompass development when in 1924 it was agreed that the Sperry was not accurate enough for fire control purposes. The French naval ordnance committee stopped its experiments. At no time were either the naval officers, the military engineers, or the civil scientists working within the navy department asked to pursue research work on gyrocompass. The compass laboratory—despite its name—remained a workshop where industrial devices were tested before their installation in a ship, and never became a place where research work was pursued. The policy of the French navy at that time was clear: if any improvement would allow the gyrocompass to become useful for fire control, it would be due to industrialists and more precisely a French industrialist.

Nevertheless, the first change in gyrocompass equipment did not come from a French industrialist. In January 1926 Navy headquarters decided to create a new technical commission at the *Centre d'Etudes de Toulon* to take charge of navigational instruments.[40] This decision

---

[38] Note from the Ordnance Commission, April 1918 [SHM, 1BB[8] 84].

[39] Note on the organisation of the Compass Laboratory, 24 December 1920 [SHM, 1BB[8] 84].

[40] Note from the CET, 14 January 1926 [SHM, 1BB[8] 139].

was probably linked to the decision to start sea trials with the gyrocompass built by the British industrial firm of Brown. Those trials seemed to convince naval officers of the technical commission and those at headquarters of the quality of the gyrocompass. In May an engineer representative of the Brown Company arrived at Toulon to have the British gyrocompass tested at the compass laboratory in preparation for a final contract. However, laboratory tests were not good. A strong debate took place between the director of compass laboratory (a naval construction engineer), and the officers of the technical commission. The director considered the poor laboratory test results of the Brown gyrocompass as proof of its inferiority to the Sperry model, while the officers argued that sea trials should prevail over laboratory tests and so far the Brown gyrocompass appeared as good as the Sperry model. Although each actor defended their own method of testing to guarantee the quality of the device, other arguments helped headquarters to make up its mind. Not only did a Brown gyrocompass cost less than the Sperry Gyrocompass, but Brown agreed to build plants in France to manufacture its devices. Consequently, it sounded like a good compromise to headquarters to choose the "inferior" Brown, given they were very anxious to being national production. Unfortunately for Brown, the debate was not over: in February 1928 the French industrialist Carpentier informed the French navy that a gyrocompass built by the French physicist Henri Béghin and an ex-military engineer, Lucien Monfray, was ready for trial. At that time officers at the headquarters asked all experiments with foreign gyrocompass to be stopped until the French model was ready for comparative trials.

Building a French Gyrocompass

The first Béghin-Monfray gyrocompass was delivered to Toulon in July 1928. Laboratory tests were started immediately. Unfortunately the device failed to resist the rolling table test and quite literally fell to pieces. However, in the report of the technical commission it was stated that, unlike other devices tested so far by the navy, the French gyrocompass was still a prototype. Consequently, these trials helped Béghin and Monfray to improve their device through feedback from the technical commission and the compass laboratory. The navy also gave a financial award to the inventors to encourage them to pursue their research. A new device was ready for testing in October

1929, less than one year after the initial trials. Although laboratory trials were once again not good, it was decided to pursue sea trials—which gave poor results as well. It is important to notice that both laboratory and sea trials were pursued in poor conditions: lack of materials appropriate for testing, bad weather, and breakdowns of various testing machines. The reports' conclusions were the identical: although the results were poor the gyrocompass was still a prototype and should be pursued. A new step was taken by headquarters to encourage the building of a French device and it was decided to buy the Carpentier/Béghin-Monfray gyrocompass. With that choice headquarters unofficially abandoned their own development of a fire control gyrocompass. The industrialists submitted a third model to the navy in May 1930. For the first time laboratory tests were pursued as planned and results were judged unanimously much better than the previous ones. Headquarters stated that a French solution has been found in the building of a gyrocompass for navigational purposes even though the device could not yet be fixed onboard a ship without previous technical arrangements.[41]

This new stage in the development of a French gyrocompass was followed by a vigorous debate on the best way for the Navy to collaborate in the building of this device. On the one hand, the director of the compass laboratory thought that the French industrialists were not able to build a gyrocompass useful for the Navy by themselves. Consequently he suggested that his department should be deeply involved in determining the final design. In other words, without the knowledge of the Compass department, Carpentier would not be able to build a gyrocompass that could be fixed on board a ship of the French fleet. On the other hand, the naval officers of the technical commission thought that in order to reach the final step of development it was necessary to buy few models from the industrial firms with the knowledge that they should not be prototypes but devices ready for use. Still, while the compass department kept an eye on the building of the gyrocompasses, they felt that their development should be kept only in the hands of the industrialists, not in the military engineers'. Headquarters decided to approve the statement of the technical commission and four gyrocompasses were ordered from Carpentier, suggesting that in 1930 headquarters still valued the expertise of the technical commission attached to the

[41] Note from the CEPIN, 13 February 1929 [SHM, 1BB^{8} 66].

Toulon centre over that of the military engineering departments. However, the "fall" of the *Centre d'Etudes de Toulon* as an innovative centre also began in 1930. Indeed, from 1927 to 1931, tensions grew stronger within headquarters because of the existence of this technical centre where naval officers could pursue military careers without spending time on board a ship. Moreover, technical commissions were judged partial in their judgement by military engineers because they both invented and examined technical devices at the same time. Eventually, the ministry of the navy started to reform the *CET* in 1930, and particularly the role of the commission. As expected, military engineering departments took more and more control of the technical development of naval armament. Many technical commissions were closed and the rearmament in 1936 just confirmed this renewal of trust of the military engineer corps within navy headquarters.[42]

In 1932, the first French gyrocompass model was installed on a ship and its accuracy was judged excellent by the ship's commander, but very few other models were built before the Second World War; the technical directors decided to equip the French fleet with foreign models, mainly Brown and Sperry. However one of the main objectives defined by naval headquarters at the end of the First World War had been reached because at least one French industrial firm was able to build a gyrocompass. Many differences should be noted on the development of a gyrocompass by the French and the British navies. First, unlike the British navy, the French navy did not develop its own gyrocompass for fire control use, they only developed one with lower precision for navigation. Second, it is important to notice that no scientist working *at* the research laboratory was entrusted to pursue research on gyrocompass or fire control systems. In fact, research and scientific collaboration within the French navy often came not only from external scientists such as Henri Béghin mentioned previously, but three very famous French physicists (Léon Brillouin, Paul Langevin, and Jean Perrin) also worked with the navy during the interwar period.[43] Last, but not least, it is

[42] For further information on the reform of the CET see Soubiran, *De l'utilisation* (note 1), part. 3, chap. 4.

[43] On this particular point see Sebastien Soubiran, "La recherche scientifique en milieu militaire: une nouvelle pratique pour les savants-universitaires entre les deux guerres ? L'exemple des marines française et britannique," *Revue 14–18. Aujourd'hui. Today. Heute* 6 (2003): 153–167.

important to underline that the French navy's judgement of the Brown gyrocompass was better than the British judgement of the same device. This difference was clearly the result of two different political choices by both headquarters. However, the development of the gyrocompass in both navies was guided by various technical as well as social and political factors. More precisely, this development was the result of the negotiations between different actors. It is this negotiation process to which I shall turn in the final section of this essay.

Successive Arenas for Testing Ideas and Technologies

In both navies the results of sea trials were presented as a proof of the value of a device. Consequently, those trials induced technical choices and shaped the development of a device. However, this specific stage of the innovation process should be understood as the result of previous agreements between the different actors involved. In this last part I would like to give a brief account of the successive and related confrontations that sustained each step of the development of the gyrocompass in the British and the French navies. I would like to show that technical development can be understood by focusing on the determination of experimental assessments by the different actors involved in the innovation process.

Negotiating Needs

The first step in the evaluation of the technical value of the gyrocompass was the precise determination of the needs. First, those needs depended on whether the device was to be used for navigation only or also for fire control purposes. For fire control use the value of the gyrocompass can be translated into its accuracy to provide a datum in azimuth (in common parlance, a constant bearing) even after rapid manoeuvres. In other words, this accuracy would be a decisive factor in regard to the average number of hits on the intended target. This accuracy was clearly defined by naval officers only. In Great Britain, the DNO stated that the gyrocompass required by the fleet should have an accuracy of 10 arc-minutes at all times. The same accuracy was chosen by French headquarters on the advice of the naval officers of the commission of naval ordnance. In both countries experiments were pursued over the course of four years, with two sea trials per year, but without any results. Consequently, in 1925 the DNO agreed to lower the accuracy expected of the gyrocompass. This decision was suggested by A.L. Rawlings who argued that this decision was necessary if the development of a gyrocompass for gunnery purposes was to be pursued because no gyrocompass would be that accurate in the near future. Almost for the same reasons the DNO once again reduced the accuracy needed for the gyrocompass in 1931. At that time gyrocompasses required an accuracy of 20 arc-minutes at all times.

The French navy did not make the same choices because since 1924 it had not been at all clear that a gyrocompass was to be used for fire control. These different choices implied implicit expectations. The French navy did not involve any of its departments in the development of a gyrocompass; rather, if a solution for fire control was to be found, it would be found by an industrial firm. By comparison, in Great Britain both the Board of Admiralty and the ARL staff were eager to find a solution with devices from their own departments. Moreover, the legitimacy of the ARL within the navy was clearly a benefit to the fire control research pursued by their scientific officers. Their involvement in the definition of the needs was clear in the 1930s: the gyrocompass was to be mechanically perfect and this perfection was based on scientific criteria. In France there was neither such a clear definition of the needs nor did any consensus exist on a specific group who should be in charge of the development of gyrocompasses for gunnery use. On the contrary, tensions between military engineers and naval officers who worked within the *Centre d'Etudes de Toulon* prevented any agreement. It is one of the reasons why the French gyrocompass in 1939 was only sufficient for navigation purposes and just as good as any foreign models. In Great Britain one can argue that although they had a highly accurate gyrocompass it was too fragile and complicated to be use by non-scientists. Indeed, the gyrocompass built by the gyrogroup was judged useless for practical use, and this and many other factors led the DNO to stop the development of the ARL gyrocompass in 1938. For instance, at that time the accuracy needs were balanced against the urgency of needing a shipboard gyrocompass ready for the impending war. This urgency was also pushed by political and strategic pressures from the late-30s demand for anti-aircraft measures.[44] Thus, it is important to underline the temporality of those choices: there were not fixed but evolved with contingency to contemporary events. The interference between various priorities shaped each country's device. In this analysis, one should also take into account the strong hierarchical organisation of the navy administration, where asymmetric powers are conferred to individuals. Thus, agreements are never settled for good but are always liable to be modified according to the individuals in command at any given time.

Negotiating Testing Methods

Once the gyrocompass requirements were defined, the actors also had to agree on the kind of experimental testing that would be accurate and legitimate in judging the value of each device. This point was highly controversial because generally each actor held his own knowledge and methods as superior and tended to deny the value of other forms of knowledge or technique. Thus, from the industrialists' point of view each device required a specific test. Indeed, Sperry, Brown and Anchütz all defined their own particular workshop testing method to certify the value of their devices. On the contrary, the gyrogroup and the director of the French compass laboratory defined testing methods that should have guaranteed the universality of their judgement. For instance, when Rawlings was nominated in 1925 as head of the gyrogroup and put in charge of sea trials, his first step was to change testing methods invented by Henderson. The British industrialist Brown criticized the method used by both navies to test its device, methods which he claimed were inspired by Sperry. Indeed, the swinging tests done by the French compass laboratory *were* invented in Sperry workshop. In the British case it was striking how little trust was put into industrial testing methods as industrialists were considered to be interested primarily with profit. They completely lost control of their devices in both laboratories and sea trials and they had no choice but to accept testing methods adopted by the navies. The debate in both navies, therefore, mainly took place between laboratory tests and sea trials.

[44] See for instance G.A.H. Gordon, *British Seapower and Procurement between the Wars: a re-appraisal of rearmament* (London, 1988).

This debate was particularly important in France where laboratory trials were directed by military engineers and sea trials by naval officers.

The struggle of both groups to control technical development made compromises difficult to reach. In Great Britain the scientists at the ARL also deployed all their knowledge to adapt their experimental assessments to new devices. Moreover, they tried to invent testing devices that would simulate ships' movements. Their testing method became very accurate indeed and consequently the DNO agreed more and more with their scientific judgement and asked for more scientific methods. However the gyrogroup never claimed that laboratory testing was superior to sea trials but step by step they took control of the sea trials which themselves became more and more scientific. No wonder that this trust then led to the building of a scientific device. Thus, it is plainly evident that the actors in charge of sea trials had strong control over technical development. Indeed, the decision strength of sea trials was very important and this event played a key role in the innovation process of both navies. The main reason is that however hard one would try in the laboratory, sea trials remained the best way to play war.

Negotiating Results

The final stage that shaped the judgement on the technical value of a device was the report in which results were presented. It was striking how trials whether at sea or in a laboratory were expected to report the steadiness of the effects of known constraints so as to make them possible to repeat. Indeed, particular attention was paid to trial conditions by each actor. A protocol was always written which stated ships' manoeuvre, specific testing criteria, and often precise hand movements involved in the trial. Any change with this protocol would interfere with the interpretation of the results. For instance, Brown simply contested the entire result of 1928 British sea trials because he claimed that the vibrations were higher than the average level defined in advance. The DNO agreed to cancel the results and to proceed to sea trials with suspension gears so as to avoid vibrations. In general, the first step after each trial consisted in identifying and eliminating all unusual behaviour and readings. During sea trials many of the results were not taken into consideration because either weather conditions were not "good enough", or a hitch occurred during the trial. Thus it is always possible to deny the results of a trial when they were not found to be adequate in terms of the proof required by the standards. Certainly the most striking example was the results obtained with the Brown gyrocompass by both the French and the British navy: bad laboratory tests were balanced by the fact that the British company agreed to build plants in France and consequently the test result were completely neglected. On the contrary, the British navy gave no special advantage to Brown or to Sperry or to its own model.

Finally, the negotiation process seemed to be incessantly on the move, only eventually brought to conclusion by the outbreak of actual warfare. Consequently, not only did research work need to be adapted to a different time scale, but one also needed to adapt to a reality ("real" war) about which one could only guess. In other words, the starting point would always be the naval battle that should be translated into different methods of approximation. The Battle of Jutland was crucial in term of technical choices for fire control. More broadly, the First World War provided the testing ground for many devices of which the gyrocompass was only one example. But of course, in so many ways, the First World War was quite unlike the Second.

These various criteria and observations demonstrate that the military environment is particularly well adapted to study the process of knowledge transfer. Indeed, each actor relies on specific criteria to define a context and tends to recreate artificially, with different experimental devices, the perceived or desired environment. The problem was precisely to make these different contexts link up. Sea trials appeared as an important stage during which different knowledges were confronted, and they also gave information about who would win the negotiations to ascertain the technical value of a device and thus framed that device's technical development. Indeed, the final judgement of the value of a device and its development results more of the domination of one set of knowledge over the others, rather than a consensus between each actor involved, or even the domination of any particular actor.

Conclusion

Differences between the French and the British gyrocompass can be correlated to the trust in different kind of practical knowledge. More broadly they reflect the different choices in terms of research and development made by both countries and in particular the military use of scientists and military collaboration with industry. Scientific knowledge as it was promoted therefore had different forms of expression in both countries. The comparative study allowed me to underline the particular status of the Polytechniciens in France since there was no such group of individuals in Great Britain that could play the same role. This French exception linked to the existence of the *Grande Ecole* gave civil scientists working for the French navy a specific and group identity and social networks radically different from their counterparts in Britain. For instance, the Polytechniciens claimed superiority of their scientific methods and their practices compared to those of the other academics. In Great Britain, the debate lay more on what should have been the place of the scientists within the innovation process than on their scientific legitimacy. The understanding of science was thus certainly different in both navies. In France, civil scientists were used by the military as an interface with the academics; the collaboration with the industry was left to the Polytechniciens. British civil scientists, on the other hand, became experts in charge of the negotiations with the industry. Finally, the whole policy defined by the Admiralty and supported by the British government helped the rise of scientific officers, advancing more than two hundred into the navy on the eve of the Second World War.

However, the most interesting point to underline is that scientific knowledge appeared to be useful only associated with other forms of practical knowledge. Indeed, the complexity of the actors involved in the innovation process remains striking. One can find scientists with an academic background working within the navy's research establishments, others used as experts working in their laboratory at universities, technical officers and military engineers working in various technical establishments, and military people at the headquarters evaluating the needs of the department. Last but not least, the collaboration with the industry also added various sorts of actors with different backgrounds and alternate forms of knowledge. I showed how the development of a technical object was a result of the interaction of these various people and know-how. Consequently, if one wants to understand the production of knowledge within a military environment it is necessary to look outside laboratories and consider more than academic knowledge. Only then would it be possible to identify the continuity of the links between the scientific and the military community and stop understanding academics and laboratories as the only technically innovative providers of instruments and materiel for the military.

CHAPTER TEN

WASHOUTS: ELECTROENCEPHALOGRAPHY, EPILEPSY AND EMOTIONS IN THE SELECTION OF AMERICAN AVIATORS DURING THE SECOND WORLD WAR

Kenton Kroker

Introduction

The idea that scientific expertise and testing of recruits, rather than socioeconomic status or the caprice of officers, should govern the selection and placement of military personnel was an innovation of twentieth-century warfare. Determining the fitness of recruits for combat had initially been a minor part of nineteenth-century military medical practice. But by the end of the First World War, a battery of psychological, psychiatric, and medical tests had evolved in the effort to make more efficient use of human capital in warfare. By mid-century, the need for scientifically-based military selection had become so engrained in the practice of war that it had even become part of the American lexicon of Cold War democracy. "We confront an adversary," declared one Department of Defence researcher at a 1951 symposium, "who, by reason of sheer control of a vast segment of humanity, can afford to make some errors in personnel utilization. Up to a point he can be wasteful; we cannot. He may also, with apparent impunity, commit all manner of offences against human dignity. We may not and yet remain a free society."[1]

The problem of matching men to tasks became more acute as the technological sophistication of warfare increased. Intelligence testing of American recruits during the First World War was applied by psychologists not only to eliminate unsatisfactory candidates, but also

[1] Frank A. Geldard, "Human Resources Aspects of Selection and Classification of Military Manpower," in Leonard Carmichael (ed.), *The Selection of Military Manpower* (Washington, D.C., 1951), pp. 16–28, at p. 16.

to place successful ones in positions suited to their abilities. The growth of air power in the early twentieth century is one of the most striking examples of how technological developments influenced military selection. The dramatic and sudden rise in American air power late in 1939 intensified the need for the rapid creation of a large group of highly skilled pilots working within a physiological and psychological environment that was both challenging and novel. Early and reliable identification of potential failures, or "washouts," was essential. The role of instruments, ranging in material complexity from pencil-and-paper intelligence tests to tilt tables and flight simulators, played an indispensable role in the selection process. Yet military historians' examination of wartime technology has traditionally overlooked the use of such instruments, and instead analyzed relationships between strategy and the transformation of war machines.[2] Historians of psychology, on the other hand, treat the rise of military testing as an integral part of psychology's intellectual and professional development. Yet they have also tended to ignore aviation psychology.[3] The historical field is thus inherited by researchers currently engaged in either human factors engineering or aerospace medicine.[4] This literature provides some valuable information and useful chronologies, but, like most disciplinary history, tends to suffer from a deficit of historical imagination. Narratives are organized around an irresistible march of progress culminating in the current

[2] Personnel testing, for example, goes unmentioned in Williamson Murray and Allan R. Millett (eds.), *Military Innovation in the Interwar Period* (Cambridge, 1996). It fares no better in some recent histories of military aviation, including Charles J. Gross, *American Military Aviation: the indispensable arm* (College Station, Tex., 2002) and Tami Davis Biddle, *Rhetoric and Reality in Air Warfare: the evolution of British and American ideas about strategic bombing, 1914–1945* (Princeton, 2002).

[3] James H. Capshew, *Psychologists on the March: science, practice, and professional identity in America, 1929–1969* (Cambridge, 1999), provides a brief discussion of aviation psychology (pp. 44–8), but this is an exception. For general histories of psychology that pay considerable attention to intelligence testing the in military, see: Kurt Danziger, *Constructing the Subject: historical origins of psychological research* (Cambridge, 1990); Ellen Herman, *The Romance of American Psychology* (Los Angeles, 1995); Graham Richards, *Putting Psychology in its Place: an introduction from a critical historical perspective* (New York, 1996); Michael M. Sokal (ed.), *Psychological Testing and American Society, 1890–1930* (New Brunswick, N.J., 1987).

[4] See Jefferson M. Koonce, "A Brief History of Aviation Psychology," *Human Factors* 26 (1984): 499–508; David Meister, *The History of Human Factors and Ergonomics* (Mahwah, N.J., 1999), especially pp. 146–170; Douglas H. Robinson, *The Dangerous Sky: a history of aviation medicine* (Oxford, 1973); and T.M. Gibson and M.H. Harrison, *Into Thin Air: a history of aviation medicine in the RAF* (London, 1984).

state of knowledge in the field. Any social, intellectual, or technological context to these developments goes largely unnoticed; failures remain invisible.

What follows is a preliminary attempt to examine the early history of aviation medicine and human factors engineering by focussing on the use of a single instrument—the electroencephalograph (EEG)—in the screening of American military aviators. Since the mid-1930s, EEG had generated considerable excitement among a wide range of medical and psychological researchers, all of whom anticipated that a record of the brain's electrical activity could help resolve long-standing questions about psychopathology, personality, and perhaps even thought itself. Some of these questions were shared by aviation psychologists, who had long been interested in developing efficient ways of identifying those individuals particularly susceptible to emotional breakdown under stress. Following the pattern of earlier electrographic instruments such as the electrocardiograph (EKG), the EEG's early successes were primarily in medical diagnosis, not psychological investigation. Thus it was a group of prominent neurophysiologists, not psychologists, who first argued in 1940 that the EEG could identify failures or 'washouts' through the early diagnosis of 'cerebral dysrhythmia', an epileptoid tendency then linked to emotional instability. Working under the auspices of the National Research Council's Committee on Selection and Training of Aircraft Pilots, these researchers claimed that the EEG could help eliminate those cadets lacking in the physiological and psychological poise required to fly in combat before they had even set foot in an aircraft.

Despite this initial enthusiasm, the EEG had itself become a washout by the end of the war—at least as far as its use in aviator screening was concerned.[5] In June of 1945, the Chairman of the NRC Committee pronounced the endeavour "not sufficiently promising" to pursue for either military or civilian applications.[6] This taciturn

[5] The term "washout" was not, to my knowledge, ever applied to the EEG. It is nonetheless an appropriate description, given that the term was already used within the military to describe both men and machines as "disappointing failures" or "wrecks" by the end of the First World War (see citations under *s.b.* "washout" in the *Oxford English Dictionary*).

[6] Morris S. Viteles, "The Aircraft Pilot: 5 Years of Research. A Summary of Outcomes," *Psychological Bulletin* 42 (1945): 489–526, at p. 498.

assessment, however, did not fully communicate the complexity of the EEG's failure. As an instrument that could purportedly depict and record the physiological basis of emotional instability, the EEG's promise as a screening device was grounded in its ability to reveal secrets of the brain that would otherwise have remained hidden. But by war's end, intellectual and organizational changes in wartime psychiatry began to promote a new analysis of emotion as a function of group behaviour, rather than the product of individual physiology.[7] The group, not the individual, became the target of expert intervention, and the EEG proved unable to adapt. Moreover, the very nature of psychologists' efforts to fit military men to their machines had also begun to shift, as they started to emphasize the 'human factor' in weapons control engineering. It would be more effective, they argued, to redesign cockpits and fire control systems to better integrate human cognition and response, than to develop screening to identify the ideal soldiers and eliminate the rest.[8] In both instances, a new analytic approach was emerging, one which supplanted the traditional boundaries between an individual soldier and his weapon or unit with an analysis of the soldier as part of a cybernetic or social system. This new interpretive frame left little room for the EEG as a screening device that depicted an individual human brain as the static, physiological expression of temperament and character that was isolated from the social or technological context of human performance.

Novel instruments of war & peace

The assessment of recruits by psychological tests and an increase in the military use of aircraft arose almost simultaneously during the First World War. Though the performance and utility of each was

[7] On the changing significance of emotions in psychiatry and psychology during the Second World War, see Ben Shephard, *A War of Nerves: soldiers and psychiatrists in the twentieth century* (Cambridge, Mass., 2001), and Joanna Burke, *An Intimate History of Killing: face-to-face killing in twentieth century warfare* (New York, 1999), especially pp. 73–90.

[8] Meister, *History* (note 4). A similar shift is described in David A. Mindell, *Between Human and Machine: feedback, control, and computing before cybernetics* (Baltimore, 2002), pp. 276–86. I am indebted to Edward Jones-Imhotep for this reference, as well as for a succinct account of his own work on the history of risk and human factors engineering, which I found helpful in my revision of this paper.

widely criticized, psychological testing and air power ultimately carried the day, both in the military sphere and in the popular imagination. During the 1920s, psychological testing was rapidly adopted as an integral part of American industrial and educational reform.[9] Calls for reform within the American military were equally strong following Billy Mitchell's dramatic demonstrations in 1921 that aircraft alone could sink any of the great ships upon which American defense was based.[10] In many ways, air power and psychological testing epitomized the application of efficient, economical, and technologically sophisticated solutions to problems of American peace and prosperity during the progressive era.[11]

These two endeavours were united under the rubric of emotion—the traditional crossroads of mind and body. The psychological tests that had been applied to nearly two million men by the end of the First World War had limited relevance to the selection of military aviators. These pencil-and-paper tests ranked recruits according to their "intelligence," a mental property that, many psychologists argued, was indicative of a candidate's capacity for military service. The tests purportedly matched soldiers to tasks by providing a measure of any related vocational abilities they might have. But so few candidates had experience with aviation that testable selection criteria remained at the generalized level of "high mental ability," even as late as the early 1940s.[12] In the 1920s, selection officers placed far greater emphasis on judgements of character. A good aviator was assumed to be intelligent, but he also needed courage, stamina, and a "good set of hands," all of which were notably similar to those traits demanded of a good cavalryman.[13] But despite the seeming distance of such characteristics from intelligence, psychologists nonetheless began to bring these qualities within the domain of their own expertise during the First World War.

---

[9] On the expansion of mental testing, see Danziger, *Constructing the Subject* (note 3), pp. 101–117.

[10] On Mitchell, see Gross, *American Military Aviation* (note 2), pp. 50ff.

[11] The inevitability of the Air Corps' emphasis on technological solutions to strategic problems is effectively examined in Timothy Moy, *War Machines: transforming technologies in the U.S. military, 1920–1940* (College Station, Tex., 2001). For a classic analysis of progressivism and psychology, see John C. Burnham, *Paths into American Culture: psychology, medicine, and morals* (Philadelphia, 1988).

[12] Capshew, *Psychologists on the March* (note 3), p. 101.

[13] Robinson, *Dangerous Sky* (note 4) observes that many British officers felt that familiarity with horses was an "outstanding recommendation" for aviation cadets (pp. 82–3).

Vivian A.C. Henmon, an educational psychologist at the University of Wisconsin at Madison, conducted some of the earliest psychological tests of aviators in the U.S. at Rockwell Field in San Diego during the summer of 1917.[14] Henmon glorified the aviator in romantic terms as a "twentieth century cavalry officer mounted on Pegasus," who was possessed of "good education and high character," even as he helped create a series of quantifiable tests that would render such traditional categories meaningless.[15] Henmon's investigations found that tests of emotional stability, mental alertness, and perception of tilt showed the strongest correlation to performance in flight training. But, as one commentator has pointed out, many psychologists at this time understood "emotional stability" as a set of measurable physiological reactions to any number of excitatory stimuli, not as an aspect of personality.[16] We might today think that succumbing to road rage or crying while watching *Bambi* signifies a lack of emotional stability, but these qualitative and complex responses are very different from the short, sharp shocks elicited by psychologists in their laboratories during the 1920s. At that time, the emphasis was inevitably on the sudden and the threatening, a cue which psychologists had borrowed from physiologists, who were then attempting to understand how emotion affected the results gleaned from their experimental animals.[17] Firing a pistol, sounding an automobile horn, or shining a glaring light in the eyes were some of the most popular means to evoke a reaction, which were then measured and recorded as hand stability, heart rate, respiratory rate, or abnormal changes in reaction times (such as that required to move a joystick in response to a light).[18] Not all investigators agreed with Henmon's

---

[14] Koonce, "Brief History" (note 4), p. 500.

[15] V.A.C. Henmon, "Air Service Tests of Aptitude for Flying," *Journal of Applied Psychology* 3 (1919): 103–109, p. 103. Henmon reported that the tests were adopted by the personnel section of the Air Service.

[16] Koonce, "Brief History", p. 500.

[17] Otniel E. Dror, "The Affect of Experiment: The turn to emotions in Anglo-American physiology, 1900–1940," *Isis* 90 (1999): 205–237. See also Dror, "The Scientific Image of Emotion: Experience and Technologies of Inscription," *Configurations* 7 (1999): 355–401. For a general overview of the quantification of emotion, see ibid., "Counting the Affects: Discoursing in Numbers," *Social Research* 68 (2001): 357–78.

[18] These examples are all taken from F.C. Dockeray and S. Isaacs, "Psychological Research in Aviation in Italy, France, England, and the American Expeditionary Forces," *Journal of Comparative Psychology* 1 (1921): 115–148. See also Robinson, *Dangerous Sky* (note 4), p. 85.

conclusions regarding a positive relationship between emotional stability and success in aviation. Some Italian researchers, for example, argued that some of the most successful pursuit pilots (in contrast to bomber pilots) showed a very low resistance to emotional stimuli.[19] But in any event, emotions were rapidly becoming part of the psychological inroad to aviator selection. Along with air power and intelligence testing, measures of emotion were emerging as a novel military technology.

Health takes flight: the rise of aviation medicine

Military interest in intelligence testing declined precipitously following demobilization from WWI. Dedication to air power within the American military also levelled off. The most important roles for the biplanes and dirigibles used in the conflict had turned out to be tactical, not strategic.[20] The primary task of the aviator had been reconnaissance and artillery sighting, not bombing or pursuit, and hence the army generals and navy admirals were dubious of aviation's overall utility. It was left to air power advocates like Billy Mitchell to convince the military hierarchy throughout the 1920s and '30s that flight had fundamentally transformed the nature of armed conflict. The novelty of the physiological problems generated by aviation, however, were quickly grasped. The ability to keep one's bearings in thick cloud cover, land in any situation, fly at high altitudes for long periods of time, or quickly recover from a steep bank or dive often proved difficult for even the healthiest officers. The vertigo, airsickness, fatigue, fainting, outbursts of anger or "staleness" that resulted could effectively disable the most stable pilots. Some solutions involved reducing or relocating the work of pilots with mechanical devices adapted from existing naval control technologies, such as gyroscopic artificial horizons or automatic pilots.[21] But the novelty of the pilot's embodied experience also seemed to demand an entirely new conception of health—one that was intimately linked not to the absence of disease, but to the maximization of the performance of this hybrid of man and machine.

---

[19] Dockeray & Isaacs "Psychological Research", p. 126.
[20] Gross, *American Military Aviation* (note 2), pp. 48–94.
[21] Mindell, *Between Human and Machine* (note 8), pp. 76–82.

American aviation medicine formally came into existence in 1921, when the War Department recognized the Medical Research Laboratory and School for Flight Surgeons at Mitchel Field, Long Island, as a special service school.[22] This official action merely acknowledged developments that had taken place during the preceding decade, first across Europe and later in the U.S. During the 1920s, pilot error emerged as a key problem in the maintenance of air power. According to one commentator, the physical condition of the pilot accounted for seventy percent of flying accidents during peacetime, with another twenty-five percent due to carelessness, overconfidence, recklessness, or ignorance of the principles of aeronautics.[23] Structural defects, on the other hand, accounted for a mere five percent of accidents. In terms of selecting suitable aviators, some problems, such as deficits in vision, coordination, balance, or reaction time, could be easily solved through existing medical or psychometric examinations. Others required considerable innovation. Flying at high altitudes emerged as one of the most pressing concerns, and the ability to identify those individuals who could tolerate a reduced partial pressure of oxygen became a priority. Flight surgeons and physiologists developed numerous means of testing responses to such conditions in pilots; the development of pressurized suits and cabins, in contrast, lagged far behind.[24]

Hypoxia caused by high altitude flight was frequently blamed for adversely affecting the cognition and judgement of even the best pilots. Poor landings, failing to recognize or engage the enemy, and an overall mental and physical deterioration were oft-cited conse-

[22] By 1931, the school had relocated to Randolph Field, Texas. See Harry G. Armstrong, *Principles and Practice of Aviation Medicine*, 2d ed. (Baltimore, 1943), pp. 1–19 and 30–62. An Aviation Medical Section was established in the Medical Department of the navy in 1921, but naval flight surgeons continued to train at the Army School of Aviation Medicine until 1940, when a similar institution was set up at Naval Air Station (NAS) Pensacola, Florida. See Frederick Ceres, "Aviation Medicine in the United States Navy," *War Medicine* 1 (1941): 43–49.

[23] J.F. Neuberger, "Aviation Medicine in the United States Navy. Physical Examination of Aviators," *United States Naval Medical Bulletin* 16 (1922): 983–1014.

[24] For a description of the various techniques of selection, see Armstrong, *Principles and Practice*, pp. 39–44. Armstrong, who later became Surgeon General of the USAF, led the American research team that developed the first pressurized cabin. See Seymour L. Chapin, "An Active Interface Between Medical Science and Aeronautical Technology: The Physiological Investigations for the XC-35," *History & Philosophy of the Life Sciences* 13 (1991): 235–248.

quences of flying at altitudes around or above 15,000 feet.[25] Physiological testing on the ground tended to reveal only the pilot's ability to adjust to reduced oxygen levels in the short-term. The long-term effects of flight, particularly upon emotional stability and fatigue resistance, were largely unknown. As a group, aviation candidates were generally perceived as highly desirable recruits in terms of physical fitness, intelligence and character. Yet they experienced a remarkably high level of nervous troubles during their training. Authorities in England reported that about half of all candidates developed a neurosis before they graduated.[26] The American situation was equally problematic. One commentator reported in 1925 that around eighty percent of all groundings of graduate pilots had been for "nervous instability."[27] The vagaries of fatigue and emotion were clearly eluding the tests for appropriate mental and physical "types" that had been established by psychologists and physiologists during the war. To the extent that they were amenable to testing, strength, courage, and intelligence were not enough to ensure that a candidate would make a good aviator. The extraordinary new environment of flight demanded that an equally new kind of soldier be discovered.

The notion that new and stressful environments could create neuroses in soldiers who were previously asymptomatic was a well-established psychiatric principle by the end of the First World War, so it was entirely reasonable for aviation medicine to turn to neuropsychiatry in the attempt to understand the effects of flight on pilots.[28] The extreme conditions of trench warfare, for example, had created "shell shock," originally described as a form of concussion from a nearby shell explosion. By the end of the war, psychiatrists had transformed this physiological account into a distinctly psychological one, and shell shock came to mean an emotional reaction to a fearful event or series of events. The condition might have physical manifestations—including fatigue, insomnia, motor or facial tics, and paralysis—but its origins were distinctly psychological. Neither a predisposition to mental illness, nor exposure to exploding shells were necessary preconditions for the diagnosis. Similar disorders in

[25] Gibson & Harrison, *Into Thin Air* (note 4), p. 35.
[26] Armstrong, *Principles and Practice*, p. 50.
[27] Louis H. Bauer, *Aviation Medicine* (Baltimore, 1926), p. 56.
[28] Shephard, *War of Nerves* (note 7). See also Allan Young, *The Harmony of Illusions: inventing post-traumatic stress disorder* (Princeton, 1995).

labourers, often in support of the patient's claims for injury compensation, began to appear in neurological clinics around the same time.[29] Students at the School of Aviation Medicine likely encountered such cases during their 14 weeks of instruction, which included lectures in neuropsychiatry, neurology and psychology, as well as clinical rounds.[30]

Flight surgeons were expected to be able to immediately recognize such neuroses in airmen. But they also hoped to be able to predict the likelihood of such events ever taking place by subjecting aviation candidates to simulated conditions of flight and measuring changes in their subsequent physiological reactions. Instability in those reactions that had been linked to pilot error represented a potential obstacle to the applicant's candidacy. Investigators in aviation medicine tended to use the language of emotions, not the language of risk, to describe such obstacles. Medical analysis based on "risk factors" was still well in the future, and physicians tended in this period to speak of patients' predisposition to disease in terms of character, of which emotional disposition was an important aspect.[31]

Unlike psychiatry, which relied on patient interviews to assess character, aviation medicine tended to depict its subjects (they were hardly 'patients' in the sense that most of them suffered from no obvious disease) in quantitative terms derived from physiological tests. Laboratory physiologists had already begun to depict their experimental animals in emotional terms through their graphical representations of vital functions,[32] so it was not particularly difficult for investigators to apply a similar approach to aviators, who were subjected to equally artificial, if less extreme, environments. Following an earlier

---

[29] See, for example, Charles Loomis Dana, "The somatic causes of the psychoneuroses," *Journal of the American Medical Association* 74 (1920): 479–483.

[30] A description of the School's curriculum can be found in "The School of Aviation Medicine," *Report of the Surgeon General of the Army* (Washington, D.C., 1924), pp. 264–269.

[31] The "risk factor approach," which calculates predisposition to chronic diseases according to discrete social, physiological and behavioral factors, has been traditionally linked to the "Framingham study," a long-term epidemiological study of coronary heart disease among 5,000 residents of Framingham, Massachusetts. The study began in 1948, but the term "risk factor" was not used by the Framingham investigators until 1961. On the Framingham study and its precedents, see Robert A. Aronowitz, *Making Sense of Illness: science, society, and disease* (Cambridge, 1998), pp. 111–144.

[32] Dror, "The Affect of Experiment" (note 17).

generation of physiologists, researchers in aviation medicine linked emotional instability to the tendency to quickly succumb to fatigue, which was in turn perceived as a potential sign of neuroticism.[33] What was needed was a reliable inscription device to evaluate the potential aviators.

Instruments & emotions

Graphical instruments played a crucial role in linking emotions and fatigue. Working in Turin, the Italian physiologist Angelo Mosso had made this link visible back in the 1890s. Mosso was best known for inventing the ergograph—a portable device that measured fatigue by providing a graphic record of a how a subject's ability to lift a small weight attached by a string to the middle finger decreased over time. Mosso acknowledged multiple causes of fatigue, and was as likely to use his ergograph to measure a bricklayer's performance before and after a day's labour as he was to demonstrate the profound physical effects of sitting emotionally-charged university exams.[34]

Such traffic between fatigue and emotion was becoming codified in aviation medicine by the end of the 1920s. The Schneider index, for example, was originally developed in 1920 as a cardiovascular rating of fatigue, and it was soon adopted by aviation medicine as a means of determining the fitness of aviation cadets.[35] But this rating was implicitly linked to the emotions and subjective experience through the concept of "neurocirculatory asthenia," a clinical concept that allowed physicians to diagnose functional heart disease in patients lacking a demonstrable pathological basis for their chest pains.[36] By the late 1930s, some investigators considered neurocirculatory asthenia (also called "effort syndrome") as a sort of nervous

[33] Gibson & Harrison, *Into Thin Air* (note 4), p. 238. On the history of fatigue, see Anson Rabinbach, *The Human Motor: energy, fatigue, and the origins of modernity* (Berkeley, Cal., 1990).

[34] For these and other examples, see Angelo Mosso, *Fatigue* (London, 1906).

[35] E.C. Schneider, "A cardiovascular rating as a measure of physical fatigue and efficiency," *Journal of the American Medical Association* 74 (1920): 1507–1510; see also Armstrong, *Principles and Practice* (note 22), pp. 44–6.

[36] On the history of neurocirculatory asthenia, see Joel Howell, "'Soldier's Heart: The Redefinition of Heart Disease and Specialty Formation in Early Twentieth-century Great Britain," *Medical History*, suppl. 5 (1985): 34–52.

predisposition to fatigue signified by exaggerated emotional responses to stimuli that raised pulse rate, and attempts were soon made to transform the Schneider index into an "emotional index."[37]

Instruments that could measure and record sudden physiological changes and display their results in terms of the ability to do work were integral to this exchange between fatigue and emotion. Instruments became an indispensable means of bridging the gap between the ideals of the physiological laboratory—a neutral place in which to examine the normal state of the living body—and the immediacy and novelty of the cockpit environment. As the invention and development of flight simulators, panel instrumentation and automatic pilots made the cockpit more regulated and predictable, graphical instruments that monitored changes in blood pressure, pulse rate, respiration and other vital functions played a similar role in the process of selecting aviators. Flight simulators that attempted to incorporate an aircraft's responses to the trainee's actions began to appear in the early 1920s, but they were little more than curios until the development of panel instrumentation fundamentally changed the phenomenology of flight itself.[38] As "flying blind," or instrument flying, became more common in the late 1920s, trainers equipped with instruments that simulated changes in attitude, altitude, and the airspeed of the aircraft emerged as a useful means of testing a pilot's ability. Automatic pilots were also emerging during this period, and the performance of aviators, like that of gunners and naval pilots, increasingly began to rely upon their manipulation of instrumental representations of the world, rather than direct control of their machines themselves.[39] The embodied experience of flight, once encapsulated by such terms as "flying by the seat of one's pants" and analogies between pilots and cavalrymen, had found its surrogate in instrumental mediation. Just as the hybrid, psychophysiological nature of the emotions served as an intermediary between mind and body, so too did instruments negotiate a sort of middle ground between machines and their human operators.

[37] See Medical Division, Training Group, *Report of the Surgeon-General of the Army* (Washington, D.C., 1939), pp. 259–63; see also Ross A. McFarland, *Human Factors in Air Transportation* (New York, 1953), pp. 101–3.

[38] On flight simulation, see M.A. Fischetti and C. Truxal, "Simulating 'The Right Stuff'," *IEEE Spectrum* 22 (1985): 38–47; and Royal Aeronautical Society, *50 Years of Flight Simulation* (London, 1979).

[39] On the emergence of automatic pilots in a variety of contexts, see Mindell, *Between Human and Machine* (note 8), pp. 69–104.

Enter electroencephalography

This tendency to intervene by manipulating mechanically generated representations of things rather than the things themselves had also come to play an increasingly important role in medical practice. Diagnosing diseases through batteries of tests and formalized laboratory procedures became the hallmark of science-based, technologically-sophisticated hospital medicine during the 1920s.[40] Following the earlier and often stunning successes of diagnostic blood assays, x-rays, and electrocardiograms (EKGs), electroencephalography (EEG) found its first institutional beachhead in the hospital during the late 1930s. From there, it quickly moved into aviation medicine, bringing with it a new diagnostic category that purportedly linked emotional instability to the psychophysiological travails of epilepsy.

Electric currents had been detected from the brains of vivisected animals as early as the mid-1870s, but it was only with the rise of vacuum-tube amplification in the 1920s that it became possible to record these weak currents from the intact scalp.[41] The first recordings of spontaneous currents from a human scalp were published in 1929 by Hans Berger, director of the psychiatry clinic of the University of Jena hospital.[42] Paradoxically, Berger's work received considerable public exposure, but was generally ignored until several years later, when German, English and American investigators began to study the phenomenon.[43] The most prominent (but not the most persistent) of these was the Cambridge neurophysiologist and Nobel laureate Edgar Adrian, who first published on the EEG in 1934. In his U.S.

[40] The success of these diagnostic technologies was not intrinsic to the instruments themselves; rather, it depended heavily upon the adoption and adaptation of parallel technological systems, including the use of standardized forms and cost accounting. See Joel D. Howell, *Technology in the Hospital* (Baltimore, 1995).

[41] For a general history of the EEG, see David Millett, "Wiring the Brain: From the Excitable Cortex to the EEG, 1870–1940" (Ph.D., University of Chicago, 2001). On the EEG's diagnostic antecedents, see Cornelius Borck, "Electricity as a Medium of Psychic Life: Electrotechnological Adventures into Psychodiagnosis in Weimar Germany," *Science in Context* 14 (2001): 565–590. On the EEG as an agent in subjective transformation, see Rhodri Hayward, "The Tortoise and the Love-Machine: Grey Walter and the Politics of Electroencephalography," *Science in Context* 14 (2001): 615–641.

[42] The paper is translated and republished in Pierre Gloor, "Hans Berger on the Electroencephalogram of Man. The Fourteen Original Reports on the Human Electroencephalogram," *Electroencephalography and Clinical Neurophysiology*, Suppl. 28 (1969).

[43] Borck, "Electricity as a Medium".

tour that same year, Adrian repeatedly offered public demonstrations of the curious appearance and disappearance of the low-voltage, 10 Hz "brainwaves" ("alpha") in subjects as they relaxed or performed various mental tasks. Whether or not the EEG really was a "thought recorder" (as one journalist described it), one thing was abundantly clear: the EEG was no artefact, and its field of signification was wide open.[44]

Hallowell Davis (1896–1992), a physiologist at Harvard University, led one of the most important of these early EEG research groups in the United States. Davis's own career began studying blood chemistry at Harvard under Lawrence J. Henderson. While Henderson went on to establish a fatigue laboratory at Soldier's Field in Boston, Davis decided that little progress could be made by a young researcher on this topic. He opted instead to take his MD while at the same time studying nerve conductivity with Henderson's colleague at Harvard, the electrophysiologist Alexander Forbes.[45] Electroencephalography began to consume Davis's research almost as soon as it was brought to his attention by a graduate student late in 1933. The EEG provided a record of the brain's invisible electrical activity with a minimum of intervention, and so it seemed to Davis and some of his colleagues that this new device suggested a way to apply physiological experimental systems to psychiatric and psychological investigations.

The time was ripe for such a transformation. The preceding generation of neurophysiologists, including Charles Sherrington, Edgar Adrian, and Forbes himself, had built their research upon meticulous micro-studies of the mechanics of the nerve impulse and the reflex arc in animal preparations. Thought, mind and personality were taken to be beyond the pale of this sort of experimental practice. By the early 1930s, however, the landscape of neurophysiological research had begun to shift. Experimentation began to move into the clinic, but clinical evidence of recovery from neurological illnesses fit poorly with animal-based models of functional destruc-

---

[44] William L. Laurence, "Electricity in the Brain Records Picture of Action of Thought," *New York Times*, 14 April 1935, p. 1.

[45] Hallowell Davis, "Crossroads on the Pathways to Discovery," in Frederic G. Worden & Judith P. Swazey (eds.), *The Neurosciences: paths of discovery* (Cambridge, Mass., 1975), pp. 311–21.

tion by cortical ablation or nerve sectioning. The brain, insofar as it was accessible to experimental knowledge, began to appear as less a sum of its individual reflexes and more a complex system with a logic of its own. The EEG afforded the opportunity to study the intact human brain in a variety of situations, even though very little was actually understood about what, exactly, was being recorded. In the words of the late historian of psychiatry, Jack Pressman, the EEG was emerging as "a medical technology in search of a function."[46]

After some initial reservations, Davis quickly moved to the forefront of EEG research. Like several other investigators, Davis was convinced that the EEG was not, like the EKG, a mere summation of an organ's electrical activity. In a 1934 letter to Forbes, Davis went so far as to ascribe distinctly human qualities to the rhythms he was recording, noting that the "big waves" had "an individuality" and even "a personality" that went against Adrian's claim that the alpha rhythm merely signified the ability of millions of neurons to fire in synchrony, and that larger waves were little more than desynchronized noise.[47]

Tuxedo Park

Davis's approach to the EEG was hardly unique, but his interests in its potential value for the study of psychodynamics did move him closer to the margins of experimental neurophysiology. This move was as physical as it was figurative: much of his EEG research took place not at Harvard, but at a private laboratory in Tuxedo Park, a wealthy enclave just north of New York City. The laboratory was owned by Alfred Lee Loomis, a wealthy financier and amateur scientist best-known for his contributions to another World War II technological development, microwave radar.[48] In the mid-1930s, however,

[46] The phrase comes from the title of an unpublished paper. See Jack D. Pressman, *Last Resort: psychosurgery and the limits of medicine* (Cambridge, 1998), pp. 456–7, n. 10.

[47] Letter, Hallowell Davis to Alexander Forbes, 9 June 1934, box 5, folder 245, Alexander Forbes Papers (AFP), Countway Library of Medicine, Harvard Medical School. For some equally expansive interpretations of the EEG, see Borck, "Electricity as a Medium" (note 41) and Hayward, "Tortoise and Love Machine" (note 41).

[48] Loomis helped organize the Radiation Laboratory at the Massachusetts Institute of Technology, as well as directing the radar research conducted through the National Defence Research Council (NDRC) and Office of Strategic Research and Development

radar was only one of many research projects at Tuxedo Park, all of which revolved around the creation and use of large and expensive precision instruments. Loomis, an adept businessman well aware of the technological transformation of the American hospital system, had recently developed an eight-foot long kymographic drum—which he claimed to be the largest in the world—to record multiple physiological functions over extended periods of time. Shortly after installing the kymograph in 1934, Loomis and a colleague hooked it up to an electrocardiograph and began tinkering with electrode locations in the hope of improving the signal. Loomis noted that when he placed an electrode on a subject's head, the baseline oscillated. He approached his old friend Forbes for advice on the phenomenon, and Forbes immediately referred him to Davis, who had started studying the EEG only a few months earlier.

By the summer of 1935, Loomis and Davis were working together at Tuxedo Park, where they were joined by Hallowell's wife and colleague, Pauline, and a number of other physiologists with an interest in EEG. Over the next several years, Tuxedo Park became a veritable hive of EEG research: the first international conference on EEG was held here in December 1935. For their part, the Davises managed to gain funding from the Josiah Macy, Jr. Foundation to investigate the use of EEG as a tool of psychological classification and psychiatric diagnosis. This particular direction of research seems to have been built into the couples' relationship. Pauline Davis had trained as a psychiatrist, and published on diagnostic issues in the field; Hallowell had begun to dabble in psychoanalysis, teaming up with a Chicago analyst to use EEG TO study emotional changes in patients during psychotherapy.[49] The Davises attempted to demonstrate that an individual's brain wave patterns were stable over long periods of time, and could therefore be usefully correlated with an

(OSRD). See Luis W. Alvarez, "Alfred Lee Loomis: November 4, 1887–August 11, 1975," *Biographical Memoirs of the National Academy of Sciences* 51 (1980): 309–341; and Hallowell Davis, "Alfred Lee Loomis: American discoverer of the EEG", box 21, folder 60, item 2 in Hallowell Davis Papers (HDP-SL), Washington University at St. Louis.

[49] Davis had also presented some of his ideas about the interface between psychoanalysis and neurophysiology to a rather hostile audience of neurological clinicians and researchers in December of 1937. See Hallowell Davis to Walter Cannon, 21 Dec 1937, box 13, folder 158, Walter B. Cannon Archives (WBCA), Countway Library of Medicine, Harvard Medical Schools.

individual's psychological profile and used as a diagnostic tool.[50] But there was no standardized account of how to decipher EEG patterns at the time, so it proved somewhat chimerical to associate such patterns to psychological types or to most psychiatric diseases. In spite of such efforts, the EEG generally resisted such calibration to existing categories of mental health and disease.

Rhythmic metaphysic—the hidden natures of epilepsy

There was, however, one important exception to the EEG's somewhat disappointing track record: epilepsy. By 1941, a raft of archaic problems, from dreams and delinquency to homosexuality and stuttering, were now afloat on a sea of brain waves.[51] The most prominent of these, however, was epilepsy, a disorder particularly amenable to a temporal phenomenology measurable by a technology like the EEG. Late in 1934, Fred and Erna Gibbs, another husband and wife team, joined the Davises at Harvard. The Gibbses had studied epilepsy under W.L. Lennox and Stanley Cobb at Massachusetts General Hospital in Boston, and, by the end of the year, they and the Davises had successfully recorded the classic three-per-second "spike and wave" pattern that characterized an attack of *petit mal* epilepsy.[52] More importantly, they were able to detect these waveforms in epileptics in a normal state, which suggested that the disease could now be diagnosed without exclusive reliance on either patient history or clinical observation of a seizure.

---

[50] This was effectively identical to what Mosso, *Fatigue* (note 34) had said about fatigue tracings. See Hallowell Davis & Pauline Davis, "The Electrical Activity of the Brain: its relation to physiological states and to states of impaired consciousness," *The Inter-relationship of Mind and Body. The Proceedings of the Association for Research in Nervous and Mental Disease, New York, December 27th and 28th, 1938, vol. XIX* (Baltimore, 1939), pp. 50–80; see also Lee Edward Travis & Abraham Gottlober, "Do Brain Waves Have Individuality?" *Science* 84 (1936): 532–533.

[51] I borrow this metaphor from the neurologist, Herbert H. Jasper. See the reprint of his 1948 article, "Charting the Sea of Brain Waves," *Journal of Clinical Neurophysiology* 14 (1997): 464–469. The EEG was also used to diagnose certain classes of brain tumours.

[52] F.A. Gibbs, H. Davis, & W.G. Lennox, "The electro-encephalogram in epilepsy and in conditions of impaired consciousness," *Archives of Neurology and Psychiatry* 34 (1935): 1133–1148.

The discovery of these "larval" forms of *petit mal* epilepsy helped determine the direction of EEG research. Unlike the spectacular bursts of uncontrolled motor activity that characterized *grand mal* epilepsy, *petit mal* seizures or "psychomotor" attacks were depicted as mental lapses that typically included inattentiveness and fits of moodiness or frustration. The most mundane of these episodes, argued EEG researchers, frequently passed unnoticed, and physicians and psychiatrists consequently ignored or misdiagnosed such "epileptoid" behaviour. The Davises called for a complete re-evaluation of psychiatric diagnosis based on the evidence of the EEG.[53] Lennox and the Gibbses went even further, claiming that this "cerebral dysrhythmia" affected some ten million Americans, which made for a prevalence rate nearly twenty times higher than our contemporary estimates for epilepsy.[54] One newspaper article noted that "momentary queer behavior" and "fleeting loss of memory" could be indicative of this new and sinister form of epilepsy, and even noted eugenic aspirations to use the EEG to advise "carriers" against "marrying their kind."[55] As one group of authors observed, if an unrecognized seizure or lapse caused an accountant to incorrectly record a number, the problem would eventually be discovered, and the employee dismissed. But if a military pilot suffered a similar episode, the consequences could be catastrophic.[56]

Given the supposedly widespread nature of epileptoid behaviour, the use of the EEG as tool for screening aviation candidates seemed a reasonable response to the risks involved in letting such secretly dysfunctional brains pilot valuable machines indispensable to the nation's defence. But Hallowell Davis had already made the link

[53] Davis & Davis, "Electrical Activity" (note 50).

[54] F.A. Gibbs, E.L. Gibbs, W.G. Lennox, "Various cerebral dysrhythmias of epilespy, and measures for their control," *Archives of Neurology and Psychiatry* 39 (1938): 298–314; see also W.G. Lennox, E.L. Gibbs, & F.A. Gibbs, "Inheritance of Cerebral Disrhythmia and Epilepsy," *Archives of Neurology and Psychiatry* 44 (1940): 1155–83. As epilepsy is not a reportable disease, prevalence is difficult to determine. Current epidemiological work suggests the crude prevalence rate is between 3.6 and 5.5 per 1,000 population. An estimate of 500,000 epileptics in the U.S. in 1940 (pop. 132 million) is equivalent to a rate of almost 3.8 per 1,000, while "cerebral dysrhythmia" estimates were at about 75 per 1,000. The news that epilepsy might be at the bottom of major psychiatric disease also made the popular press. See, for example, George Gray, "The attack on brainstorms," *Harper's* 183 (1941): 366–376.

[55] "Triumph in Epilepsy," *New York Times*, 3 June 1940, p. 12.

[56] Melvin Thorner, Frederic A. Gibbs & Erna L. Gibbs, "Relation between the electroencephalogram and flying ability," *War Medicine* 2 (1942): 255–262.

between EEG and aviation medicine in another area. In 1938 he initiated a series of experiments in which he used EEG as a means of graphically recording the effects of lowered oxygen levels on cognitive performance. His juxtaposition of brain waves to an experimental subject's introspective, handwritten accounts of oxygen deprivation indicate the degree to which Davis linked internal sensations or feelings with performance and EEG.[57] (figure 1) As the orderliness of the subject's handwriting began to deteriorate, so, too, did the EEG begin to lose its organized appearance. Hypoxia, one of the most important problems of aviation, seemed to have its own distinctive signature—one that could be linked to subjective feelings as much as to physiological effects.

The perceived need for such a device as an instrument of selection, however, did not exist until the sudden and massive expansion of air power inundated training fields with thousands of young men who, despite their health, talent and intelligence, were nonetheless utterly inexperienced when it came to flying—and yet all of whom wanted to be fighter aces. American rearmament began in earnest in January of 1938, after Japan renounced all naval arms limitation treaties in 1936 and invaded China the following year. The emphasis of U.S. efforts was on naval force that could counter Japanese expansion through a blockade of the home islands. Early rearmament thus included only a modest increase in the navy's air power, and provided nothing at all for the army Air Corps.[58] It was only after the Nazis threatened to deploy the *Luftwaffe* during the Munich crisis in September 1938 that the U.S. began to pay much attention to the depleted state of their air forces. By January 1939 Congress had authorized the Air Corps to spend $300 million to bring its inventory of aircraft closer to that of Germany, although the number of American personnel was still far behind that of England or the Axis powers. The Nazi invasion of Poland and their rapid conquest of France by the spring of 1940 conclusively demonstrated the strategic role air power now played in modern warfare. That summer, Congress approved spending $4 billion to fully mechanize and modernize the navy, a figure that included increasing the number of naval aircraft from just over 1,700 to 14,500. Any future increases

[57] Davis & Davis, "Electrical Activity" (note 50).
[58] Gross, *American Military Aviation* (note 2), pp. 80–94.

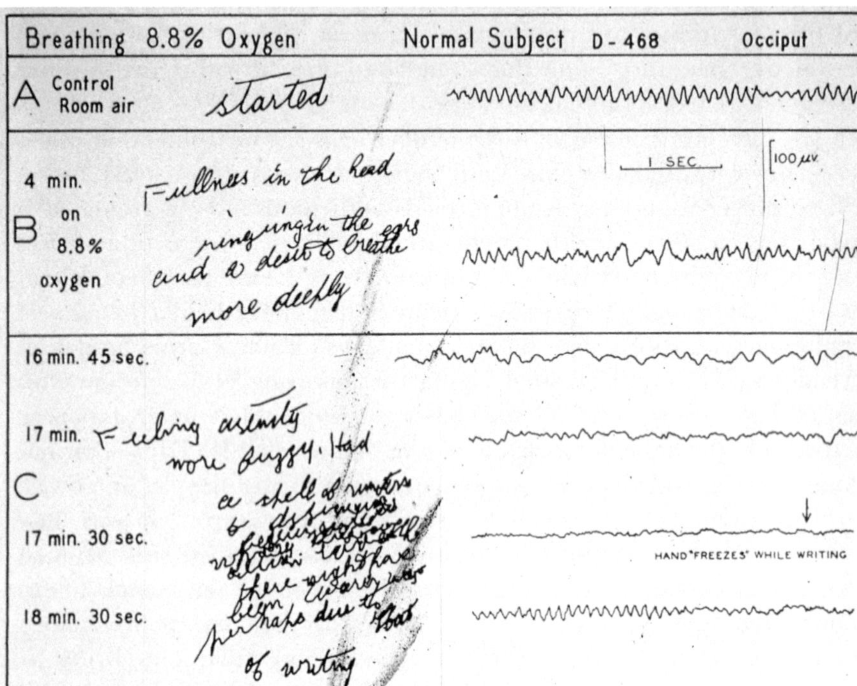

Figure 1. Hallowell Davis's juxtaposition of handwriting and brain waves during oxygen deprivation. From Davis & Davis [1939—see n. 50].

to the number of aircraft in the army Air Corps or in the navy required only the approval of the President.

It was in this context of rapid expansion of a specialized form of military manpower that the EEG came into its own as diagnostic instrument. The Canadian military had been confronted with this problem since 1939, and had already turned to the EEG to help rationalize their selection of aviators by the time the U.S. entered the war in 1941.[59] Both forces were confronted with similar problems that were framed by the spectre of epilepsy. In civilian life, a history of seizures would typically bring an epileptic to seek treatment; this history could also be confirmed by family and friends. In contrast, a military candidate's medical history, which was extracted to determine fitness for service rather than treatment, was often difficult, if not impossible to confirm.[60] Worse, the stigma of psychiatric illness encouraged candidates to dissimulate. This created the paradoxical situation in which a candidate's eagerness to fly, a well-recognized component of success in aviation, actually created an impediment to determining his fitness for the task by encouraging him to deceive.[61] In such a situation, the expert judgement of instructors and flight surgeons seemed to require the additional authority of the EEG, an instrument that could convincingly and objectively reveal the secret and personal nature of the brain's rhythmical functioning.

Numerous studies demonstrated that EEG enabled the identification of epileptics from those applicants who either admitted to suffering fits, fainting, dizziness, sleepwalking, temper outbursts or mood swings,

[59] A *New York Times* article noted in 1940 that "[s]o successful is this new way of testing the human cortex that Canada will probably use it to weed out pilots who may be subject to momentary lapses" ("Triumph of Epilepsy" [note 55]). Three months later, the *Toronto Daily Star* reported that "[b]rain wave readings have already been taken of 1,500 men and with this help have been weeded out men with potential epilepsy or other mental disturbances which would unfit them for the air service, although not as much for normal life" (23 September 1940, p. 6). Abstracts of such studies can be found in Defense Research Board. Department of National Defense, Canada. *Bibliography of Canadian Reports in Aviation Medicine*, 1939–1945 (Ottawa, 1962). EEG research in the Canadian military will be the subject of a future paper.

[60] Robert S. Schwab, "Application of Electroencephalography in the Navy in Wartime," *War Medicine* 4 (1943): 404–409.

[61] R. Barry Bigelow, "Psychiatric Problems in Military Aviation," *War Medicine* 2 (1942): 381–402.

or who were suspected of lying about such problems in their preliminary psychiatric exams, which were notoriously unreliable.[62] The high cost and frequency of neuropsychiatric discharges during the First World War had led psychiatrists to design a series of lengthy, sophisticated interviews with candidates to determine their predisposition to mental illness. But the pressures of getting a large number of inductees through the process soon reduced the examinations to a series of rote questions that took about a minute to complete, and were usually administered by physicians with little experience in psychiatry. Assessments of a "psychoneurosis" or "constitutional psychopathology" were thus regarded as costly mistakes that needed to be avoided, if possible. Epilepsy, which was only just beginning to emerge as a neurological rather than a psychiatric disease, was a common problem among recruits, and appeared to be rife in the navy.[63] In this context, electroencephalography seemed a potentially valuable way of definitively separating true epileptics, who would inevitably be discharged, from those simulators, who might (many hoped) be cured by psychotherapy or similar moral persuasion. The new device could help determine whether a seizure or head injury experienced while on active duty was simulated or organic, and it could thus help determine the length of sick leave provided. As the second or third tier of a screening process aimed at identifying an established disease, the EEG's utility in military medicine was assured.

Selecting *against* established psychiatric disease had long been an integral part of military medicine, and the EEG fit easily within this context, in which problems of simulation specific to military service had been grafted onto practices then current in civilian medical diagnosis. Despite the differences, the procedure was largely the same: collect a patient history, record his symptoms, make a preliminary diagnosis, confirm with a test. The EEG simply verified the suspi-

[62] On World War II psychiatric exams, see Herman, *Romance of American Psychology* (note 3), pp. 83–90. For wartime EEG usage in diagnosis, see Ceres, "Aviation Medicine" (note 22); John E. Harty, Erna L. Gibbs & Frederic A. Gibbs, "Electroencephalographic Study of Two Hundred and Seventy-Five Candidates for Military Service," *War Medicine* 2 (1942): 923–930; Philip Solomon, Herbert I. Harris, Cecil L. Wittson & William A. Hunt, "Electroencephalography in the Selection of Naval Recruits," *U.S. Naval Medical Bulletin* 41 (1943): 1310–1317; and Daniel Silverman, "Electroencephalography in the Army General Hospital," *War Medicine* 5 (1944): 163–168.

[63] Forest M. Harrison, "Psychiatry in the Navy," *War Medicine* 3 (1943): 113–138.

cion of the diagnosis of epilepsy. But many EEG investigators—particularly high-profile ones likes the Gibbses and the Davises—hoped for something more. They wanted to use the EEG as a creative instrument that could develop new diagnoses and select *for* characteristics, as defined by an EEG record, rather than merely identifying and eliminating the unfit. Against this backdrop of searching for the optimal form of "alpha," these investigators used the categories of "epileptoid" behaviours and "cerebral dysrhythmias," to extend the boundaries of neuropsychiatry to include anomalous phenomena (such as sleepwalking, or outbursts of emotion) that were not, in themselves, considered or treated as mental illnesses in civilian life. This particular expansion was ultimately unsuccessful, but it presaged the "normalization" of mental illness that took place in the following decades, as psychoanalytic concepts and therapies, community psychiatry, and "mental health" movements spread across the United States following demobilization.[64]

EEG and selection at NAS Pensacola

The earliest American efforts to use the EEG to sort out the good from the bad pilots emerged at the Naval Air Station at Pensacola, Florida, where Hallowell Davis and Alexander Forbes were stationed. Naval aviators had trained at NAS Pensacola since the First World War, but the introduction of the air cadet training program in 1935, the Naval Expansion Acts of 1940, and the Selective Service and Training Act of September of the same year, turned the base into a factory for churning out cadets. The previous decade had seen 100 students pass through Pensacola each year. By 1940, 1,100 new cadets were arriving for training each *month*, thus increasing the pressure to identify washouts quickly and accurately.

The Pensacola research station was one of several psychological laboratories that opened at training grounds across the U.S. between 1940 and 1942. Although Pensacola operated under the auspices of the navy, its very existence was a testament to the growth of air power in general, as well as to the expanding authority of psychology

[64] Herman, *Romance of American Psychology* (note 3), pp. 238–275.

in both civilian and military life. In fact, most of these aviation laboratories actually began as a civilian, rather than a military, initiative. Soon after Congress appropriated nearly $6 million for the training of approximately fifteen thousand civilian pilots, psychologists successfully lobbied the National Research Council to establish a Committee on Selection and Training of Civilian Pilots to fund psychological research in this field. Following the outbreak of hostilities in 1939, the term "Civilian Pilots" was replaced by "Aircraft Pilots," and its contract with the Civilian Aeronautics Authority became one of the largest wartime projects administered by the NRC.[65] Much of this money went towards the Aviation Psychology Program (APP), which had been created through the Office of the Air Surgeon shortly after the army Air Corps was reorganized as the United States Army Air Forces (USAAF) in June 1941. By the time the army officially recognized the USAAF as independent and equal to army ground and service forces the following March, there were five well-staffed psychological laboratories at air training centres at Santa Ana, Nashville, San Antonio, Forth Worth and Randolph Field.[66]

Electroencephalography had been part of the NRC Committee's plans from the start, and Hallowell Davis, who had established himself as both a pioneer and a great popularizer of the EEG, was a natural choice to head research in this field.[67] His interest in low-oxygen physiology was well-known, his Macy Foundation funding had run out, and his research at Tuxedo Park had come to an end by mid-1940, when Loomis became deeply involved in helping to organize the British-American efforts to develop a viable form of microwave radar.[68] After leaving Tuxedo Park, Davis spent the summer of 1940 with a number of fellow physiologists and psychologists

[65] On the NRC's Committee, see Capshew, *Psychologists on the March* (note 3), pp. 44–8.

[66] On the APP, see *ibid.*, pp. 107–10; on the USAAF reorganization, see Gross, *American Military Aviation* (note 2), p. 92.

[67] On Davis's efforts at publicizing the EEG, see Kenton Kroker, *The Sleep of Others: rapid eye movement and the creation of modern sleep research* (Toronto, forthcoming), chapter seven.

[68] On Davis's funding, see his letter requesting financial support to Walter B. Cannon, 14 June 1940, box 19, folder 223, WBCA (note 49). On his interest in low-oxygen physiology, see his progress report to the Josiah Macy, Jr. Foundation of 17 January 1938, box 15, folder 179, WBCA; and his letter to Cannon of 5 July 1939, box 17, folder 201, WBCA.

setting up the laboratory at Pensacola. The equipment, much of which Davis borrowed from Henderson's Fatigue Laboratory in Boston, indicated the nature of his expectations. The EEG was not simply another diagnostic tool. It was an instrument for engineering a superior kind of aviator. In a letter to Walter B. Cannon, the famed physiologist of "homeostasis" and departmental head at Harvard, Davis enthused about high calibre of the candidates:

> The study of the aviation cadets is most interesting and we hope will have some fruitful results, particularly as the program of expansion in aviation puts a premium on picking the promising candidates early and correctly. They are a wonderful group of physical specimens. We simply have to adopt different mental standards of exhaustion on such things as vital capacity.[69]

Davis and Forbes's early work at Pensacola, however, was still governed by expectations that the EEG could provide "a means of eliminating from flight training those individuals possessing latent epileptoid trends and emotional instabilities not revealed by other forms of examination."[70] Despite the exceptional quality of the applicants at Pensacola, attrition rates were high: in 1940, roughly thirty percent of cadets never finished their training. Their reasons for failure varied, but Forbes and Davis noted that instructors frequently described washouts as "emotionally unstable," "tense," having "poor motor coordination," "poor judgement," or lacking in "poise, military bearing, or the ability to command."[71] Such problems typically appeared only once training had begun, and the costs of overlooking these traits were potentially high. Besides the unnecessary expense of actually training these failures, the possibility that they might eventually washout during active service raised the spectre of a neuropsychiatric discharge and pension, routinely pegged at around $30,000.[72]

---

[69] Letter, Hallowell Davis to Walter Cannon, 5 August 1940, box 19, folder 223, WBCA.

[70] "Editorial Forward" of Alexander Forbes & Hallowell Davis. *Electroencephalography of Naval Aviators.* With a supplement by Pauline A. Davis: EEG Analysis of 79 Selected C.A.A. Subjects. Civil Aeronautics Administration, Division of Research, Report no. 13 (Washington, D.C.). April 1943 (box 2, Hallowell Davis Papers, Countway Library of Medicine, Harvard Medical Schools [HDP-H]).

[71] Alexander Forbes & Hallowell Davis. *The Selection of Naval Aviators: Pensacola Project.* Preliminary Report. May 1941, box 2, HDP-H, pp. 1–2.

[72] On the oft-cited average pension of $30,000, see Solomon *et al.*, "Electroencephalography" (note 62); see also Herman, *Romance of American Psychology* (note 3), p. 86. The British context was similar: Sir Arthur Harris, the Royal Air Force Bomber

Measurement was crucial to the identification of washouts, just as it had been to the assessment of intelligence among First World War recruits. But by 1940, extensive criticism of a narrow measure of "intelligence" to signify mere "fitness" had generated considerable interest in the "whole soldier" as an object of research, much as the "whole patient" had become an integral part of the biomedical reaction to bacteriological reductionism a decade earlier.[73] But such an approach created problems of its own. As two New York City psychologists observed in an early review on psychology and aviation medicine, "it is really difficult to think of another field of such practical importance as that of the selection of aircraft pilots in which so much confusion reigns and in which research has been attempted and interpreted by investigators of such varying background and training."[74]

Nonetheless, the laboratory at Pensacola was physically organized along precisely these sorts of interdisciplinary lines. (figure 2) An EEG unit for recording brain waves was set up next to physiological and psychomotor tests for determining metabolic rate, eye-hand coordination, reaction time, startle response, and eye movement. Despite the epileptic origins of the EEG's authority, the laboratory at Pensacola was not set up as a neuropsychiatric clinic. Rather, it was arranged like a series of stations in a circuit through which candidates would proceed in a prescribed order. In isolation, the meaning of any of the circuit's components was ambiguous. Did blacking out on a tilt table mean you were more likely to pass out during a steep bank or dive? Was the ability to stand perfectly still for minutes on end truly relevant in an age of flying by instruments? Was a rapid reaction time beneficial in combat and flight, or did it indicate a jittery subject who might be easily startled? When put together, however, these components seemed to offer an objective measure analogous to the emotional stability and poise that made for a good pilot. The ideal aviator, like the ideal aircraft, showed stable and uniform responses under all circumstances.

Commander, suggested that the cost of training a single airman was about £10,000, or the equivalent to the cost of sending ten men to Oxford or Cambridge for three years. See Shephard, *War of Nerves* (note 7), p. 283.

[73] On biomedical holism, see Christopher Lawrence & George Weisz (eds.), *Greater Than the Parts: holism in biomedicine, 1920–1950* (New York, 1998).

[74] G.H.S. Razran & H.C. Brown, "Aviation," *Psychological Bulletin* 38 (1941): 322–30, at p. 326.

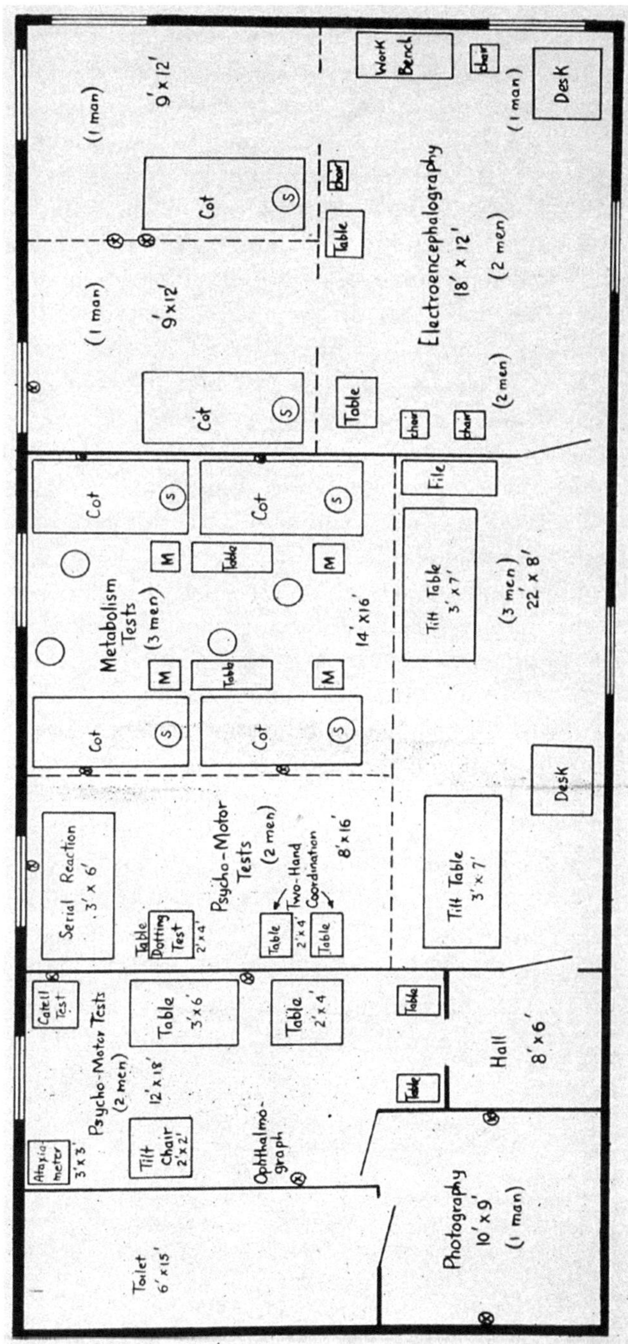

Figure 2. Floor plan of the laboratory at NAS Pensacola. From Forbes & Davis [1943—see n. 70].

This notion of "stability" was a dominant metaphor at Pensacola.[75] This virtue could be recorded graphically, through such measures such as the "dotting test," which recorded eye tremor while the subject fixed his gaze on a dot of light (figure 3), or the "ataxiameter," which recorded all movements a subject made while standing as still as possible, with eyes closed, for one minute. (figure 4) Stability on a subject's EEG record was no less crucial in the minds of the Pensacola investigators. Anything less than consistency and stability on such tests indicated a potential washout. But at Pensacola, the optimal pilot was not merely a composite of reactions: even the candidate's physical appearance was indicative of his skills as a pilot. According to Davis's colleague at Harvard and Pensacola, Ross A. McFarland, bombardiers or transport pilots would likely have "massive or corpulent" bodies, while the more aggressive and agile fighter pilots would tend to be leaner and more muscular.[76]

For Davis, the EEG seemed well-adapted to this atmosphere, as it offered accounts of stability in a familiar graphical form. Returning to the physiology of fatigue, Davis revived one of the oldest theorems of the field; namely, that the changes in physiological states that influenced performance were actually measures of emotion.[77] Not all emotions, of course, were of equal relevance. At Pensacola, love and desire (with their potential relevance to motivation) passed without mention, while "presence of mind under stress, courage and emotional stability," the investigators argued, were the characteristics "most likely to be revealed by the EEG."[78]

The nature of a candidate's EEG was determined by a "stability scale" originally created by Pauline Davis. The scale, which classified records according to the number of "episodes" or bursts of EEG and the relative prominence of waves outside of the alpha range (defined as 8.5–15 Hz), had originated in her work with psychiatric patients at the McLean hospital, near Boston. A low score (1) indi-

[75] It was an equally important ideal for systems of fire control, and signal and power transmission. See Mindell, *Between Human and Machine* (note 8), pp. 121–3 & 142–5. On fire control, see Soubiran, in this volume.

[76] McFarland had worked in Henderson's Fatigue Laboratory during the 1930s and '40s, before joining the Harvard School of Public Health in 1947. He went on to became one of the leading postwar researchers on pilot selection. On his "somatotyping," see Forbes & Davis, *Selection* (note 71), pp. 72–80; and McFarland, *Human Factors* (note 37), pp. 98–101.

[77] Dror, "The Affect of Experiment" (note 17).

[78] Forbes & Davis, *Electroencephalography* (note 70), p. 55.

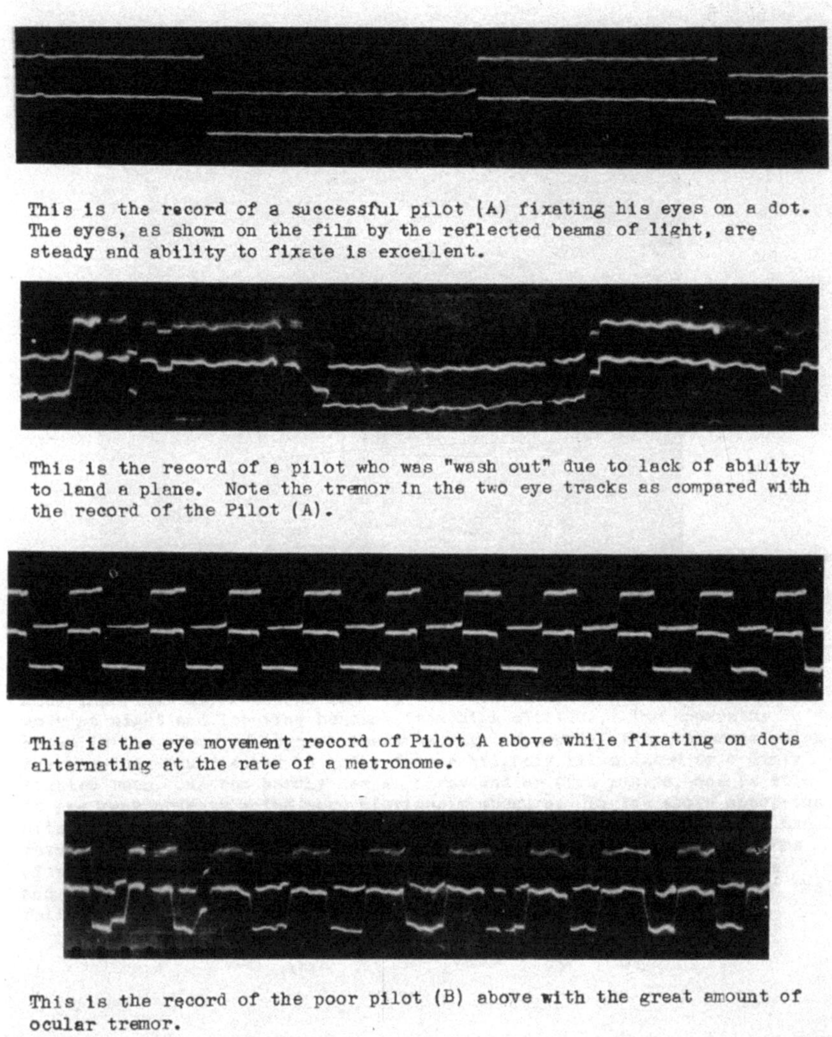

Figure 3. Records from "dotting" tests that measured ocular tremor. From Forbes & Davis [1943—see n. 70].

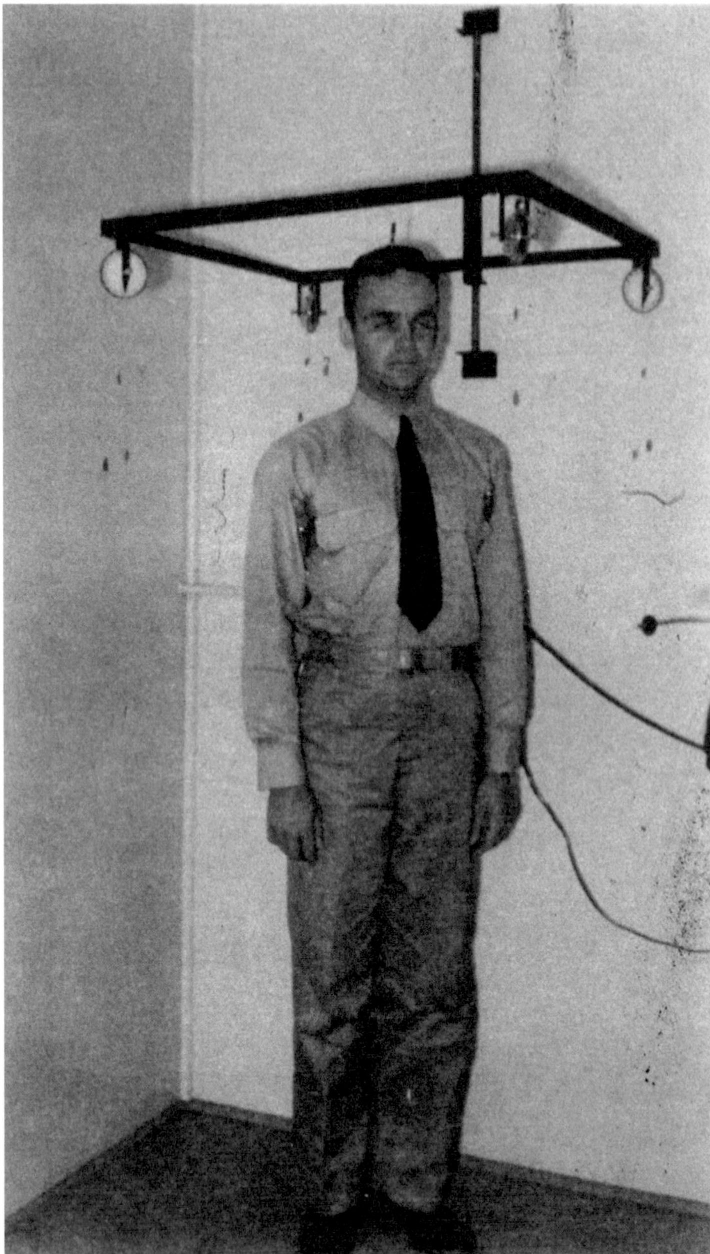

Figure 4. An ataxiameter. Note the pulley system for recording movement on a kymograph (not shown). From Forbes & Davis [1943—see n. 70].

cated a normal EEG, while a high score (5) was taken to be clearly abnormal. This was an explicitly clinical system, and the analogy between assessing abnormal records and diagnosing epilepsy was not lost on the investigators, who noted that they were able to distinguish "relatively orderly records which have been found to have the highest incidence among healthy well-balanced persons" from "the more irregular patterns with waves resembling, in greater or less degree, those clearly abnormal patterns associated with epilepsy."[79] The candidates were then divided up into four classes of success in flight training, ranging from those who had been subjected to a performance review by an advisory board to those who were dropped from the course. The distribution of stability scores were compared across these classes.

At an anecdotal level, Davis and Forbes were particularly pleased with one result: the only EEG record that scored a "5" (extremely abnormal) belonged to a candidate (referred to as "US 1160") that had successfully completed a Civil Aeronautics Authority course, and had even passed his solo flight check at Pensacola. (figure 5) But he suffered from airsickness, and his performance had been judged "erratic" and his personality "infantile" by his instructors. The cadet subsequently dropped on his own request, an outcome that hardly surprised Davis and Forbes, who noted that his EEG was "practically diagnostic of an epileptoid condition," and "showed episodes of high-voltage, low-frequency waves, at times closely resembling those of a petit-mal epileptic seizure."[80] The investigators duly used this example to illustrate the EEG's potential value for positively identifying the emotionally unstable candidates hidden among the normals.

At a statistical level, however, the investigators' conclusions were more ambiguous. Davis and Forbes discovered that the least successful candidates did have a larger percentage of poor EEG scores, but only a small portion of those with distinctly abnormal EEG records (16%) had been rejected by the advisory board. Their scale was clearly not specific enough to independently pick out failures. More problematic was the fact that many candidates whose records indicated "borderline degrees of instability" still passed the course.

---

[79] *Ibid.*, p. 47.
[80] *Ibid.*, pp. 52, 56.

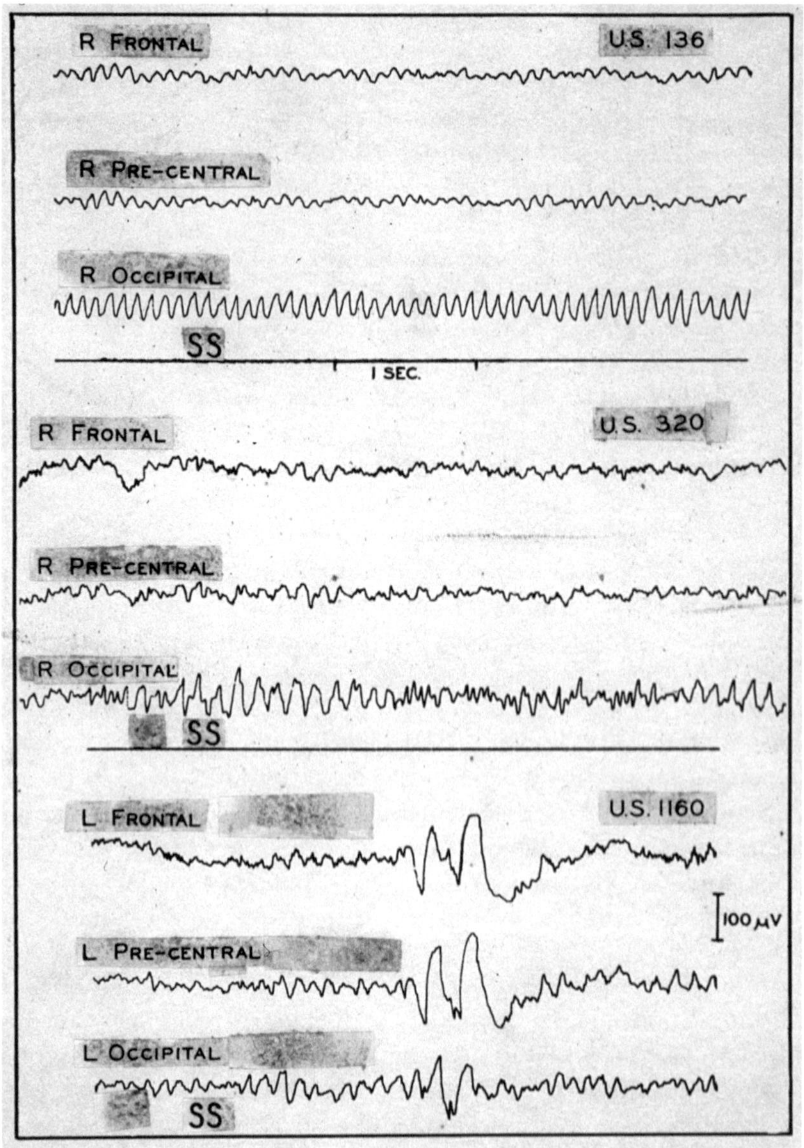

Figure 5. EEG records from three Pensacola aviation candidates. The first (136), who passed, showed a normal, stable EEG, while the second (320), who failed, showed predominant rhythms both faster and slower than alpha. The third (1160), who had successfully completed flight training, but later withdrew, had been judged to be suffering from several psychophysiological defects. His record, the authors suggested, was "practically diagnostic of an epileptoid condition." From Forbes & Davis [1943—see n. 70].

Confronted with this conflict between military authority and instrumental testimony, Davis and Forbes yielded to the former. They concluded that, although the EEG seemed to detect levels of emotional stability, many other factors not measurable by EEG, including physical fitness, intelligence, and eagerness to fly, went into making a good pilot. At most, they suggested, an EEG record might serve as a "court of appeal" when an instructor doubted a candidate's capabilities; or perhaps, the authors mused, the EEG might be capable of differentiating between different types of aviators, such as fighter and bomber pilots.

The final 1943 report included an important modification to the original stability scale—one which took Davis and Forbes further from the original clinical applications of the EEG, and closer to its development as a psychological instrument of selection rather than diagnosis. Instead of scoring records in terms of their similarity to an epileptic's EEG, the authors adapted a calculation of the normal distribution within a population of aviation candidates of two categories: 1) alpha (8–13 Hz) and beta (19–42 Hz) waves; and 2) dysrhythmic or episodic activity (waves slower than 7 Hz or between 14–17.5 Hz).[81] They then rated records according to their deviation from the normal, and correlated them to success in completing flight training. "No preconceptions," argued the authors, "as to resemblance to 'epileptic' patterns entered into the construction of this scale, but simply the principle of deviation from the mean for the group."[82] In preparing the report, Davis told Forbes he had decided to "omit the figure from Pauline's measurement paper that shows normal and abnormal brain waves, inasmuch as the emphasis of the present report is against epileptoid waves and emphasizes a new feature,—the presence of prominent activity in the range between alpha and beta," although he did insist on "including our famous case 1160" as a reminder that crypto-epileptics could be detected by the EEG.[83] (figure 5)

---

[81] The scale was developed by Mary A.B. Brazier at NAS Squantum, near Boston. Brazier, who relocated to the Brain Research Institute at UCLA in the early 1950s, became one of the most prominent American EEG researchers in the postwar period.

[82] Forbes & Davis, *Electroencephalography*, p. 4.

[83] Letter, Hallowell Davis to Alexander Forbes, 12 February 1942, box 5, folder 247, AFP (note 47).

Davis and Forbes were attempting to move the EEG's field of application from the neurological to the psychological, a shift which they emphasized by ranking the cause of failure according to the percentage of highly deviant EEG scores for each cause. While the incidence of such scores among the general population of aviators was around 30%, these scores were higher among those who failed or were dropped for reasons such as their own request (58%), excessive nervousness or anxiety (47%), adverse physical reactions, such as airsickness (67%), or being "psychologically" (50%) or "temperamentally" (53%) unsuited for military aviation.[84] Reasons detectable by conventional physiological assessments or psychophysical and intelligence tests, in contrast, had an incidence of deviant EEG scores closer to that of the general population. This seemed to confirm that, despite the paucity of clinically diagnosable epileptics among the candidates, the EEG might still be valuable in detecting psychological, temperamental or psychosomatic defects.

The Pensacola researchers, however, ultimately remained unable to find any significant correlation between brain wave patterns and flying ability.[85] After eliminating what few crypto-epileptics did appear, they found too many instances of good aviators—including some instructors—with abnormal EEGs to warrant using the device as a screening tool. One central difficulty was that, in the absence of a mathematical analysis of brainwaves, EEG classifications were rather crude, and could only be definitively calibrated to clinical observation of gross behavioural change, such as seizure, coma, or sleep. In contrast, "emotional stability"—the perennial target of EEG investigators—remained a determination of character or personality which

[84] Forbes & Davis, *Electroencephalography*, pp. 1 & 13–15.

[85] Standard postwar textbooks in the field of aviation medicine and pilot selection either ignored or rejected the use of EEG in pilot selection. McFarland, *Human Factors* (note 37) noted in 1953 that the Pensacola experiments demonstrated that "no significant relationships existed between [EEG] ratings and flight performance, as judged by success or failure in the training course" (p. 122). Armstrong was initially enthusiastic about using the EEG to detect "epilepsy, epileptic personalities, and other personality changes" in 1943, noting that "the electroencephalograph and the interpretations of the electroencephalogram hold forth almost unlimited possibilities in the selection of candidates for flying training" (*Principles and Practice* [note 22], p. 59). But in later editions, all references to EEG were unceremoniously dropped (although Armstrong continued to classify epilepsy as a "psychic," rather than a "neurologic" condition). See Harry G. Armstrong (ed.), *Aerospace Medicine* (Baltimore, 1961).

was made, at least in the case of aviation candidates, exclusively within the structure of military hierarchy.

Emotion's transformation and the EEG as washout

Assessments of emotional "tone" resisted laboratory-based attempts at measurement, and they did not yield to calibration by the EEG. But this was not a simple failure of the instrument or of its operators. Emotional stability and success in military aviation themselves turned out to be moving targets that ultimately transcended physiological measurement, The meaning of emotional stability was narrow, task-driven, and highly contingent, as the changing dynamics of military aviation influenced the definition of what it was to be an ideal pilot. At the outset of the war, many believed an "eager" pilot to be "a young fighting gamecock, without serious life responsibilities, reckless of danger, who enjoys the thrill of flying and fighting and shooting down the enemy."[86] Such an image closely corresponded to the heroic vision of the lone pursuit pilot so carefully cultivated by proponents of military aviation since the First World War. But a few years into the Second World War, this image began to change. The allies experienced massive losses in their bombing campaigns over Germany between 1942–44. At its height, the RAF estimated that aviators had a one in five chance of surviving two tours in a heavy or medium bomber, and only a one in ten chance of making it through a full thirty-sortie tour.[87] The fate of many experienced flyers was succinctly summed up as "coffins or crackers." Aviators in the USAAF often fared just as poorly. An infamous daylight bombing raid on the Messerschmidt factory at Regensburg and the ball-bearings works at Schweinfurt led to a loss of 32 of the 230 bombers that left England on the morning of 17 August 1943. Over 400 men were lost or captured.[88] Under such extraordinary conditions, psychiatrists began to argue that pilots with the once-celebrated anti-disciplinarian and individualist streaks often broke with formations,

---

[86] E.S.C. Ford, "Principles and Problems of Maintenance of Fighter-Bomber Pilots," *War Medicine* 8 (1945): 26–31, at p. 31.
[87] Shephard, *War of Nerves* (note 7), p. 283.
[88] *Ibid.*, pp. 280–1.

were quickly shot down, or succumbed to a crippling anxiety. The new vision of an ideal pilot called for a careful blend of aggression and deference to authority that more closely matched "the needs of contemporary social morality."[89]

But how could EEG possibly determine which candidates conformed to social mores and which did not? The fact was that only psychodynamic practice, which had come to assume a prominent, if controversial, position in wartime psychiatry, could detect such traits. Military psychiatrists and clinical psychologists, most of whom had acquired their training not in asylums or in classrooms, but by treating traumatized soldiers in "community" therapy sessions, were rapidly turning away from psycho-physiological concepts, and moving towards talk therapy and so-called "brief psychoanalysis."[90] Their "environmental" explanations of the psychoneuroses no longer resolved on the individual soldier's unconscious strategies to defend his ego, but rather focussed on the more immediate problem of "adaptation," a necessary skill for the aviators whose skies had became the new trenches of the Second World War. Biologized accounts of how "instinct" could be tapped to teach soldiers to kill gave way to considerations of how socialization could instil appropriate combat behaviour.[91] The group or squadron, not the individual aviator, emerged as the key unit of analysis,[92] and there was little room here for the discourse of electroencephalography, based, as it was, on the performance of the individual's brain. The analysis of emotions had become the exclusive domain of these investigators, just as EEG researchers had earlier claimed authority over the study of epilepsy.

Historians traditionally depict the electroencephalograph as a biomedical or psychological technology, not an instrument of war. But aviator selection during the Second World War created a unique new field of application for the EEG that was not clearly medical or psychological. "Cerebral dysrhythmia" was not quite a disease; nor was it a character or personality trait. It was a visual account

[89] Ford, "Principles and Problems". The shift was paralleled by changes in the technology and tactics of combat flight. Ford's analysis was of a bomber group who had shifted from flying twin engine light bombers (A-20s) to dive bombers (A-36s) in the spring of 1943.

[90] Herman, *Romance of American Psychology* (note 3).

[91] Bourke, *Intimate History* (note 7), pp. 73–9.

[92] Shephard, *War of Nerves* (note 7), pp. 279–97.

of how a brain performed, but this performance signified little until it could be linked to existing categories, like emotional instability and success in flight training. This context disappeared once it became apparent that the psychology of the air crew might be as much or even more relevant than that of the individual that the EEG purportedly analyzed.

For his part, Hallowell Davis, the principal architect behind the Pensacola project, never returned to the problem of EEG screening after the war. He abandoned his attempts to link the EEG with personality, and an autobiographical account of his career does not mention such projects, despite its often candid recollection of failures and oversights.[93] A central reason for this outcome was the shifting ground upon which the EEG was operating. Not only were the individual aviator's emotions no longer the central object of psychiatric or psychological interest, but the very qualities that aviators were supposed to possess had been redefined in such a way as to make EEG data about cerebral dysrhythmia irrelevant. Social interaction and teamwork became more highly valued than an individual pilot's steady nerves. The latter might be measurable by the EEG, but the former required psychodynamic technology to be properly understood. The investigation of psychophysiological capacities, on the other hand, had began to change its focus of intervention. Rather than select the ideal pilot based upon psychophysiological tests, human factors engineers began to create an ideal cockpit that featured instruments that meshed more closely with human cognition and performance.[94]

Another possible explanation—one equally fraught with emotion—might be given for Davis's abandonment of EEG screening following the war. Davis had never drawn particularly clear boundaries between his research and his personal life. Archival records, for example, indicate that his interest in auditory neurophysiology was motivated, at least in part, by his determination to build better hearing aids for Forbes, his mentor and colleague at Harvard.[95] Exchanges

[93] Davis, "Crossroads" (note 45).

[94] Meister, *History of Human Factors* (note 4) identifies the work of Paul M. Fitts as particularly crucial in this regard (pp. 151–3). See, for example, Paul M. Fitts, "Psychological Requirements in Aviation Equipment Design," *Journal of Aviation Medicine* 17 (1946): 270–275.

[95] Letter, Hallowell Davis to Alexander Forbes, 9 June 1934, box 5, folder 245,

between Davis and his colleagues also include several pleas to improve his wife's second-rate position as a department technician.[96] Moreover, some early American EEG research had been (at least around Boston and Pensacola) more or less family affairs. The Gibbses worked together as a couple, and Pauline and Hallowell Davis had even involved their son in some of their experiments at Tuxedo Park. But Pauline's crucial contributions to the Pensacola project were tragically cut short on 11 July 1942, when she died from a concussion after falling and hitting her head on a wastepaper basket in her study.[97] Eager to sensationalize the event, some newspapers noted that the toxicology report indicated she had taken sedatives shortly before her death, and speculated that her death was a suicide. Hallowell acknowledged that his wife did, indeed, use sleeping pills to combat her insomnia; but he insisted that her death had been accidental, and the coroner concurred. He did not return to EEG research in any significant way after this, and he left Harvard for Washington University at St. Louis shortly after the war ended, despite Forbes's attempts to persuade him to stay.[98] Perhaps Davis's own confluence of emotions had now made this particular investigative path beyond the pale of electroencephalographic measure.

AFP (note 47); Letter, Alexander Forbes to Hallowell Davis, 8 April 1941, box 5, folder 247, AFP.

[96] Letter, Hallowell Davis to Alexander Forbes, 15 January 1939, box 5, folder 247, AFP.

[97] See the collection of newspaper clippings in the "Hallowell Davis" vertical file at the Countway Library, Harvard Medical Schools.

[98] Letter, Alexander Forbes to Hallowell Davis, 4 September 1945, box 5, folder 247, AFP.

CHAPTER ELEVEN

A MATTER OF GRAVITY: MILITARY SUPPORT FOR GRAVIMETRY DURING THE COLD WAR

Deborah J. Warner

Scientists had long known that the earth's gravity varies from one point to another, but on their own could not command the resources needed to measure this force with very much precision or in very many places. When gravity came to be seen as practical, however, money became available. In the early twentieth century, having learned that gravitational anomalies might indicate deposits hidden underground, petroleum prospectors began developing new gravimetric instruments and using them in specific areas. The U.S. military establishment conducted similar but much larger programs during the Cold War in order to develop ever more precise maps of the earth and charts of the seas, and improve the possibility that missiles might hit their intended targets.[1]

While militaries have long sought the services of cartographers, the introduction of long-range weapons and communications technologies forced them to recognize the importance of the foundational geodesy for maps and charts. Civilian scientists who had worked on military projects during World War II, and thereby gained first-hand knowledge of these new geodetic challenges, told the military about gravimetric geodesy in the immediate post-war period. Their motives were clearly mixed: they wanted to solve a military problem, demonstrate the benefits of this relatively new science, and

[1] "Press Release for Committee on Geophysical Sciences" (Draft), 9 December 1946, National Archives and Records Administration (hereafter NARA), RG330, Entry 341, Box 227, Folder 11; and "Application of Gravity Measurements to Geodetic Problems," 25 April 1951, NARA, RG330, entry 341, box 389, folder 34. See also Helmut E. Landsberg, *Geophysics and Warfare* (Washington, D.C., 1954). Landsberg wrote this in 1948 when, as executive director of the Committee on Geophysical Sciences of the Research and Development Board, he attempted to answer "the frequently heard question of why the military departments should be sponsoring research work in the field of geophysical sciences."

obtain funding for their own research projects. Military support was soon forthcoming—but it came with strings. As the civilian scientists soon learned, they could only work on parts of the world of interest to the military, and they could only publish results cleared by military authorities. As the military services became convinced of the practical benefits of this science, they developed in-house gravimetric programs. Some of the results of these programs—most notably the instrumentation and the techniques for analysis—were shared with the wider scientific community.[2]

Army Programs

The army's involvement with geodesy began in the immediate postwar period when the Army Map Service (AMS) worked with Allied counterparts to create the precise map of Europe they had needed during the recent conflict. As part of this project, the AMS gave the U.S. Coast & Geodetic Survey $54,000 to wrap up a pilot project on the reliability of the gravimetric method for determining the deflection of the vertical. Walter Lambert, Chief of the Gravity and Astronomy Branch of the Coast & Geodetic Survey, had started this project in the early 1940s, after having learned of similar work being done by the Soviet Institute for Geodesy and Cartography.[3] Lambert

[2] For those not familiar with geodetic terms, a few definitions are in order. The *geoid* is the irregular surface on which the earth's gravitational field is everywhere the same. The *ellipsoid* is the regular geometrical surface that best approximates the earth's size and shape. The relation between the two is known variously as the *deflection of the vertical* or the *elevation of the geoid.* Because a plumb bob points to the center of the geoid, positions determined by astronomical means often differ markedly from those determined by horizontal measurements.

[3] Walter Lambert, "Deflections of the Vertical from Gravity Anomalies," *Transactions of the American Geophysical Union* 28 (1947): 153–156; and "Deflections of the Vertical from Gravity Anomalies," *Army Map Service Technical Report* 2 (November 1949). Donald Rice, "Gravimetric Deflections by the Method of Condensation," *Transactions of the American Geophysical Union* 30 (1949): 323–327. See also Lambert, "The Use of Values of Gravity in the Adjustment of the Triangulation of Europe," NARA, RG77, Box 2/3, Folder 914; this was later published in *Bulletin Géodésique* 3 (1947): 19–22. Floyd Hough to Acting Executive Office (of the Army Map Service), 2 April 1947, NARA, RG77, Box 2/3, Folder 914. "Digest on the Determination of Deflections of the Vertical from Gravity Anomalies," NARA, RG330, Entry 341, Box 457, Folder 6. The Research and Development Board kept track of the extent and importance of gravimetric research in the U.S.S.R. See for instance M.S. Molodenskiy and V.V. Fedynakiy, "Thirty Years of Soviet Gravimetry," *Reports of*

served as president of the International Association of Geodesy in the 1940s, and as an advisor to the U.S. Research and Development Board, and his opinions were highly regarded.

In 1947 the Research and Development Board asked the Army Map Service to develop a "Geodesy for Guided Missiles" (as long range missiles were then often referred to).[4] This would eventually morph into the World Geodetic System, a massive multi-year project to which the several military services contributed, and which produced the geodetic information needed for missile defense. It would be sponsored in turn by the Defense Intelligence Agency, the Defense Mapping Agency, and the National Imagery and Mapping Agency.[5]

One aspect of the "Geodesy for Guided Missiles" project might be seen as an extension of the European readjustment to the entire world, placing the various national and regional geodetic surveys on a common basis.[6] Another aspect of the project involved an analysis of the effect that the deflection of the vertical at the launch site and the target, and over the path in between, might have on missile trajectories.[7] By 1958 the Army Ballistic Missile Agency understood that "improved gravity information would increase accuracy in the employment of longer range missiles."[8] Military reports on this subject were classified, but their gist was well known. An article in *Surveying and Mapping* in 1959 noted, obliquely yet transparently, that,

the Academic [sic] *of Sciences, U.S.S.R.* 11.5, Series Geographic and Geodesic (1947), English translation prepared by the Board's Technical Intelligence Branch, NARA, RG330, Entry 341, Box 457, Folder 6.

[4] "Press Release for Committee of Geophysical Sciences" (Draft), 9 December 1946, NARA, RG330, Entry 341, Box 227, Folder 11. "Geodesy for Navigation of Long Range Guided Missiles," 27 April 1948, NARA, RG330, Entry 341, Folder 31.

[5] D.J. Warner, "Political Geodesy: the Army, the Air Force, and the World Geodetic System of 1960," *Annals of Science* 59 (2002): 363–389.

[6] J.D. Abell to The Commanding Officer, Army Map Service, 10 March 1950, NARA, RG456-91-3079, Box 3/4, Folder AMS Post Hostilities. See also Central Intelligence Agency, *Probable United States and USSR Geodetic Accuracies Between ICBM Launch Sites and Targets* (CIA/RR-GR-166), November 1957.

[7] Bernard Chovitz, "The Effect of the Undulations of the Geoid on Guided Missile Paths," *Army Map Service Technical Report* 9 (1951); Chovitz, "A Statistical Analysis of the Effect of Deflections of the Vertical on Inertial Guidance Systems," *Army Map Service Technical Report* 16 (1954). See also Walter Lambert, "Geodetic Controls for the Flight of Guided Missiles" *OSURF Report* #504 (April 1953), done on Air Force contract 19 (604)–287.

[8] V.P. Bauer, memo for the record, 23 October 1958, NARA, RG77, Box 2/3, Folder 914.

"A guidance system which relies on the earth's gravity field en route would of course benefit from a detailed knowledge of the changes in that field from place to place."[9]

In addition to this theoretical work, the army conducted gravimetric surveys in numerous third world countries. At the height of the Cold War, the AMS had some 125 gravity crews going "everywhere except behind the Iron Curtain."[10] This project began with the Inter-American Geodetic Survey (IAGS), a branch of the Army Engineers that was established in 1947 and charged with helping engineers in Central and South America produce maps that were up to American standards and available for American military and commercial purposes. *National Geographic* ran a lengthy piece on the IAGS in 1956 noting that, together with the Venezuelan government, it was spending $1,000,000 on a project to find out "which way is down." A photograph of a Worden gravimeter carried a caption stating that this very delicate instrument cost $8,000 and "can detect variations in gravity equivalent to the adding of 1/10,000,000 gram to a one-gram weight."[11] The Worden gravimeter had been designed by Sam Worden of Houston (see below) and was manufactured by Texas Instruments.

The LaCoste & Romberg geodetic gravimeter was more precise than the Worden, and substantially more expensive. Lucien LaCoste had developed the central element of these instruments (a zero length spring) in the early 1930s while he was a graduate student in physics at the University of Texas, and went into business with his professor, Arnold Romberg, in 1939. LaCoste & Romberg thought that petroleum prospectors would buy most of their instruments, but found the military to be their major customer. The AMS eventually bought at least 67 LaCoste & Romberg gravimeters.[12] Some were used in

---

[9] F.W. Diercks "The Role of the Topographic Map in the Missile Age," *Surveying and Mapping* 19 (1959): 349–354. See also the paper on the "Geodetic Interests and Activities of the Army Map Service" read at the meeting of the International Union of Geodesy and Geophysics held in Rome in 1954.

[10] D.J. Warner interviews with Robert M. "Ivy" Iverson, who studied geodesy with Woollard at the University of Wisconsin, and who helped organize and equip the AMS gravity program.

[11] R. Conly, "Men who Measure the Earth," *National Geographic* 109 (March 1956): 335–362, on 348 and 356–357. See also "Measurement of Gravity," *Transactions. American Geophysical Union* 44 (1963): 319.

[12] Records at LaCoste & Romberg in Austin, Texas. The Naval Hydrographic Office bought 12 of these instruments, the Geological Survey bought 7, and the

the western United States, some by army personnel, and others by academic scientists with army contracts. Many were used by army personnel and contractors in Cambodia, Laos, Thailand, and Vietnam.[13]

Office of Naval Research Programs

In the 1920s, a Dutch geodesist named F.A. Vening-Meinesz developed a complicated pendulum apparatus that gave good results when taken by submarine to the calmer waters below the surface. The U.S. Navy, working in collaboration with the Coast & Geodetic Survey, allowed Vening-Meinesz to bring his apparatus aboard a U.S. submarine in 1928. That cruise and the two that followed in the 1930s are good examples of the navy's traditional, occasional, and partial support for science at sea. The navy supplied the submarine, but the Carnegie Institution of Washington picked up the tab for the scientists' food and other expenses.[14]

In the late 1940s, the newly-established Office of Naval Research (ONR) provided tens of thousands of dollars so that Maurice Ewing, a geophysicist at Columbia University, could measure gravity at sea with an instrument of this sort. Over the course of the next dozen or so years, Ewing and his graduate students would make thousands of observations of gravity at sea. The navy clearly understood the military purpose of the project. So too did Ewing, who noted in 1952 that the Naval Air Missile Test Center at Point Mugu, California, was concerned with the deflection of the vertical "in connection with problems of range instrumentation and missile guidance." And although

Air Force and the Coast & Geodetic Survey bought 3 each. For early observations see "Gravity Observations and Reductions," *Transactions of the American Geophysical Union* 41 (1960): 139; and "Measurement of Gravity," *Transactions of the American Geophysical Union* 44 (1963): 318–319.

[13] T.H. Nilsen, "The Army Topographic Command Gravity Program in the Western United States," *EOS* 50 (1969): 528–529; N.I. Christensen, R.C. Bostrom and R.S. Crosson, "The Gravity Program of the University of Washington," p. 548; Z. Dane, "Gravity Results in Western Washington," pp. 548–550; and K.L. Cook, T.H. Nilsen, and J.F. Lambert, *Gravity Base Station Network in Utah-1967* (Salt Lake City, 1971). W.H. Van Atta, "Mapping of Southeast Asia," *Surveying and Mapping* 28 (1968): 41–44. For budgets see OCE-AMS Conference on DOD Gravity Survey Requirements, 5 October 1962, NARA, RG77, Box 2/3, Folder 914.

[14] Naomi Oreskes, "Weighing the Earth from a Submarine: The Gravity Measurement Cruise of the U.S.S. S-21," in G. Good (ed.), *The Earth, the Heavens, and the Carnegie Institution of Washington* (Washington, D.C., 1994), pp. 53–68.

Ewing often extolled the disinterested nature of science, he was admittedly "eager to do whatever is possible to further surveys of the oceans and to make our results and technical experience in this field available to the armed forces."[15] After the Navy Hydrographic Office noted that "detailed gravity surveys of strategic ocean areas must be made," the ONR informed Ewing that "interest in accelerating the present gravity program at sea has increased" and suggested that proposals for work of this nature would be warmly received. The scientists, however, would make observations "from U.S. navy submarines in the course of normal submarine operations," not where and when they wished.[16]

While money was easy to come by, publication proved problematic. Ewing learned in the summer of 1954 that the path of the submarines and the location of their observations could no longer be disclosed.[17] By November, heightened security considerations had led the navy and air force to declare that the gravity data itself should remain secret. Clair Ewing (no relation to Maurice), Director of Range Development at the Air Force Missile Test Center in Florida, explained that this decision was "probably based on the premise that as additional geophysical data are amassed that may contribute to world wide use of new weapons, the overall knowledge constitutes a national advantage which should be maintained, if possible, by resort to classification." A 1958 Department of Defense memo formalized this decision: "Any gravimetric data which may be of value to a potential enemy" had to "be considered for security classification." Oceanographic data was generally confidential. Land data was usually unclassified, but most "gravimetric data from the U.S.S.R. which would be of military or scientific value are not made available."[18]

---

[15] J.H. Russell to M. Ewing, 27 February 1952; and Ewing to Russell, 2 March 1952, Maurice Ewing papers, Box 81, Center for American History, University of Texas at Austin (hereafter CAH).

[16] Survey Branch, U.S. Navy Hydrographic Office, "Application of Gravity to Problems of the Navy," prepared for the 10 March 1952 meeting of the RDB Panel on Cartography and Geodesy, NARA, RG330, Entry 341, Box 389, Folder 36. Gordon Lill to M. Ewing, 26 November 1952, Maurice Ewing papers, Box 81, CAH. Panel on Cartography and Geodesy, "1953 Technical Estimate," 5 March 1953, NARA, RG330, Entry 341, Folder 26; this refers to contract NR-081-018. Office of Naval Research, *Quarterly Project Summary* (1 July 1949), p. 38.

[17] J.L Worzel to Chief of Naval Research, 2 August 1954, Maurice Ewing papers, Box 126, CAH.

[18] Clair Ewing to J.W. Worzel, 10 July 1956, Maurice Ewing papers, Box 81,

Soviet scientists faced similar security concerns. At the triennial meeting of the International Union of Geodesy and Geology held in Toronto in September 1957, Ewing and his colleague, J.L. Worzel, discussed gravitational instruments and observations with a Professor Boulanger of the Russian delegation. When Boulenger promised to send "a mimeographed list containing five to six thousand gravity values from the U.S.S.R.," the American scientists opined that "it is just possible that this may be the turning point which will open up to the western countries the great mass of data which has been withheld in the past. If so, the meeting in Toronto, and the 'trade' in gravity data will have been one of the most significant scientific interchanges of geophysical data in the last 30 years."[19] It would have been. But, in the context of the Cold War, it was not to be.

A second geodetic project supported by the Office of Naval Research was led by George P. Woollard, a geophysicist with a Princeton Ph.D. who hoped to correlate geological features and gravitational anomalies in order to better understand the structure of the earth.[20] Later in life, in response to military concerns, Woollard would invert his analysis, using geological and geophysical parameters to predict gravity "in unsurveyed areas."[21]

Woollard began this work in the late 1930s, using gravimeters that had recently been developed for petroleum prospecting. These instruments were incredibly efficient. Even the Coast & Geodetic Survey, which had been measuring gravity since the 1870s, acknowledged that the gravimeter "revolutionizes the observations for gravity, as from 20 to 30 observations can easily be made in one day, whereas only a single observation can be made with the pendulum equipment."[22] Unlike pendulums, however, which gave absolute results, gravimeters simply measured gravimetric changes between one place and another, and between one moment and another.

CAH. F.O. Diercks to Chief of Engineers, U.S. Army, re: Iron Curtain Gravity Meter Survey Data, 27 September 1960, NARA, RG77, Box 2/3, Folder 914.

[19] M. Ewing to the Commanding Officer at the Air Force Cambridge Research Center, 13 December 1957, Maurice Ewing papers, Box 66, CAH.

[20] V. Godley, "Memorial to George Prior Woollard, 1908–1979," *Geological Society of America Memorials* 15 (1985): 1–6.

[21] W.E. Strange and G. Woollard, *The Use of Geological and Geophysical Parameters in the Evaluation, Interpolation and Prediction of Gravity* (November 1964); this is the final report of contract AF23(601)3879, a continuation and expansion of AF23(601)3455, a project sponsored by the Aeronautical Chart and Information Center.

[22] Lansing Simmons, "Geodesy," in U.S. Coast and Geodetic Survey, Staff Meeting, 4 April 1951, NARA, NN-370–96–3, 79–36, Box 1.

The first gravimeters were large, heavy, and unwieldy. With the return of peace, Woollard asked several instrument makers "if they would make a 'good' high range, lightweight gravimeter suitable for use with air transport." The challenge was accepted by Sam Worden, a graduate of Rice Institute who had begun in business in his garage in 1945 and who made lightweight gravimeters equipped with a quartz spring. Woollard would later recall that when he first approached Worden he admitted that he had "no backing or grant to foot the bill."[23] This statement may have been literally true, but it was slightly misleading. In October 1947 the Research and Development Board authorized funds so that Worden could develop a "gravity meter of extreme simplicity, portability, and accuracy."[24] Worden applied for a patent on this instrument in December of that year. Texas Instruments bought Worden out in December 1953, "Worden" became a TI trademark, and Sam Worden became a consultant to the larger firm.[25]

Woollard had spent the war years at the Woods Hole Oceanographic Institute on Cape Cod working with Maurice Ewing on acoustic problems relating to undersea warfare.[26] In the course of this work Woollard learned that the navy faced the problem of determining the positions of islands that held monitoring stations for LORAN (a land-based radionavigation system) and SOFAR (an acoustic method of measuring long distances under water) and that were located beyond the limits of conventional geodetic ties. Woollard would later recall that, when he heard that "uncertainties as to the actual positions of the proposed monitoring stations" forced the navy to scrap several

---

[23] G.P. Woollard, "Honorary Membership Citation for Sam P. Worden," *Program of the 48th Annual International Meeting of the Society of Exploration Geophysicists* (30 October 1978).

[24] "The National Military Establishment Digest of Current Research Projects, Research and Development Board, Prepared by Committee on Geophysics and Geography," 1 August 1948, NARA, RG330, entry 341, box 173, folder 5. The listing is for the Houston Technical Laboratory, which was the name of Worden's firm.

[25] S.P. Worden, "Gravity Meter," U.S. patent #2,674,887. Worden and Boyd Cornelison, "Large Range Gravity Sensitive Instrument," U.S. patent #2,738,676. Cornelison, "Gravity Meter," U.S. patent #2,732,718. All three patents were assigned to Texas Instruments. "Texas Instruments Expanding," *New York Times* (10 December 1953), p. 83:7.

[26] Woollard *curriculum vita* in National Academy of Sciences (NAS) Archives, Washington, DC, IGY papers, Projects 5.6 and 5.7.

sets of charts "entailing thousands of man hours work," he suggested that gravity might "offer a solution to this geodetic problem."[27]

In the summer of 1948 Woollard knew that the navy was still "encountering some difficulties with the deviation of the vertical in the Pacific," and noted that he was working "in collaboration with the navy's guided missile program."[28] It is thus not surprising that ONR funds enabled Woollard to take his Worden gravimeter around the world, establish the value of gravity at outlying islands and at ports that had been or might be used by submarines, and tie together the major gravity base stations "so that there will be no question about gravity difference between the locations."[29]

Like Ewing and many other American scientists active during the Cold War, Woollard considered himself an academic who took advantage of opportunities that military funding made available. According to Lambert, he traveled by military air transport and dressed "in the uniform of a naval technician," but believed himself to be "the same sort of scientist as ever."[30] The military offices that supported his work, however, saw him as a servant of the state. One military document stated that "Woollard's activities will be directed toward performing gravity surveys in areas where such information is considered most urgent to the unified program."[31]

Military concerns were a two-edged sword, facilitating some observations and making others impossible. Speaking to the American Geophysical Union in April 1949, Woollard mentioned that Russian authorities had blocked his attempt to measure the acceleration of gravity in Potsdam, a city that had been the reference point for

[27] G.P. Woollard, "The Status of Gravimetric Control for Global Geodetic Studies," in S.H. Laurila and W.A. Heiskanen (eds.), *Geodesy in the Space Age* (Columbus, Ohio, 1961), p. 98.

[28] Woollard's comments are in John O'Keefe to Chief, Geodetic Division, Army Map Service, 30 June 1948, NARA, RG77, Box 2/3, Folder 914.

[29] Office of Naval Research, *Quarterly Project Summary* (1 April 1949), p. 243; *Quarterly Project Summary* (1 July 1949), pp. 53–4; and *Project Summary Quarterly Supplement* (1 October 1949), p. 23. G.P. Woollard, "Report on Field Tests on Special Worden Gravity Meter" (16 August 1948), Woods Hole Oceanographic Institution 48–31; "World Wide Gravity Measurements with a Gravity Meter" (1949), Woods Hole Oceanographic Institution 49–33; and "The Gravity Meter as a Geodetic Instrument," *Geophysics* 15 (1950): 1–29.

[30] W. Lambert to P. Tardi, 25 June 1948, NARA, RG77, Box 2/3, Folder 914.

[31] Ralph J. Ford, "Geodesy and Gravimetry," *Air Force Surveys in Geophysics*, No. 22 (December 1952), p. 26. This was declassified in 1964.

gravimetric observations for the previous half century and that was now in the Russian-occupied zone of Germany. The headline in the *New York Times* declared: "Russia Rejects Attempt to Gauge Gravity's Pull."[32] The *Times* did not explain the reasons for this rejection, but it must be imagined that the Soviets understood the connection between Woollard's science and the Pentagon's desire to gain target information for sites "within the impenetrable land mass of Eurasia."[33] Throughout the rest of his career Woollard remained eager to obtain information about gravity anomalies in areas behind the Iron Curtain. He hoped, for instance, to use the International Geophysical Year to connect areas that had been "politically inaccessible" to his growing network of gravity stations.[34] When international cooperation was not forthcoming, Woollard may have resorted to stealth: he once mentioned that he had disappeared "for months at a time with no word as to my being behind the communist lines in China or stranded out in the Libyan desert."[35]

ONR's third gravimetric contract was with Louis B. Slichter, head of the Institute of Geophysics at UCLA. Slichter was interested in detecting earth tides, and by 1955 had received some $53,000 from ONR for this purpose. The navy was probably not particularly interested in earth tides but, as explained below, it did appreciate academic interest—especially when those academics would be able to evaluate other gravimetric instruments as they became available. With ONR money Slichter purchased an extremely precise, and extremely expensive ($25,000), geodetic gravimeter made by LaCoste & Romberg.[36]

[32] "Russia Rejects Attempt to Gauge Gravity's Pull," *New York Times*, 21 April 1949, p. 5 See also "D.C. Meetings Told of Radar Wonders," *Washington Post*, 21 April 1949), p. 8.

[33] Science Advisory Board recommendations, 11 March 1949, Panel: Explosives and Armament. Subject: Recommendations on Target Marking Techniques for Target Identification, NARA RG341 (Hq USAF), entry 165, box 23, folder 21.

[34] G. Woollard, "World Gravity Connections," in W.A. Heiskanen (ed.), *Size and Shape of the Earth* (Columbus, Ohio, 1957), pp. 73–75.

[35] G. Woollard, Acceptance of the William Bowie Medal, *Transactions of the American Geophysical Union* 54 (1973): 708–709.

[36] L.B. Slichter's 1955 proposal to NSF for "Mean Rigidity of the Earth" and other documents in NAS Archives, IGY Papers, File: Gravity. UCLA, Slichter, 1955–July 1957.

Navy Programs

The Navy Hydrographic Office (NHO) contracted with the Coast Geodetic Survey in the late 1940s for gravimetric observations and analysis. One Survey scientist noted that most of their gravimetric work at that time was done "primarily for the national defense people who are interested in the method of determining deflections of the vertical in out-lying islands several hundred miles from other control schemes." This scientist went on to explain that "If you can determine the deflection of the vertical at these outlying islands, you will be able to relate your surveys to geodetic control at other places."[37]

The NHO established its own gravity office in 1950, acquired several Worden and LaCoste & Romberg gravimeters, and used them to determine the position of Clipperton Island, a French atoll in the Pacific Ocean that the U.S. Navy had occupied during World War II. It also conducted gravity surveys around Wake, Eniwetok, and the Midway Islands, in order to obtain precise geodetic positions "as required for the Pacific Missile Range."[38]

While the islands were important, the navy was primarily interested in the ocean itself. And while the Vening-Meinesz pendulum apparatus was reliable, it was fairly inefficient. In 1951, after the Chief of Naval Research informed the Research and Development Board that ocean gravity data was "virtually nonexistent," the Board approved the procurement of a submarine gravimeter.[39] Several individuals and organizations considered having a whack at the problem. The Department of Terrestrial Magnetism of the Carnegie Institution of Washington proposed using building an instrument with

[37] Captain Hemple, "Division of Geodesy," in U.S. Coast & Geodetic Survey, Staff Meeting, 15 November 1950, NARA, NN-370-96-3, 79–36, Box 1.

[38] A.L. McCallan, "The Navy Ocean Gravity Program," in *Proceedings of Military Geodesy Seminar, December 1958, Air Force Cambridge Research Center* (1959), pp. 26–32. See also Don Parker's manuscript history of the gravity program of NHO/NAVOCEANO. See also W. Kaula, Memorandum for Record, 10 December 1958, NARA, RG77, Box 2/3, Folder 914. This concerned a meeting to revise and bring up to date the Joint Chiefs of Staff statement of ocean gravity survey requirements.

[39] "Tentative Submarine Gravity Meter Technical Specifications," and Memo from Chief of Naval Operations, 10 October 1951, in NARA, RG330, Entry 341, Box 457, Folder 6. This Navy Project, NR-081-116, was mentioned in Research and Development Board, "1953 Technical Estimates on IO-13," NARA, RG330, Entry 341, Box 389, Folder 26.

funds provided by the IGY.[40] Ewing and Worzel at Columbia University designed a submarine gravimeter, using some of their ONR money for this purpose.[41] Sam Worden was apparently put off when he learned that he would need a security clearance to test an instrument at sea.[42] Lucien LaCoste developed another but, allegedly "touchy" about military contracts, did so on his own dime and then sold the finished instrument to the navy.[43]

While the submarine gravimeter was under development, the navy urged Slichter to submit a proposal for the acquisition and use of a Vening-Meinesz pendulum apparatus. Ewing objected that the navy was not providing enough submarine time for his projects, let alone for two projects of this nature, but Slichter believed that, if the navy "wishes to support *two* groups, that seems a step in the right direction." This was, Slichter noted, in accordance with navy policy of "having two sources for everything."[44] When the LaCoste & Romberg submarine gravimeter became available in 1955, Slichter's team used it in the Pacific Missile Test Range between the launch site at Vandenberg AFB and target islands in the ocean, and they compared the results with those they had obtained in this same area with the Vening-Meinesz apparatus. The two instruments gave similar results, but the gravimeter was easier to manipulate. The navy used it on three under-ice submarine cruises in the Arctic in early 1959, and on the USS *Triton* during its historic submerged circumnavigation of the globe.[45]

Since submarines were expensive to operate and often not available for scientific purposes, the navy was also interested in gravi-

[40] M. Ewing to Merle Tuve or Howard Tatel, 18 May 1955, Maurice Ewing papers, Box 81, CAH.

[41] T.W. Yerzley to Commanding Office of ONR, 18 September 1956, Maurice Ewing papers, Box 81, CAH.

[42] Gordon Lill to S.P. Worden, 17 August 1951, Maurice Ewing papers, Box 94, Folder U.S. Navy/ONR, CAH.

[43] D.J. Warner telephone interview with H.N. Clarkson, 18 July 2003. See also Lucien LaCoste to M. Ewing, 26 May 1955, Maurice Ewing papers, Box 181, CAH.

[44] L.B. Slichter to M. Ewing, 26 March 1954; Ewing to Slichter 30 March 1954; and J. Worzel to M. Ewing, 2 April 1954, Maurice Ewing papers, Box 123, CAH. See also J.C. Harrison, G.L. Brown, and F.N. Spiess, "Gravity Measurements in the Northeastern Pacific Ocean," *Transactions of the American Geophysical Union* 38 (1957): 835–840.

[45] F.N. Spiess and G.L. Brown, "Tests of a New Submarine Gravity Meter," *Transactions of the American Geophysical Union* 39 (1958): 391–396. "Gravity Measurements," *Transactions of the American Geophysical Union* 44 (1963): 321.

metric instruments that could function on surface ships. The first instrument of this sort was designed by Anton Graf of Munich, and produced by Askania-Werke AG. Its first test in the open sea was conducted in November 1957 by J. Lamar Worzel, Assistant Director of the Lamont Geological Observatory at Columbia University. The following spring, Worzel took the Graf gravimeter on a five-week cruise from New York to the Mediterranean.[46] As with many projects of this nature, this one had civilian and military aspects. The National Science Foundation provided some funds under its IGY budget, while the navy provided the *Compass Island*, an 18,000 ton vessel that had been acquired expressly for the purpose of developing and evaluating "a navigation system independent of shore-based and celestial aids, a necessary adjunct of the ballistic missile program."[47]

The first American instrument of this sort was a LaCoste & Romberg submarine gravimeter equipped with a gimbal mount and accelerometers so that it could withstand the substantial horizontal accelerations encountered on a surface ship. The first tests of this instrument were conducted on the *Hidalgo*, a ship owned by Texas A&M University. A somewhat modified instrument was tested on the *Horizon*, a research vessel operated by the Scripps Institution of Oceanography. This voyage along a 300 mile track passing close to 13 submarine gravity stations off the coast of southern California showed that it was possible to obtain relatively accurate measurements "in calm and slight swell conditions on board a surface vessel of less than 1,000 tons total laden displacement without the aid of a stabilized platform."[48]

These tests led one geodesist to enthuse that it was now "possible to measure gravity profiles at sea, instead of point values as has been the case in the past."[49] Although the celebration was somewhat

[46] [J. Lamar Worzel], "First Sea Surface Gravimeter," *IGY Bulletin* 39 (February 1958): 175–178; "Gravity Measurements on a Surface Ship at Sea," *Transactions of the American Geophysical Union* 41 (1960): 701–706; and "Continuous Gravity Measurements on a Surface Ship with the Graf Sea Gravimeter," *Journal of Geophysical Research* 64 (1959): 1299–1315.

[47] Online at <www.multied.com/navy/MISC/CompassIsland.html> (accessed 15 May 2004).

[48] L. LaCoste, "Surface Ship Gravity Measurements on the Texas A. and M. College Ship, The 'Hidalgo,'" *Geophysics* 24 (1959): 309–322. J.C. Harrison, "Tests of the LaCoste-Romberg Surface Ship Gravity Meter I," *Journal of Geophysical Research* 64 (1959): 1875–1881.

[49] W.A. Heiskanen, "The Latest Achievements of Physical Geodesy," *Journal of Geophysical Research* 65 (1960): 2827–2836.

premature, the ocean gravity program was well underway. By March 1960, the NHO's continuous-reading sea gravimeters had produced more than 1,000,000 observations, and the NHO's gravimetric staff numbered more than 20.[50] In addition, civilian scientists at Oregon State University used a LaCoste & Romberg surface ship gravimeter owned by the Office of Naval Research; and scientists at UCLA, working with the Naval Ordnance Test Station at China Lake, California, used a similar instrument in the Santa Barbara Channel.[51]

LaCoste & Romberg produced a gravimeter suitable for use on a smaller surface ship in 1963.[52] The NHO, now renamed the Naval Oceanographic Office (NOO), soon had seven sea gravimeters in constant use, as well as a gravity processing system that used electronic accounting equipment and high-speed computers. In 1965, the NOO acquired the *Silas Bent*, a medium-sized ship that could collect and process geophysical, acoustic, and meteorological as well as gravimetric information, even while underway. In 1969 the Commander of the Naval Oceanographic Office noted that gravity "will receive the highest priority among the geophysical parameters being studied" under the navy's Geophysical Survey System.[53]

After the NOO acquired three large ships that had been mothballed at the end of WW II, LaCoste "tuned" a gravimeter to each ship's particular characteristics. With further improvements, NOO could take gravity measurements, along with related observations for latitude and longitude, while traveling at 18 knots. These ships remained at sea until the late 1980s, two of them amassing more mileage than any other vessels. They were replaced by two other large ships, designed and built for the purpose in the late 1980s.[54] The results of the navy's Ocean Gravity Survey Program were delivered to the Strategic Submarine Program Office in Crystal City, Virginia, an office complex near the Pentagon, and remain classified.

---

[50] D.L. Mills addendum to J.A. Bernard, Memo for Record, 2 March 1960, NARA, RG77, Box 2/3, Folder 914.

[51] P. Dehlinger and S.H. Yungul, "Experimental Determination of the Reliability of the LaCoste and Romberg Surface-Ship Gravity Meter S-9," *Journal of Geophysical Research* 67 (1962): 4389–94. Roland von Huene and J.B. Ridlon, "Offshore Gravity and Magnetic Anomalies in the Santa Barbara Channel, California," *Naval Ordnance Test Station Technical Paper* 3917 (*ca.* 1965).

[52] Naval Oceanographic Office, *Annual Report of the Commander* (1963): 6.

[53] Naval Oceanographic Office, *Annual Report of the Commander* (1964): 9–10 and 30 and *Annual Report* (1965): 11–12, and *Annual Report* (1969): 59. See also Naval Oceanographic Office, *USNS Silas Bent T-AGS 26* (Washington, D.C., 1966).

[54] D.J. Warner, telephone interview with John Hankins.

The rapid growth in extent and expense of the navy's gravity program was directly linked with the rapid expansion of America's missile and nuclear submarine programs. A December 1958 assembly of geodesists "concurred that ICBM guidance capabilities after 1962 was the primary military reason for obtaining gravity data" at sea. Another geodetic meeting reported that the navy's program of hydrographic surveying was "predicated on the needs of the Armed Services missile programs and on general geodetic and navigational applications."[55] Some of this gravity data was used "for improvement to dimensions used for earth models"—that is, for the World Geodetic System, mentioned below.[56]

The navy also used gravity data to determine the locations of places where nuclear submarines might hide and from which they might launch their missiles. Major General Daley, Director of Special Weapons in the Office of the Deputy Chief of Staff for Research and Development, raised this issue in April 1958, asking "how the Navy planned to get any geodetic accuracy in firing the Polaris missiles from submarines." John O'Keefe, a geodesist with the Army Map Service, responded in a once-classified memo that the navy would probably use seamounts as reference points for firing, that these seamounts would be referenced to nearby islands, and that the islands would be referenced to the continents.[57] In 1969 the navy acknowledged using gravity data "as an interim input to inertial guidance and navigation systems."[58]

Navy support led to the development of yet other gravimetric instruments. In March 1964 the Naval Oceanographic Office issued a request for proposals for a gravity meter that incorporated an accelerometer of the sort that had been developed for the guidance systems of submarines and long range missiles. The Coast & Geodetic Survey was also involved with this project, providing some 10% of the funds.[59] Bell Aerosystems received a $490,000 contract to develop an instrument based on its Model VII accelerometer. The first Bell

---

[55] W. Kaula, Memorandum for Record, 10 December 1958, NARA, RG77, Box 2/3, Folder 914; this concerned a meeting to revise and bring up to date the Joint Chiefs of Staff statement of ocean gravity survey requirements. A.L. McCahan, "The Navy Ocean Gravity Program" in *Proceedings of Military Geodesy Seminar, December 1958, Air Force Cambridge Research Center* (1959), pp. 27–32.

[56] Naval Oceanographic Office, *Annual Report of the Commander* (1969): 17.

[57] B. Chovitz, Memorandum for Record, 3 April 1958, NARA, RG77, Box 2/3, Folder 914.

[58] Naval Oceanographic Office, *Annual Report of the Commander* (1969): 17.

[59] Wilbur Kattner (ed.), *Advances in Dynamic Gravimetry* (Pittsburgh, 1970).

Gravity Meter (BGM-1) is said to have "performed well within specifications under a wide variety of sea conditions" and was soon accepted for operational use.[60] Bell introduced the BGM-2 in 1968 and the BGM-3 in the early 1980s, selling them for commercial as well as military use.[61]

While the BGM project was underway, William von Arx, a professor of oceanography at MIT, suggested that the Vibrating String Accelerometer (VSA) that the American Bosch Arma Corporation had built for the guidance system of the Atlas Agena rocket could also be used for a gravimeter. The ONR then provided funds so that academic scientists could pick up on this suggestion. Charles Wing, a graduate student at MIT, built a VSA gravimeter that could operate on the ocean bottom in 1967 and another that could operate on a surface ship in 1969.[62] Carl Bowin, head of the gravity program at Woods Hole, developed a similar instrument and used it on surface ships for several years.[63]

Since the movements of a ship at sea affect the readings of gravimeters on board, the navy asked Bell Aerosystems to develop a moving base gravity gradiometer—an assemblage of accelerometers that measured the gravity gradient between one point and another.[64] This instrument was a key component of the navy's Gravity Sensors System, a silent alternative to SONAR designed to prevent Trident II submarines from bumping into underwater ridges or mountains. Gravity gradiometers were partially declassified at the end of the Cold War and made available for geological purposes.[65] Lockheed Martin acquired Bell Aerosystems in 1996 and continued the gradiometer program. With funding from the Defense Threat Reduction

---

[60] "Report on Prototype Gravity Measuring System," in *Proceedings of First Marine Geodesy Symposium, 1966*, pp. 189–199.

[61] Don Norton, "Measuring Earth's Gravity," *Rendezvous* 7 (1968): 13–15; this is a monthly magazine of Bell Aerosystems. Bell's advertisement in *Geophysics* (June 1968): A-65. Hyman Orlin, "Gravity Sensing Instruments," *Transactions of the American Geophysical Union* 48 (1967): 353. Robin Bell and A.B. Watts, "Evaluation of the BGM-3 Sea Gravity System onboard R/V Conrad," *Geophysics* 51 (1986): 1480–1493.

[62] C.G. Wing, "An Experimental Deep-Sea-Bottom Gravimeter," *Journal of Geophysical Research* 72 (1967): 1249–1257. Wing, "MIT Vibrating String Surfaceship Gravimeter," *Journal of Geophysical Research* 74 (1969): 5882–5894.

[63] Carl Bowin, T.C. Aldrich, and R.A. Folinsbee, "VSA Gravity Meter System: Tests and Recent Developments," *Journal of Geophysical Research* 77 (1972): 2018–2033.

[64] Mark A. Gerber, "Gravity Gradiometry. Something New in Inertial Navigation," *Astronautics and Aeronautics* 16 (May 1978): 18–26.

[65] Robin Bell, "Gravity Gradiometry," *Scientific American* 278 (June 1998): 74–79.

Agency, Lockheed Martin developed an Arms Control Verification Gravity Gradiometer. A civilian version of this instrument was known as the Airborne Gravity Gradiometer.[66]

Air Force Programs

U.S. Air Force support of geodesy began shortly after that service was established in 1947 and increased substantially as missile defense gained prominence and as it came under air force control. A good way to access the air force story is through the Mapping, Charting and Research Laboratory (MCRL) of the Ohio State University (OSU). The MCRL was an off-campus organization established in 1947 that did contract research, primarily for the Pentagon.[67] In November 1949, shortly after the Soviet Union tested its first nuclear bomb, the MCRL informed the air force that Russian scientists claimed to have developed a "much more exact figure of the earth than the International Spheroid" (the figure used in the West), that two-thirds of all the gravity measures in the world were made in Russia, and that the Russians "make much use of the combination of gravimetric and astronomic measures to obtain the deflection of the vertical." Unstated but understood by all was the fact that this science would enable Soviet missiles to hit American targets.[68]

Taking the bait, the air force provided funds so that Weikko Heiskanen could come to the U.S. and "continue on a world-wide scale his studies on the size and shape of the earth which were started in Finland twenty-five years ago." Heiskanen was Director of the International Isostatic Institute, an organization that gathered and analyzed the gravimetric observations needed to support the theory that the crust of the earth was in isostatic equilibrium—a

[66] D. DiFrancesco, "Gravity Gradiometry Developments at Lockheed Martin," online at <www.cosis.net/abstracts/EAE03/1069/EAE03-A-01069> (accessed 15 May 2004).

[67] John Cloud, "Crossing the Olentangy River. The Figure of the Earth and the Military-Industrial-Academic-Complex, 1947–1972," *Studies in the History and Philosophy of Science* 31 (2000): 371–404.

[68] N.T. Bobrovnikoff, *Russian Science* (15 December 1949). Interest in Soviet work in this area continued over the years; see, for instance, Yu D. Bulanzhe, *et al.*, *Report on the Research in the Field of Gravimetry Conducted in the Soviet Union in 1962–1965*, English translation by Linguistic Section, Western Area Branch, Chart Research Division, ACIC, December 1965.

theory now discarded, but then very much in vogue. When talking to scientists, Heiskanen emphasized the connections between gravity and the theory of isostasy; but when talking to military men he stressed the connections between gravity and geodesy.

Heiskanen arrived in Columbus, Ohio in the late summer of 1950 and was immediately asked to study "to what practical geodetic purposes gravity anomalies can be used."[69] By early November, when the Panel on Cartography and Geodesy of the Research and Development Board convened at the MCRL, Heiskanen discussed the geodetic significance of gravity studies, while another colleague referred to the lack of an adequate world-wide gravity survey project as "a serious gap in our geodetic program."[70] The MCRL hosted another symposium for the air force in November 1952 on "The Importance of Gravity for Guided Missile Operational Requirements," and all participants needed a security clearance. Heiskanen made the summary remarks.[71]

Heiskanen also played an important role in the Institute of Geodesy, Photogrammetry, and Cartography (IGPC). Founded at OSU in late 1950, the IGPC was the first graduate program of its sort in the western hemisphere. Like other American science programs of the period, this one was well funded: students had, for instance, access to geodetic, gravimetric, and other instruments valued at some $90,000.[72] Unlike most other programs, however, most of the students here were military officers or civilian employees of a military service. After graduation, whether they went into academia or stayed in the military, these students remained effective advocates of gravimetric geodesy. Clair Ewing, the first Ph.D. graduate from this program, went

---

[69] W. Heiskanen, "The World-Wide Gravity Program of the Mapping and Charting Research Laboratory of Ohio State University," *Geophysics* 16 (1951): 697–700.

[70] "Suggested Program, Technical Papers, Afternoon, November 3, 1950," NARA, RG330, Entry 341, Box 456, Folder 6. W.A. Heiskanen, "World Gravity Needs for Geodetic Purposes," *MCRL Technical Paper* 118 (1950); this work was supported by Air Materiel Command contract AF 33 (038)-379. See also Heiskanen, "On the World Geodetic System," *Publications of the Isostatic Institute* 26 (1951); this also appeared as *Publication* #1 of the Institute for Geodesy, Photogrammetry and Cartography at Ohio State University. W.O. Byrd, "A Serious Gap in Our Geodetic Program," *MCRL Technical Paper* 125 (1950).

[71] W.O. Byrd to Roger Prior, 10 October 1951, NARA, RG330, Entry 341, Box 388, Folder 200.

[72] John Cloud, op. cit., p. 385. For the gravimeter see *Gravity Survey of the State of Ohio* (Columbus: Ohio Dept of Natural Resources, Division of Geological Survey, 1956).

on to become the first Director of Range Development at the Air Force Missile Test Center at Patrick Air Force Base. Richard Burkhard, who received an M.Sc. from OSU, went on to write *Geodesy for the Layman*, a wonderfully intelligent and intelligible publication that was used for decades to educate congressional and military decision makers about the need for geodetic research in general, and gravimetry in particular.

The air force had recognized as early as 1952 that "it would be desirable to conduct a training program for assigned personnel in gravity survey and data reduction work." Following the launch of Sputnik in October 1957 it worked with the IGPC to develop an intensive program in which students could obtain a masters degree in one year. During its first two years this program attracted 68 students, all from the Aeronautical Chart and Information Center (ACIC) in St. Louis, the air force counterpart of the Army Map Service and the Navy Hydrographic Office. After their stint at OSU, most of these students returned to work at the ACIC.[73]

Heiskanen had been attracted to the U.S. by the promise of funds for his research. Accordingly, he asked the air force for $150,000 (at that time an exorbitant sum for non-physical sciences) to acquire and analyze gravity data from around the world. The Air Force Scientific Advisory Board appreciated the scientific merits of this project. But, believing that Heiskanen should focus on the "basic military problem," it endorsed those phases of the proposal that related to the "determination of deflections of the vertical at stations in the U.S.S.R. east of the line Leningrad-Moscow-Stalingrad." As was said at that time: "Our map knowledge stops at Stalingrad—so did the Germans."[74] In other words, Heiskanen should work on "the reduction to a specified ellipsoid of points in the U.S.S.R. for which only astronomical position data is known."[75] After the successful completion of this practical task, Heiskanen and his colleagues received additional funding for similar work in other parts of the world. The

---

[73] Memorandum for Record, "Air Force Gravity Program," 7 October 1952, NARA, RG330, Entry 341, Box 389, Folder 34. C.H. Frey comments in S.H. Laurila and W.A. Heiskanen (eds.), *Geodesy in the Space Age* (Columbus, Ohio, 1961), pp. 7–8.

[74] Report of Working Group on Utilization of World Gravity Data, 18 October 1951, NARA, RG330, entry 341, box 389, folder 34.

[75] Norman Haskell to C.H. Harding, 10 August 1951, NARA, RG77, Box 2/3, Folder 914.

air force would continue supporting gravimetric and other geodetic research at OSU for the next four decades. With the end of the Cold War, the OSU geodetic program came to an end.

Funding for Heiskanen's research began under the auspices of the Air Force Cambridge Research Center (AFCRC). A secret report written for this organization in 1952 explained the military benefits of geodesy and gravimetry: if the size and shape of the earth were known, "the geodetic position of any point whose astronomical coordinates are known could be computed in reference to any other desired point." The first step towards this goal "would be the initiation of a worldwide gravity program of reasonable extent, in cooperation with the other agencies of the Department of Defense and civilian institutions."[76]

Heiskanen was not the first academic geodesist to receive support from the AFCRC. George Woollard had received a contract in 1949 to establish a network of gravity standards across the country—standards against which gravimeters, which tended to drift, could be regularly calibrated.[77] Woollard's project led to the World Gravity Base Plan, a massive international effort that produced a uniform calibration system, a first-order gravity network, and an absolute gravity system.[78] When the military brass learned about this project, they decided that the Department of Defense could not "openly or directly support an international effort as it would be politically infeasible." It could, however, support academics to do this work.[79] Woollard and his team also tested gravimeters for the AFCRC.[80]

Using funds provided by the AFCRC, Ewing and his colleagues at the Lamont Geophysical Observatory made gravimetric and other geophysical observations in the Arctic during the International Geo-

[76] Ralph J. Ford, "Geodesy and Gravimetry, Preliminary Report," *Air Force Surveys in Geophysics*, No. 11 (September 1952). This was declassified in 1964.

[77] G.P. Woollard, J.C. Rose, and W.E. Bonini, "The Establishment of an International Gravity Standard," *Transactions of the American Geophysical Union* 37 (1956/57): 143–155. Woollard and Rose, *International Gravity Measurements* (Tulsa, Okla., 1963), p. vii.

[78] *Report on Research at AFCRL, July 1962–July 1963*, pp. 239–40. See also, B. Szabo, "World Calibration Standard First-Order Gravity Net and Absolute Gravity System," *AFCRL Research Report for 1963*, pp. 19–28.

[79] F.L. Culley, "Memorandum: Conference on DOD World Gravity Base Plan," 31 May 1962, NARA, RG77, AMS, Box 2.3, Folder 914.

[80] G.P. Woollard, Bonini, *et al.*, "A Study of Methods for Measuring Large Changes in Gravity on an Intercontinental Basis," Woods Hole Oceanographic Institute, *Technical Report* 53–36 / AF 19 (122)-234; *Final Report*, 1 August 1953.

physical Year. In 1960, representatives from the air force visited Lamont "to discuss geodetic programs and possibly research for geodetic data which may be useful to USAF programs."[81] The AFCRC sponsored the first successful flights of a LaCoste & Romberg sea gravimeter in 1958–59, and it contracted with Sam Worden for an airborne gravimeter. Although the project manager looked forward to "an airborne gravity survey system which will provide data over the entire world in a few years," the project took much longer than expected and airborne gravimetry was never as successful as satellite gravimetry.[82]

The Air Force Cambridge Research Center was dissolved in 1960.[83] The reconstituted Air Force Cambridge Research Laboratory (AFCRL) provided the funds for James Faller's first laser interferometric apparatus, a device that produced very precise absolute measurements of gravity. This apparatus was used a various sites in the U.S., France, and England in 1968–1969, and provided the first transfer of absolute gravity measurements across the Atlantic.[84] AFCRL then hired James Hammond, the doctoral student who had helped Faller develop this apparatus, and assigned him the task of producing a smaller and more efficient instrument of this sort.[85] Absolute gravity measurements made by Hammond and Faller were included in the International Gravity Standardization Net adopted by the International Association of Geodesy in 1971.[86] The Air Force Geophysics Laboratory, as the organization was then known, sponsored a third and

[81] Folder "AFCRC", Maurice Ewing papers, Box 75, CAH.

[82] L.G.D. Thompson, "Airborne Gravity Meter Test," *Journal of Geophysical Research* 64 (1959): 488. L. LaCoste, "Airborne Gravity Measurements," *Journal of Geophysical Research* 64 (1959): 1127. Thompson and LaCoste, "Aerial Gravity Measurements," *Journal of Geophysical Research* 65 (1960): 305–322. "Study of Gravity from Plane Cited," *New York Times*, 9 August 1959) p. 7. AFCRL contract with S.P. Worden, AF19 (604)–5893. Uotila and Rapp, "Studies of the Earth's Gravity for Geodetic Purposes," Final Report of AFGL-TR-78-301 (December 1978). Robert M. Perry, "AFCRL's Experimental Aerial Gravimetry Program," AFCRL-ERP-366 (1971).

[83] Ruth Liebowitz, *Chronology. From the Cambridge Field Station to the Air Force Geophysics Laboratory, 1945–1985*, AFGL-TR-85-0201.

[84] AFCRL, *Report on Research, July 1967–June 1970*, pp. 112–113. "AF to Install Gravity Measuring Apparatus," *Electronic News* (14 October 1968). J.A. Hammond and J.E. Faller, "Results of Absolute Gravity Determinations at a Number of Different Sites," *Journal of Geophysical Research* 76 (1971): 7850–7854.

[85] AFGL, *Report on Research, July 1974–June 1976*, pp. 150–151. James Hammond and Robert Iliff, "The AFGL Absolute Gravity Program," in *Applications of Geodesy to Geodynamics* (Columbus, Ohio, 1978), pp. 245–249.

[86] W.E. Strange, "Land Gravimetry," *Reviews of Geophysics and Space Physics* 13 (1975): 255.

much improved version of the interferometric instrument in the mid-1980s.[87]

In the early 1970s the AFCRL began investigating the possibility of using the Bell Aerosystems model VII accelerometer for a moving base gravity gradiometer for airborne use. The resultant Air Force Gravity Gradiometer Survey System (GGSS), like its navy counterpart, was partially declassified at the end of the Cold War.[88] Two other gradiometer systems were also developed at this time. The Draper Laboratory developed a spherical floating gradiometer using a hollow beryllium sphere with two radially opposite proof masses. With funding from NASA, the Hughes Research Laboratory developed a rotating gradiometer.[89]

The Aeronautical Chart and Information Center (ACIC) began a study of "geodetic requirements for pilotless aircraft" in March 1954.[90] This soon led to an awareness of the "lack of special knowledge such as gravity and gravity anomalies which affect the size and shape of the earth and which have an undetermined effect on guidance systems" and a request that "gravity and magnetic data collection efforts be increased so that side effects of these phenomena may be eliminated or reduced as factors introducing errors to weapon guidance sub-systems.[91] By 1957 ACIC had begun work on an earth-centered mathematical model that made significant use of gravimetric data and that would lead directly to the USAF World Geodetic System.[92] ACIC became responsible for the Department of Defense gravity library in 1960. By 1968 this library had a staff of 87 who stored, cataloged and evaluated data from various sources.[93] ACIC also sup-

---

[87] Robert Iliff and Roger Sands, *The AFGL Absolute Gravity Measuring System*, AFGL-TR-83-0297.

[88] Christopher Jekeli, "The Gravity Gradiometer Survey System (GGSS)," *EOS* 69 (1988): 105, 116–117.

[89] Mark A. Gerber, "Gravity Gradiometry (note 64)". Hughes Research Laboratories, *Development of a Rotating Gravity Gradiometer for Earth Orbit Applications (AAFE)* (1973).

[90] *History of the USAF Aeronautical Chart and Information Center, 1 January–30 June 1954*, p. 20. *History of the USAF Aeronautical Chart and Information Center, 1 July–30 December 1954*, p. 59.

[91] *History of the USAF Aeronautical Chart and Information Center, January–June 1955*, classified supplement pp. 9–10.

[92] *History of the USAF Aeronautical Chart and Information Center, July–December 1957*, p. 46.

[93] D. Barnes and H.W. Oliver, "Symposium on Gravity Surveys in Western North America," *EOS* 50 (1969): 524–525.

ported a number of studies, many of them classified, concerning gravitational models of the whole earth and of those regions near missile launch sites.[94]

Finally, it should be noted the air force understood by the early 1950s that the deflection of the vertical around a launch site would affect the missile's path; and by the early 1960s the guidance systems had improved to the extent that the absolute value of gravity at the launch site was used to calibrate the accelerometers and velocity meters of long range missile.[95] Both of these factors pointed to the need for accurate gravity surveys around launch sites.[96] Although the early missile site surveys were conducted by the Coast & Geodetic Survey and by private contractors, the air force took matters into its own hands in 1959. The 1381st Geodetic Survey Squadron (which became the 1st Geodetic Survey Squadron in 1968, and simply the Geodetic Survey Squadron in 1972) eventually had a staff of about 600 working around the world.[97] While the Squadron worked largely behind the fence, some of their results were shared with the scientific community: notable in this regard were their tests of geodetic gravimeters under various environmental conditions.[98]

Conclusion

During the Cold War, the U.S. military establishment supported a number of expensive and expansive gravimetric projects. The trajectory of this engagement followed the development of long range missiles and their guidance systems. It was modest in the 1950s when gravity played but a small part in the error budget of missiles, but

---

94 See for instance the list of classified papers in the *Classified History of the USAF Aeronautical Chart and Information Center, July 1969–June 1970*, pp. 166–167.

95 Geodetic Working Group, Inter-Range Instrumentation Group, *1963 Technical Papers* (IRIG Document 105–64).

96 *History of the USAF Aeronautical Chart and Information Center, 1 January 1952–11 May 1952*, p. 19, mentions a secret conference held at Patrick Air Force Base in March 1952 to discuss the gravity surveys needed in and around the Air Force Missile Test Center.

97 "Air Photographic and Charting Service (MATS)," *The Military Engineer* 53 (1961): 137. Online at <http://164.214.2.59/GandG/control/rbshist.html> (accessed 10 March 2004). For private contractors see ad in *Sky & Telescope* 15 (Sept 1956): 508.

98 David Anthony, "Environmental Testing of Gravimeters," *Transactions of the American Geophysical Union* 46 (1965): 46.

grew substantially in the 1960s when guidance systems had so improved that geodetic factors became major considerations in their design.[99] Benefits from this military project were largely confined to the military, but not exclusively. Some gravimetric data developed for the military was classified, but some was available for academic and commercial purposes. Likewise, while some gravimetric instruments developed for the military remain under wraps, civilian scientists eagerly adopted the secret instruments as soon as they were declassified as well as those that have always been open. Moreover, many important discoveries about the earth's structure that have been made in recent decades are based on observations made with portable and precise gravity measuring instruments that the military developed for geodetic purposes.

[99] *Investigations in the Area of Geodetic, Gravitational, and Mapping Factors in Ballistic Weapons Systems.* Prepared by Ballistics Systems Division and Aerospace Corp. for USAF Military Geodesy Coordination Committee Meeting, Orlando AFB, 24–25 February 1966.

CHAPTER TWELVE

PHYSICS BETWEEN WAR AND PEACE*

Peter Galison

Introduction

Three hundred and fifty years ago, Galileo introduced the notion of mechanical relativity by invoking the experience of sea travel:

> Shut yourself up with some friend in the main cabin below decks on some large ship, and have with you there some flies, butterflies, and other small flying animals. Have a large bowl of water with some fish in it. . . . The fish swim indifferently in all directions; the drops fall into the vessel beneath; and, in throwing something to your friend, you need throw it no more strongly in one direction than another. . . . You will discover not the least change in all the effects named, nor could you tell from any of them whether the ship was moving or standing still.[1]

The imagery and dynamics of ships permeate Galileo's works, at once tying the new physics rhetorically to the modern navigational achievements of early seventeenth century Italy, and providing an effective thought-experiment laboratory for the new "World System."

Several centuries later, when Albert Einstein was struggling to overthrow Galilean-Newtonian physics, he too chose an image of contemporary transport as the vehicle for his radically new *Gedanken-experimente.* Now the railroads, symbol of the success of German technology and industry, replaced the sailing vessel in the argument. As Einstein put it, no optical experiment conducted in a constantly moving train could be distinguished from one performed at rest.[2]

---

* This essay originally appeared in Everett Mendelsohn, Merritt Roe Smith, and Peter Weingart (eds.), *Science, Technology and the Military*, Sociology of the Sciences 12 (Dordrecht; Boston: Kluwer Academic Publishers, 1988), pp. 47–86, and appears reprinted here with a new postscript with Kluwer's permission.

[1] Galileo, *Dialogue Concerning the Two Chief World Systems*, trans. Stillman Drake (Berkeley, Calif., 1967), pp. 186–187.

[2] See, for example, Einstein's popularization of relativity first published in 1916: *Relativity: the special and general theory* (New York, 1961).

Yet a third technological image of transport appealed to the American physicist Richard Feynman as he assembled a synthetic picture of quantum mechanics and relativity in the years after World War II. The positron, as he saw it, could be viewed as an electron moving backwards in time. If so, then the simultaneous creation of an electron-positron pair (ordinarily seen as involving two quite distinct paths) could instead be viewed as a single, continuous track: the positron travels backwards in time until it reaches the moment of creation, whereupon it becomes an electron moving forward in time. Feynman conveyed his vision with a vivid metaphor: "It is as though a bombardier flying low over a road suddenly sees three roads and it is only when two of them come together and disappear again that he realizes that he has simply passed over a long switchback in a single road."[3]

Every age has its cultural symbols, and Feynman's was as telling as Galileo's. Young American physicists of the 1940s and 1950s had seen their discipline recrystallize around the twin poles of radar and the atomic bomb. The venerated B-29 bomber carried both—an apt symbol, therefore, of the fruits of their labor—and served as a perfect vantage point from which to view the new theoretical and experimental physics.

Of course, the effect of the war on the development of physics in general, and of high-energy physics in particular, goes far beyond a passing metaphor. Indeed, the problem is that the effects are too great, and too varied, to be treated comprehensively in any one essay. For in a sense, the effects of the war permeate every aspect of postwar history. In the history of physics these effects include the thoroughgoing overhaul of the institutional structure of government-supported science. From the National Science Foundation to the Atomic Energy Commission, and the Office of Naval Research, no aspect of science funding remained unchanged.[4] Among the war's

[3] R.P. Feynman, "The Theory of Positrons," *Physical Review* 76 (1949): 749.

[4] On the World War II scientific organizations and their effects on research, see Irvin Stewart, *Organizing Scientific Research for War* (Boston, 1948); D. Kevles, *The Physicists* (New York, 1978); B. Hevly, "Basic Research within a Military Context: The Naval Research Laboratory and the Foundations of Extreme Ultraviolet and X-Ray Astronomy 1923–1960" (Ph.D. dissertation, Johns Hopkins University, 1987), esp. chap. 2; and A. Hunter Dupree, "The *Great Instauration* of 1940: The Organization of Scientific Research for War," in G. Holton (ed.), *The Twentieth-Century Sciences* (New York, 1972), pp. 443–467. On the postwar contributions of one important

consequences was a profound realignment of all relations between the academic, governmental, and corporate worlds, especially as physicists began contemplating the funding necessary for the construction of atomic piles, larger accelerators, and new particle detectors. Further, the war forged many collaborations and working groups among scientists that continued smoothly into the postwar epoch. And finally, the war provided astonishing quantities of surplus equipment that fed the rapidly expanding needs of postwar "nucleonics"—the study of a broadly construed nuclear physics, situated at the nodal point of research problems of cosmic rays, nuclear medicine, quantum electrodynamics, nuclear chemistry, and the practical imperatives of industry and defense.

Above all, one cannot ignore the new relation of university physics to military affairs that in a sense began, rather than ended, in the skies over Nagasaki. Suddenly academic physicists could negotiate with high-ranking officials from the Navy, the Air Force, and the Army to acquire new machines. At the same time, the military became an active participant in the shaping of postwar scientific research, through university contracts, the continuation of laboratories expanded during the war, and the establishment of new basic research programs under the aegis of individual armed services. Projects of joint civilian and military interest were lavishly funded, offering physicists the chance to think about exploring cosmic rays, not 3 or 4, but 100 miles above the earth's surface. Where a handful of technicians had once been sufficient to aid the physicists as they constructed new instruments, now the physics community began a deep new alliance with the various branches of scientifically informed engineers.

The present essay, addressing the impact of wartime research on postwar experimental and theoretical physics, can only begin to sketch some of these effects, drawing a few of the lines along which such a history of physics between war and peace might advance. I will not treat, for example, the vicissitudes in the dramatic careers of scientist-politicians such as Vannevar Bush, James Conant, or Robert

agency see S.S. Schweber's contribution to this volume, "The Mutual Embrace of Science and the Military: ONR and the Growth of Physics in the United States after World War II." [*i.e.*, in Everett Mendelsohn, Merritt Roe Smith, and Peter Weingart (eds.), *Science, Technology and the Military*, Sociology of the Sciences 12 (Dordrecht, 1988)].

Oppenheimer; the establishment and internal politics of funding organizations; or the alterations in industrial physics research policy. Instead, my goal is to peer into the effects of wartime science on the quotidian proceedings of physics itself, and into the experience of physicists in their research capacity. To do this I have chosen to focus on four exemplary physics departments: those at Harvard, Princeton, Berkeley, and Stanford. Each had its own trajectory, shaped in part by different war experiences and earlier patterns of research. Yet the four had much in common: each had to confront a sudden expansion, search for a new relationship between theorists and experimentalists, and solve the difficulties that accompanied the move into the epoch of large-scale, centralized, and cooperative research.

Together these changes transfigured the physicists' approach to research. In the most visible and dramatic fashion the war provided concrete instances of scientific accomplishments, though it remains an open question whether or not the lessons drawn from that experience were actually the ones responsible for the physicists' success. But in a sense that I will develop further below, the major weapons systems—radar, atomic bombs, rockets, and proximity fuses—formed *guiding symbols* that inspired the strategy of much postwar research. Needless to say, one can find earlier instances of one aspect or another of large-scale research: the great philanthropically funded telescopes, Ernest Lawrence's growing array of cyclotrons in the 1930s, and institutes of physics in Europe come immediately to mind—monumental telescopes cost millions of dollars, cyclotrons took several people to operate, and at certain institutes state and scientific concerns shared a common roof. Indeed, at universities such as Stanford and Berkeley, the 1930s saw the establishment of joint endeavors involving physics and electrical engineering. But despite the importance of such successes as the cyclotron and the klystron, before the war there were no physics achievements born of such large physics/engineering efforts that made the continuation of centralized big research seem either inevitable or inarguable. Beginning in the war, however, the physicists' and engineers' large-scale collaborative work on electronmagnetic and nuclear physics-based weapons systems provided just such exemplars. In part as a result of these successes, between 1943 and 1948 key segments of the American physics community came to accept a mutation in the ideal of the physicists' work and workplace. One after another, physics departments began to conceive of a style of orchestrated research that has come to dominate the character of modern investigations in high energy physics, and increasingly in other domains as well.

Expansion and the Repositioning of Physics in the University: Teaching, Surplus, and the Relation of Science to Engineering

Our first need is for a closer analysis of how the expansion affected physics departments, and to achieve this we need to dispose of the myth that changes in scientific planning began only after the guns of World War II had ceased firing. For it was *during* the period 1943–45 that physicists and administrators first debated and set in motion the coming boom. Across the country, from Berkeley to Harvard, pressure to think about postwar expansion began at the top.[5]

Harvard

At a meeting on 20 November 1944 the president and Fellows of Harvard College agreed to create a panel whose task was to direct the expansion of Physics, Chemical Physics, and Engineering. Selecting representatives from the various physical sciences at Harvard, President James Conant established a Committee on the Physical Sciences.[6] As a driving member of the principal organizations shaping scientific war research, Conant had arrived at a clear conception of the shape of postwar science. It was an image shared by several of his Harvard colleagues, as was evident at that committee's very first assembly. Edwin C. Kemble lobbied for physics in these terms": *The war* has given a great boost to physics. It has stressed the importance of physics to industry and national defense and has underscored the

[5] On the postwar work at Berkeley see R. Seidel, "Accelerating Science," *Historical Studies in the Physical Sciences* 13 (1983): 375–400.

[6] J. Conant to Kemble, 28 November 1944; summary of meeting of 20 November 1944, in Hiektaann Papers, Physics Department Historical Records, box 3, ca. 1930–65, Harvard University Archives (hereinafter abbreviated as HP). Already in the summer of 1943, Conant was deeply involved in postwar planning for military/scientific relations, though the early speculations did not lead directly to accepted policy. See the excellent book on the military's expectations for their postwar condition: M. Sherry, *Preparing for the Next War. America plans for postwar defense 1941–45* (New Haven, Conn., 1967), pp. 137–138.

usefulness of men trained in pure physics when emergency requires that they turn to applications."[7] Consequently, he argued, the university needed to expand to new fields, get the best personnel, and enlarge the group for instructional purposes. In nuclear physics, which was "the field of greatest interest," Kemble conveyed the department's desire to keep its existing staff, to add "top caliber" theoretical physicists—for example, Julian Schwinger, Hans Bethe, or Harvey Brooks—and, of course, to augment their total budget with funds for mechanical assistance and construction costs.

To justify the expansion of theoretical physics, Kemble composed a memorandum on 9 December 1944 to the Physical Sciences Panel which began, under the rubric "presuppositions," by contending that there would be a "nation-wide acceleration in the growth of the science of Physics as a result of war emphasis." As Kemble saw it, it was a growth that would take place in two areas. *Solid-state physics* ("the field of properties of matter in bulk") demanded the efforts of a new, more powerful contingent of theoretical physicists, who would be masters of quantum mechanics, statistical mechanics, and chemical thermodynamics. For Kemble, the need for more theory was illustrated by his colleague. P.W. Bridgman, whose superb investigation into the high-pressure domain had nonetheless "undoubtedly fallen short of its maximum potentialities since, to date, he has worked without steady and effective collaboration from theoretical physicists.[8]

Beyond solid-state physics lay *nuclear physics,* which was, by Kemble's lights, "the most spectacular field in physics today." It was there that "the riddle of the physicist's universe is found," amid the cosmic rays, mass spectrographs, cyclotrons, and forms of radioactivity. And as the war was making perfectly clear, medical and chemical applications appeared "manifold," and were accompanied by the even more tantalizing "possibility of unlocking stores of atomic energy [which added] urgent significance to the investigations." Exactly this combination of the purely intellectual and of hoped-for practical consequences characterized the physics community's justification of the needed expansion. In brief, the intellectual argument positioned atomic physics as a stepping-stone to nuclear physics.[9]

[7] File cards in handwriting of E.C. Kemble, HP (note 6).

[8] Kemble, "Panel: Physical Sciences. Memorandum on Proposals for the Development of the Department of Physics," 9 December 1944, HP (note 6).

[9] *Ibid.*

Speaking for his department, Kemble argued that since World War I the focus of physicists' concern had altered from understanding atomic structure and the structure of simple molecules. Previously, Harvard's own efforts had to a large extent been devoted to spectroscopy of all kinds; but now problems of this type "have largely been solved," and "work in this field operates against a law of diminishing returns," to be replaced by the more alluring problems of nuclear and solid-state physics. Both of these new domains demanded a new, deeper cooperation between experimental and theoretical workers to handle

> the increasingly abstract and complex character of present-day physical theories. This quality is the result of the intensive search for more powerful means of attack on problems of a more and more difficult character. As one consequence of the complexity of the theory, the most brilliant experimental physicist is in need of theoretical collaboration to an extent previously unknown.

At the same time, an increasing number of the "more gifted young men" were choosing theoretical physics. Oppenheimer's presence at the Berkeley Radiation Laboratory was an example for all to see of the theorist's usefulness in joining the skills of experimental and theoretical physicists; the young theorist's contribution "has been of crucial importance in the meteoric rise of that laboratory to its place as principal center for nuclear investigations in this country." Unspoken—but undoubtedly understood—was Oppenheimer's masterful guidance of the Manhattan Project. The Harvard physicists hoped Schwinger or Bethe could enliven theoretical life in Cambridge.[10]

Acquiring theorists was, however, only part of a much larger pattern of growth. By the end of December 1944 Kemble was sure enough of the expansion plans to write Conant:

> My personal acquaintance with the state of engineering arts at the time of the last war and at the present time convinces me that it is imperative for our future national safety that the scientific bases of engineering practice shall have far more intensive study than heretofore. We must have a much increased number of analytical engineers with brains and advanced training if we are to hold our own in the technological race.[11]

[10] *Ibid.*
[11] Kemble to Conant, 22 December 1944, HP (note 6).

Kemble added that the universities could best aid this cause by prosecuting work in "pure science," which would yield more "in relation to the investment" and attract the best "quality of the men . . . likely to execute it."[12] Here the Harvard physicist embraced the central justifications for expansion that would be repeated over and over again during the following decades: connections between fundamental research and teaching, industrial spin-offs, and military preparedness.

The "modest" investment Kemble had in mind was hiring Julian Schwinger and Edward M. Purcell; adding $25,000 for operating expenses, $30,000 per year for operating the cyclotron, and $30,000 for the physics of metals; a new electronics building at a cost of $100,000; a new mechanical engineering building at $500,000; an electronics research budget of $10,000 per year; further appointments in electronics, mechanical engineering, and aeronautical engineering; construction of a wind tunnel for $75,000; and building an interdisciplinary science center for between 1.5 and 2 million dollars.[13] Clearly, the expansion envisioned by plans such as these extend far beyond nuclear or "fundamental" physics and embraced a picture of a much-enlarged program for both physics and engineering. A few days later Kemble contacted his physicist colleagues Jabez Curry Street and Kenneth Bainbridge, requesting them to consider how their respective research areas might participate in the expansion.[14]

A boosted physics budget made possible both the new machines and the increased role of theoretical physics. Throughout the American university system the growth of physics was also fueled by a spectacular jump in the number of students. During the war, the armed forces had called upon the physics departments to teach thousands of conscripts the elements of physics so they could cope with a new generation of technical war apparatus, especially radar, radio, rockets, and navigational equipment. After the war the G.I. Bill funded these and other students as they came back to the university in droves. Although providing instruction to so many students often taxed their already depleted resources, wartime instruction also presented universities with an unprecedented opportunity to expand their clientele.

---

[12] *Ibid.*

[13] Kemble, "Tentative Summary of Proposals before Physical Sciences Panel," HP (note 6).

[14] Kemble to Street, 30 November 1944; Kemble to Bainbridge, 30 November 1944, HP (note 6).

Temporary physics programs had formed during the war at Harvard and Bowdoin in electronics and communication, at MIT in radar, at Los Alamos, and at other institutions as well. At Harvard alone, 5,000 students passed through the pre-radar electronics course. After demobilization the faculty expected that large numbers of veterans from that course would return to further their physics education. As E.L. Chaffee put it on 9 December 1944, the Officer War Training Courses put "Harvard in an advantageous position for attracting students after the war."[15]

Aside from introducing promising students to the university, Chaffee remarked that "the war-training courses have provided us at no expense with a very considerable amount of laboratory equipment. There is insufficient space in the Cruft Laboratory even to store this equipment, say nothing of setting it up for instruction."[16] Instructional equipment could be supplemented by war surplus apparatus suitable for research. Chaffee noted that Harvard's antiradar work and other war projects would offer

> an unusual opportunity to purchase advantageously some very valuable equipment from the OSRO [Office of Scientific Research and Development] projects and perhaps some equipment from the war training courses. . . . We should be prepared to purchase a considerable amount of this equipment. There will [be] available machine tools, obtainable at much reduced prices from the same sources, and I believe we should purchase a considerable amount of this machinery both to increase our present shop facilities and to replace some outmoded and worn machine tools.[17]

With this new equipment Chaffee expected that physicists in the postwar period would be able to exploit their newly developed capability to generate microwave signals "by methods which have worked but which are not understood." Such applications included detecting molecular resonances, pulse systems of communication, and high-frequency heating. As Purcell rather over modestly put it: you didn't have to be too smart to design an experiment with the extraordinary resources offered by the new electronics.[18]

[15] Chaffee, "Expansion of Research and Instruction in the Cruft Laboratory," HP (note 6). We still have relatively few systematic data on the effects of this teaching: were there lasting effects on the physics curriculum? Where did the students go after the war: to academic pursuits? industrial positions? military assignments?

[16] *Ibid.*

[17] *Ibid.*, p. 8.

[18] E.M. Purcell, interview 26 May 1987. Purcell discovered the 21-centimeter

Surplus equipment came from many war sources and went to a wide spectrum of users. Connections established between the scientific and defense communities grew, yielding benefits for scientists long after V-J Day. Since they were on contracts from the Office of Naval Research, as late as 1960 physicists at Brookhaven could acquire large armor plating, originally intended for cruisers, to use in neutrino experiments.[19] On the West Coast, Robert Hofstadter collected Naval gun mounts on which he could perch his magnetic spectrometer.[20] Overseas, A. Gozzini's experiments with surplus pulse-generator circuits and microwave equipment led to his development with Marcello Conversi of the "flash tubes" that played such an important role in cosmic-ray physics and in subsequent work on the spark chamber.[21] Phototubes, crucial for scintillation devices, had been much improved and exploited during the war as sources for noise generation in radar countermeasures. As we will see, this type of continuity between wartime and postwar work ran deep; beyond the formation and evolution of administrative organizations such as OSRD, an essential consequence of the war work was the carry-over of physicists' techniques, equipment, and collaborations into the late 1940s.

With the promise of new physicists, new students, and new equipment, departments could embark on major new research projects. Harvard—along with many other universities—was determined to restart its cyclotron program on a much-increased scale. Planning to accelerate both electrons and protons, the nuclear physics planning committee met for the first time in January 1946. They wrote: "From the point of view of physics this program represents a vigorous and progressive plan which should enable Harvard to compete favorably for financial support and, in addition, enhance its attractiveness as a center of research in the nuclear field."[22]

line, and won the Nobel Prize for his co-invention of Nuclear Magnetic Resonance—both exploited Rad Lab techniques.

[19] M. Schwartz, interview 20 October 1983.

[20] R. Hofstadter, H.R. Fechter, and R.H. Helm to Commandant, Mare Island Naval Shipyard, 9 February 1952, Hofstadter private papers, Stanford University.

[21] See P. Galison, "Bubbles, Sparks, and the Postwar Laboratory," in L. Brown, M. Dresden, and L. Hoddeson (eds.), *From Pions to Quarks: elementary particle physics in the 1950's* (Cambridge, forthcoming). [the volume appeared in 1989]

[22] "Proposal for Nuclear Physics Program at Harvard," submitted 2 February 1946 and enclosed with minutes of the first meeting (11 January 1946) of the

When the Harvard physicists turned from war work to accelerator work they brought with them experience from the Manhattan Project (Kenneth Bainbridge), from the Radar Project (Edward M. Purcell, Julian Schwinger), and from Radar Countermeasures (Roger Hickman, John Van Vleck). All served on Harvard's newly established Committee on Nuclear Sciences. After selling their old cyclotron to the government for $200,000, and getting a commitment of $590,000 from Harvard and $425,000 from the Navy, the committee members could begin to plan an 84-inch cyclotron.[23] Their physics program included proposals to produce 25 MeV deuterons and 50 MeV alphas in order to explore the nature of the proton, to produce high-energy neutrons, and to extend the wartime fission experiments to elements lighter than uranium and thorium. With accelerated electrons on tap, the planning committee hoped that the cyclotron could also be exploited to pursue radiation therapy, the photodisintegration of nuclei, and the formation of electromagnetic showers.[24]

Princeton

At Princeton, as at Harvard, planning for the postwar expansion began long before the euphoric crowds descended on Times Square. On 4 January 1944 a somewhat overoptimistic John Wheeler wrote to H.D. Smyth that he trusted the war would soon be over and he could return to physics shortly. He then went on to formulate a "Proposal for Research on Particle Transformations," of which the "Ultimate Purpose" was "to determine the number of elementary particles, the transformations between them, the combinations which they permit, the nature of their interactions, the relation between these particles and the existing theories of pair formation, electromagnetism, gravitation, quantum mechanics and relativity." As he was witnessing an immensely successful collaboration of theory and

Committee on Nuclear Physics [later Committee on Nuclear Sciences], file "Index and Records," box "Committee on Research and Nuclear Sciences, Records 1946–1951," Harvard University Archives.

[23] P.H. Buck to R. Hickman, 5 October 1946, file "Miscellaneous Financial Material and Correspondence," box "Committee on Research and Nuclear Sciences, Records 1946–1951," Harvard University Archives.

[24] "Proposal for Nuclear Physics Program at Harvard" (note 22).

experiment at the University of Chicago's Metallurgical Laboratory (Met Lab), culminating in the achievement of fission in December 1942, he clearly saw this as the wave of the future. Wheeler began: "*Plan.* Effective progress calls for the collaboration of experiment and theory. The interaction between the two will be most fruitful, I believe, when in addition both approaches are combined in a single institution, under the same leadership."[25] He argued that such collaboration ought to be the model for physics research at Princeton; the lesson he drew was one taken to heart at institutions across the United States.

Wheeler needed help to solve the main theoretical problems. These included the need to invoke action-at-a-distance theories to eliminate the self-energy difficulty in quantum electrodynamics, the necessity of classifying relativistic quantum field theories, and exploring the theory of positronium. There were also such "associated experimental" problems as nuclear meson capture, mass distribution of cosmic ray particles, and the gamma-ray production of mesons. Like Kemble at Harvard, Wheeler counseled building the theoretical side of the Princeton department, by hiring three theoretical assistants of "the type of Feynman or Jauch," along with some experimentalists of the "type of Luis Alvarez or Bob Wilson." Of course the thirty-three-year-old Wheeler would "[w]elcome collaboration of interested older members of staff," and, again grounding his recommendation on his Chicago experience, he suggested that experimentalists should call on experienced "electronics men" for the design and development of instruments. The program promised, as Wheeler continued to insist after the war, the possibility of sources of energy many times more powerful than all known nuclear reactions, with "obvious implications for the problem of national defense."[26]

As ambitious as young Wheeler's plans may have sounded initially, even before the war's end the Physics Department had begun to address the administration in a new, more confident tone. With the successes of proximity fuses, radar, and then the atomic bombs, physicists—who for years had occupied a decidedly secondary place

[25] J.A. Wheeler to H.D. Smyth, 4 January 1944, file "Postwar Research," in Physics Department Departmental Records, Chairman, 1934–35, 1945–46, no. 1, Princeton University Archives (hereinafter, PUA).

[26] *Ibid.*

within the university—acquired a radically improved self-image, evident as they spoke to colleagues and administrators. As the Physics Department put it in a draft of their department report,

> The end of the war finds the department in a praiseworthy but embarrassing condition. The record of the members of the department in war work is laudable, so much so that many of them, particularly in the younger group, are receiving very attractive offers from other institutions and from industry. Such offers are not only attractive in terms of salary but are usually backed by promises of large expenditures for apparatus and equipment. The university must choose between going ahead vigorously, capitalizing the fine record of this department during the war or letting its physicists drift away to such a degree that it may take a generation to restore the department. The first course will require money for men and for equipment, a great deal of money, but it offers a magnificent opportunity, completely in the tradition of the university. We have never been in a better position to push forward in the field of fundamental physical research.[27]

Indeed, their position *was* entirely unprecedented.

Physicists everywhere were attracted by the promise of the new technology and science for advancing the physics of nucleons and mesons, but developments in "fundamental" experimental physics at Princeton took on a particular cast. In part this reflected an imaginative style of work that John Wheeler had developed before the war, but the echoes of what he had seen in the Metallurgical Laboratory can clearly be heard in his ideas for postwar research. As early as June 1945 Wheeler had penned a proposal on the future of physics research that gave three goals for the postwar epoch. First, though he voiced doubts about some of their features, he advocated the development of accelerator sources for particles. Among these, he mentioned Luis Alvarez's latest plans to build a linear electron accelerator; Wheeler judged that one would want at least a 5 GeV proton accelerator in order to produce pairs of mesons.[28] Second, he wanted Princeton to establish a wide-ranging "ultranucleonics" program. Third—and here he saw the real payoff in physics—he

[27] "Department of Physics Report to the President 1944–45," Physics Department Chairman's Correspondence 1942–43, 1943–44, no. 16, PUA.

[28] J.A. Wheeler, "Three Proposals for the Promotion of Ultranucleonic Research #6: H.D.S.," 15 June 1945, copy to Smyth, in Physics Departmental Records, Chairman 1934–35, 1945–46, no. 1, PUA.

hoped that the department would seize upon cosmic rays as the primary domain in which to search for answers to basic questions, because there and only there could one find the high-energy collisions needed to probe the subnuclear domain. For the cosmic-ray project, Wheeler suggested using Flying Fortresses to hoist experiments and experimenters into the upper atmosphere.

The idea of enlisting bombers for the study of high-altitude cosmic radiation had several appealing aspects. It would alleviate the costs of such exploration for the universities, while leaving control over research apparatus entirely in the hands of the physicists. As for its physics justification, Wheeler noted that only by reaching far into the sky could one study particles with something like 10^{17} electron volts, and therefore exhibit the multiple meson production process that interested him. "This plan calls for army transportation of equipment up to 10 tons to altitudes of the order of 40,000 feet. Research money would in this way be freed for research itself, and for research of a most effective kind."[29]

Finally, Wheeler felt that a survey of the entire field of ultranucleonics was of the highest priority. From what had been learned by cosmic-ray studies, he suspected that it might be possible to transform matter directly and completely into energy on the model of protons being transformed into mesons in the upper atmosphere.

> Discovery [of] how to release the untapped energy on a reasonable scale might completely alter our economy and the basis of our military security. For this reason we owe special attention to the branches of ultranucleonics—cosmic ray phenomena, meson physics, field theory, energy production in supernovae, and particle transformation physics—where a single development may produce such far-reaching changes.[30]

To reach these dramatic goals, physicists had to make their needs known, and here the survey would play a vital role. It would offer workers in postwar physics "a prospectus of long-range objectives," and it would gain financial support for fundamental physics by making research public and by demonstrating "that scientists in free association can show more vision and judgment on research planning

[29] *Ibid.*

[30] *Ibid.*; and see J.A. Wheeler, "Elementary Particle Physics," *American Scientist* 35 (1947): 170, 172, 174, 177–193, 223.

than any centralized government authority." Not least, the ultranucleonic survey would "uncover lines of investigation of evident present or future value to the country's war power."[31] Though these functions were already far beyond the typical prewar involvement of the government in basic research, within a few months Wheeler had in mind new even more active roles for the government to play in physics.

One such role came from Nazi Germany's development of the dreaded "vengeance weapons," the V-1 "buzz bomb" and the V-2 guided missile. This had been an engineering project of immense scope, costing over $3 billion—fully one and a half times the resources put into the Manhattan Project. The V-2 was brought into the war relatively late, but starting in September 1944 the Germans successfully launched more than 3,000 V-2s, killing almost 10,000 people in England. Soon, however, the Allies began advancing across Europe, and Wernher von Braun retreated from his headquarters at Peenemunde with some 4,000 workers to the V-weapon production facility situated in the concentration-camp complex of Dora-Nordhausen in Thuringia. Installing themselves in the Harz mountains, the V-2 workers successfully evaded the approaching Russian army, eventually surrendering themselves to an American garrison. Under a secret mission code-named "Overcast," the American army shipped the Nazi scientists to the United States to continue missile development work at several sites. The Germans arrived in October 1945, and soon the Peenemunde team was split into groups, working with American industry to produce a variety of rocket types.[32]

From 16 April 1946 to 19 September 1952, 64 V-2s were launched from White Sands. The first failed three and a half miles into the air when a fin ripped off and the rocket was destroyed; the next launch, on 10 May 1946, successfully rose to 71 miles.[33] For physicists at Princeton, the capture and reinstallation of the German rocket team offered an immediate opportunity. In November 1945—just a month after von Braun and his associates were brought to White

[31] Wheeler, "Three Proposals" (note 28).

[32] F. Ordway III and M.R. Sharpe, *The Rocket Team. From the V-2 to the saturn moon rocket* (Cambridge, Mass., 1982); see also the important article by Linda Hunt, "U.S. Coverup of Nazi Scientists," *Bulletin of the Atomic Scientists* 41.3 (April 1985): 16–24.

[33] Ordway and Sharpe, *The Rocket Team*, pp. 353–354.

Sands, M.H. Nichols jotted an interoffice memo to Smyth suggesting that the department ought to propose to study optical and electrical phenomena in the upper regions of the earth's atmosphere. At the same time the Princeton group could explore cosmic rays and neutron densities. All this was made possible by "new advances in rocket technique as well as progress here at Princeton and elsewhere in the field of radio telemetering from aircraft and missiles," which would "make possible an extension of present data to regions as high as 500,000 feet."[34]

For some time Wheeler had seen cosmic-ray physics, not accelerator physics, as the primary vehicle for understanding elementary particles. In a memo of January 1946 he stated that "cosmic ray research will take on an even more important role in physics in the next few years," and he advocated immediately setting up a joint experimental and theoretical research group. "Inasmuch as the V-2 firings will not last indefinitely, and inasmuch as the experienced researchers are becoming scarcer every day, it appears that some action [should] be taken as soon as sound decisions can be made." Presumably addressing himself to the department chairman, Wheeler stressed that they would be needing four to six assistants with experience in experimental electronics, nuclear physics, or cosmic rays, as well as experienced cosmic-ray and nuclear physics experimentalists. "Research in physics is starting fresh, and . . . new techniques and new vehicles are now available"—and so it behooved the department to search out consultants from among the best universities, institutes, and weapons laboratories.[35]

Wheeler himself headed a Navy-funded project that would handle the telemetric transmission of cosmic-ray data from the German team's missiles. Begun on 1 January 1945 for other purposes, the Navy grant had been extended in March 1946 and was to cover the development of telemetry equipment, while at the same time serving to study cosmic-ray showers and the properties of "mesotrons" through the design, fabrication, and operation of cloud chambers and Geiger counters to be mounted on the V-2s. In July 1946 D.J.

---

[34] M.H. Nichols to H.D. Smyth, 26 November 1945, file "Postwar Research," Physics Department Records, Chairman 1934–35, 1945–46, no. 1, PUA.

[35] No author listed [probably Wheeler], "Program in Cosmic Rays," January 1946, file "Postwar Research," Physics Department Records, Chairman, 1934–35, 1945–46, no. 1, PUA.

Montgomery reported to the Navy that the Princeton V-2 expedition had arrived at White Sands and was making final tests for what they hoped would be the 100-mile-high Princeton Shot on 6 August. At this point $335 thousand had already been allotted, with $250 thousand more to be shared over the next two years between chemistry and physics.[36]

Cooperation with the military remained close. Military and elementary-particle problems were interspersed in planning and designing the mission. Both physicists and strategic planners needed a comparison of "Lark" and Naval Research Laboratory telemetering systems, especially with regard to the reliability, intensity of signals, and freedom from disturbances of each system. Both civilians and uniformed personnel had to study radio signal propagation in the ionosphere by transmitting and receiving signals from the missile. In addition, the physicists could use the high-altitude flight to measure cosmic-ray intensity, to distinguish primary cosmic-ray electrons from primary protons, and to measure the neutron productivity in the atmosphere as a function of altitude.[37] Such studies directly continued some of the Princeton group's wartime accomplishments in telemetry. In fact, at least one member of the staff, Dr. Walter Roberts, wanted to continue this work on guided missiles at the Johns Hopkins Applied Physics Laboratory. As Wheeler assured his readers, this would ensure a "satisfactory liaison" between the Princeton and weapons laboratories.[38]

During the period from 1945 to the early 1950s, the liaison between civilian and military nucleonics functioned well—from both parties' perspectives. The Office of Naval Research (ONR) liberally funded civilian science, and in return the scientists moved easily back and forth between nuclear physics, cosmic rays, and weapons problems.

---

[36] D.J. Montgomery, "Annual Report of Project Assisted by Outside Funds," 23 July 1946, file "A-475 Wheeler," Laboratory and Research Files, 1929–54, box I of 5, PUA (see also "Elementary Particle Projects as of 6 May 1946," in same file). Among physics goals listed in this report were: determination of total cosmic-ray intensity, meson production, neutron intensity, multiply charged particles at rocket altitudes; study of radio propagation in ionosphere; telemetry tests; pressure, temperature studies; and coordination with the Schein group's ground cloud-chamber and balloon tests.

[37] Wheeler, "Appendix IV—General Survey of the Princeton Project Program—Cosmic Rays and Telemetering," 28 August 1946, file "A-475 Wheeler," Laboratory and Research Files, 1929–54, box I of 5, PUA.

[38] *Ibid.*, see under "Guided Missile Developmental Work."

Princeton's nuclear physicist, Milton White, for example, was pleased to report on some recent Princeton instrumentation work that seemed perfectly suited for transfer to the military sector. The laboratory had perfected a new, simple, rugged, and reliable scintillation counter, and White lost no time in alerting ONR to possible defense applications of the new device:

> If the U.S. Government has need of a-particle counters, either in connection with plutonium plants, or atomic bombs, there should be set in motion a program for further engineering and quantity manufacture. I can visualize an eventual need of many thousands of counters; if this is correct then our contract with the Navy will already have given the government more than a fair return on the money thus far allocated.[39]

White added only that he hoped the Princeton researchers could be spared the engineering details.

Berkeley

On the West Coast, Stanford and Berkeley had no intention of being spared the engineering details. The style of research in the West was somewhat different from that in the East: it was more tightly bound to engineering, and it drew more liberally from philanthropic and industrial sources. Such entrepreneurial physics had brought Berkeley's E.O. Lawrence international fame for his big accelerators paid for from private coffers; as engineering accomplishments big accelerators were unrivaled, though Lawrence was less successful at drawing deep physics from them. As Robert Seidel has so nicely shown, World War II brought a substitution of federal for philanthropic funds, and the important assignment that Lawrence's laboratory direct the electromagnetic separation of U-235. Lawrence was as well prepared to begin large-scale research as anyone, and shortly was supervising a dramatically bigger laboratory with a wartime expenditure of $692 thousand *per month.* By the middle of 1944 the Radiation Laboratory held a total working population of 1,200 scientists, engi-

[39] M.G. White to Urner Liddel, Nuclear Physics Section, ONR, 11 July 1947, file "761," box V of 5, Laboratory and Research Files, 1929–54, PUA.

neers, and technicians; indeed, once their war work began in earnest the number of engineers at the laboratory never dipped below sixty.[40]

After a few months of intense discouragement because of difficulties with his electromagnetic separation facility, Lawrence began escalating his expectations for the postwar period. During the summer of 1944 he began lobbying for ten new isotope separation facilities, leading General Leslie Groves (who was in charge of the Manhattan Project) to some cautious thinking about spending $7–10 million. Just a year later Lawrence began arguing for the rapid expansion of nonweapons facilities, including Luis Alvarez's plans for a linear accelerator and Edwin McMillan's for a synchrotron. Within a few months of the end of the war, Groves authorized $250 thousand in surplus radar sets for the linear accelerator, $203 thousand in surplus capacitors for the synchrotron, $630 thousand for construction in the laboratory, and $1.6 million for six months of operating expenses. Building on earlier experience, engineering and physics had grown together at Berkeley, to make the university one of the models of postwar physical research. In fact, the Berkeley Radiation Laboratory became the pacesetter for the Atomic Energy Commission's development of regional laboratories.[41]

Stanford

Stanford, like its Berkeley neighbor, had successfully linked engineering and physics before the war. While Lawrence and his team were building ever-larger cyclotrons, the Stanford physicists were binding electrical engineering to physics as they learned to manipulate microwaves. William Hansen had set the character of that collaboration at Stanford with his stunning development of the "rhumbatron," which set electrons in an oscillatory dance by creating electromagnetic resonances within a copper cavity. Although the device was quickly superseded as a particle accelerator, it formed the core of the klystron, a powerful microwave tube that the Varian brothers

[40] Seidel, "Accelerating Science" (note 5).

[41] *Ibid.* See also R. Seidel, "A Home for Big Science: The Atomic Energy Commission's Laboratory System," *Historical Studies in the Physical Sciences* 16 (1986): 135–175.

designed and deployed in airplane navigation and locating systems. Soon the Sperry Gyroscope Company was underwriting a good deal of the joint physics/electrical-engineering efforts. With the help of their electrical engineer leader, Frederick Terman, Stanford's electrical engineering department built a myriad of radio communications systems around their jewel, the klystron. Gradually, the Stanford engineers transformed the klystron from a fascinating, isolated tube to a standardized component within a whole gamut of microwave circuits.[42]

On a technical level, the microwave klystron-based research continued unabated into the early years of World War II: weapon innovations included instrument landing systems and doppler radar. But dramatic changes quickly accompanied the increased pace, scope, and funding of laboratory work. Already in April 1942, Paul Davis (Stanford's general secretary) was writing Terman that "there are many things that could be done under the pressure of the present war situation that will be more difficult to achieve in peace times"—including ambitious plans for electrical engineering. In August 1942, Stanford issued its "Proposal to Organize the Stanford Resources for Public Service," focusing on how to organize "a vastly augmented program of service on a contractual basis."[43]

After listing suggested projects (from surveys of mineral and industrial resources to the creation of psychological warfare tunes, such as "Marching Civilization"), the August proposal turned to the effects of war work on Stanford. A radical increase in contractual research would provide an opportunity to reorganize faculty administration and to improve the physical plant for more effective war and postwar work. Substantial contracts would bring federal war priority, keeping faculty on campus, creating interdisciplinary research, and engaging a new cadre of talented students who would stay on after the war. Moreover, working on contracts would make Stanford stu-

[42] On Stanford's early combination of physics and engineering see the excellent article by S.W. Leslie and B. Hevly, "Steeple Building at Stanford: Electrical Engineering, Physics, and Microwave Research," *Proceedings of the IEEE* 73 (1985): 1169–80. [see more recently Stuart W. Leslie, *The Cold War and American Science* (New York, 1994).]

[43] P. Davis to F. Terman, 18 April 1942, Terman Papers, SC 160, 1:1:2, Stanford University Archives (hereinafter SUA); unsigned typescript, "A Proposal to Organize the Stanford Resources for Public Service," 24 August 1942, Terman Papers, SC 160, 1:1:2, SUA.

dents known to public and private agencies, and might contribute to the long-term development of the West. But above all, government-sponsored research would rocket Stanford to a position comparable to Harvard, Chicago, Caltech, the University of California, and Columbia by bringing in "substantial additional income."[44]

Ironically, while Stanford did greatly expand during the war, its great advocate, Frederick Terman, spent the war years on the East Coast as director of the Radio Research Laboratory in Cambridge (RRL), a facility built to produce radar countermeasures. As the new laboratory grew into a powerful organization, Terman became ever more conscious of the models that Harvard, MIT, and RRL itself presented for the postwar situation at Stanford. He was also deeply impressed with many of the administrators from Harvard—especially with his neighbor, William Henry Claflin, Jr., Harvard's treasurer.[45] Terman was quite keen for Stanford's general secretary to speak with Claflin, and the Stanford engineer-administrator soon sought and arranged a meeting between Donald Tresidder, Stanford's president, and Claflin. By December 1943, Terman had concluded that

> the years after the war are going to be very important and also *very critical ones* for Stanford. I believe that we will either consolidate our potential strength, and create a foundation for a position in the West somewhat analogous to that of Harvard in the East, or we will drop so a level somewhat similar to that of Dartmouth, a well thought of institution having about 2 per cent as much influence on national life as Harvard.[46]

Terman went on to set out a plan to "lick" Caltech by equaling it in the physical sciences—since "after all they are only a specialized school, and Stanford is a complete university." In part, Terman wanted a "technical institute" that would create a joint identity among scientific and engineering fields. This alliance would aid in attracting and placing students, raising money, and creating an identity based on special Western areas of strength. One such field was the characteristically Western oil industry, which would link geology, heat transfer, and chemical engineering with the radio industry and accompanying research. Moreover, Terman argued, the competition was

---

[44] *Ibid.*

[45] Terman to Davis, 23 August 1943, Terman Papers, SC 160, 1:1:2, SUA.

[46] Terman to Davis, 29 December 1943, SC 160, 1:1:2, SUA; partially cited in Leslie and Hevly, "Steeple Building" (note 42).

softening: Caltech had become "smug," leaving "cracks in its armor" by not developing electrical engineering, and Harvard, Yale, Columbia, and Princeton had thus far slighted the applied sciences in favor of natural philosophy and the humanities.[47] The codevelopment of electrical engineering and physics was the hallmark of Stanford physics, as it passed from the Microwave Laboratory to the High Energy Physics Laboratory, and eventually to the two-mile-long Stanford Linear Accelerator.

The Continuity of Technique and Discontinuity of Results

On many levels, then, physics began to change *during*, not after, the war at institutions like Harvard, Princeton, Stanford, and Berkeley. Nonetheless, there is a natural tendency among physicists and historians to overlook the continuity between wartime weapons development work and postwar research, and to reach back before the war to points of common peacetime research. The difficulty may stem from an understandable focus solely on *results*, ignoring the techniques and practices of the discipline. Contributing to the physicists' inclination to elide the effects of war on research is the preponderance of theorists among those who have narrated the discipline's history.

It may also be that war/postwar continuities are slighted because after the war the physics community found itself divided over the opportunities and hazards of the links to weapons research. Physicists walked a tightrope, using government funds to build the machines and teams they needed, but at the same time trying to reestablish a domain of work free from the constraints of a too-closely directed and supervised research. The struggle to maintain that independence also contributed to a vision of the history of physics as, in a sense, skipping lightly over the war years.

But whatever the source of this hesitancy in tracing the continuity of wartime to postwar research, we have inherited a broken narrative. Let me illustrate this point by focusing on the work of the long-productive cosmic-ray physicist Bruno Rossi.

Rossi, an important contributor to cosmic-ray physics before the war, to the war effort itself, and subsequently to high-energy physics, offered the following recollection:

[47] *Ibid.*

> In 1939, a systematic investigation of air showers was initiated by Auger and his collaborators. Their work, still carried out by means of Geiger-Muller counters, produced results of very great significance. However when, in the late 40's, air shower work was resumed, it became clear that, in order to substantially advance and refine these studies, more sophisticated kinds of detectors were needed.[48]

If attention is paid only to the specific results of air-shower research, Rossi's comment makes perfect sense. But instead of halting our historical inquiry at that point, let us descend to a "lower level" of analysis—that is, let us focus on the instruments and techniques of the work in question.

Many of the instruments developed after 1945 to detect air showers were fundamentally linked to war work. In Rossi's particular case, this link was abundantly clear since he, with H. Staub, literally wrote the book on the subject. Their volume, *Ionization Chambers and Counters* (1949), was produced for the National Nuclear Energy Series, Manhattan Project Technical Section. It summarized the advances in electronics and detectors that issued from the radar and bomb projects. Starting in July and August of 1943 Staub had directed a Los Alamos team in charge of improving counters, and Rossi led a group to improve electronic techniques. In September 1943, the two groups were merged into a single Experimental Physics Division group P-6, the detector group, under Rossi.[49]

Roughly speaking, their task was to design and implement detector systems that could determine the type, energy, and number of particles emerging from a variety of interactions. Their principal mission was to develop ionization counter systems that functioned in four stages: a first device detected the particle by producing a small current; a second amplified the current; a third separated the signal from unwanted noise; and a final instrument counted and recorded the total number of pulses. Physicists from the two big war projects had improved electronic instrumentation in all four areas—detection, amplification, discrimination, and counting.

An ionization counter works as follows: A charged particle passes through a gas that is contained between two parallel plates at different

---

[48] Bruno Rossi, "Development of the Cosmic Ray Techniques," in A. Berthelot (ed.), *Colloque International sur l'Histoire de la Physique des Particules*, in *Journal de Physique*, no. 12, vol. 43 (1982), pp. C8–82. [*sic*].

[49] D. Hawkins. "Part I: Inward Trinity," in D. Hawkins, E.C. Truslow, and R.C. Smith, *Project Y: the Los Alamos story* (Los Angeles, 1983), p. 90.

voltages. Along the particle's track it ionizes gas atoms; the electrons wander toward the positive plate, and the ions toward the negative plate. If the field is not too strong, the charge deposited on one of the plates is equal to the number of ions produced. When these charges arrive at the collecting plate, the current that they produce can be amplified; the shape and height of this current can then be used to determine the charge and energy of the incoming particle.

Rossi's immediate postwar contribution to physics involved the development of fast timing circuits for cosmic rays. His work of the late 1940s built directly on the wartime timing circuits that he had used to link ionization chambers in tests of the Los Alamos "Water Boiler" reactor. For that "Rossi Experiment," as it became known, the Italian physicist set a neutron detector to register the presence of a chain reaction inside the reactor. Using a fast coincidence circuit, the experimenter could count the number of other neutrons emitted during a brief period after the start of fission. In this way Rossi and his coworkers determined the period between the emission of prompt neutrons (those simultaneous with the fission event) and the delayed neutrons.[50]

By 1947 military authorities had declassified not only Rossi's electronic contributions, but a compendious batch of 270 Los Alamos technical reports. Immediately, journals on instrumentation brimmed with the new information. Even a cursory perusal of the 1947 volume of *Reviews of Scientific Instruments* indicates the depth of interest in the instrumentation that had been developed in the weapons projects.

Consider just one example from each of the four stages of measurement mentioned above. One way to find a neutron's energy was to scatter it from a hydrogen nucleus inside an ionization chamber. The recoiling hydrogen nucleus, since it is charged, ionizes other particles in the gas; these, in turn, cascade toward the negative plate. The pulse is then proportional to the number of ions, which is proportional to the energy of the recoiling proton, which is proportional to the energy of the original fast neutron. A variety of such "proportional counters" issued from the Manhattan Project, including sensitive ones that could measure the energy of neutrons traveling in a particular direction.[51] Signals from devices like these could then

[50] *Ibid.*, pp. 104–107.

[51] J.H. Coon and R.A. Nobles, "Hydrogen Recoil Proportional Counter for Neutron Detection," *Reviews of Scientific Instruments* 18 (1947): 44–47.

be analyzed. The simplest possible device only registered a pulse if its height came above a certain set level. In more sophisticated instruments developed at other laboratories (e.g., by the Chalk River group), separate channels were activated by pulses of varying energy. Such "pulse-height analyzers" gave an immediate energy spectrum; they were (and are) essential instruments in postwar nuclear physics.[52] Finally, once the pulses emerged from the discriminator they needed to be counted. Here too a great deal of progress was made during the war. One such device that was designed at Los Alamos to be used with a wide variety of detectors was the "Model 200 Pulse Counter," which, like the other devices just described, was first made public in 1947.[53]

Physicists from the Manhattan and Radar projects disseminated their work widely. One, William C. Elmore, prepared a series of Saturday lectures that he delivered at Princeton in the spring of 1947. The Department of Physics mailed nearly 300 copies of the lectures to physicists all over the United States, and many of the Princeton physicists made quick application of the techniques;[54] Robert Hofstadter, to offer one example, recalled that his own work on the inorganic scintillation detector was strongly shaped by Elmore's talks. The next year Elmore published an expanded version of his lectures in the journal *Nucleonics* as a four-part article, "Electronics for the Nuclear Physicist." According to the author, the series constituted in part a "commentary on electronic instruments designed at the Los Alamos Scientific Laboratory, and now employed extensively at various university laboratories."[55]

---

[52] Before the war there were essentially two ways to obtain an energy distribution. One could record the pulses photographically with an oscillograph, which was cumbersome and required large amounts of film. Or one could employ a counting circuit with a discriminator that would record the number of counts N above an energy amplitude E; the resulting "bias curve" (N versus E) then had to be reduced after the experiment by taking $N(E) = dN/dE$, and plotting this quantity against E. See H.F. Freundlich, E.P. Hincks, and W.J. Ozeroff, "A Pulse Analyser for Nuclear Research," *Reviews of Scientific Instruments* 18 (1947): 90–100. As of February 1947, descriptions of the other devices still had not been published: *e.g.*, E.A. Sayle, British Project Report, January 1944 (cited in Freundlich, *et al.*).

[53] W.A. Higinbotham, J. Gallagher, and M. Sands, "The Model 200 Pulse Counter," *Reviews of Scientific Instruments* 18 (1947): 706–714.

[54] Committee on Project Research and Inventions, Princeton University, "Proposal for Continuation of Research Projects in Proton-Nuclear Reactions for the Year 1948–49," 5 March 1948, loose in box V of 5, Princeton Physics Department Laboratory and Research Files, 1929–54, PUA.

[55] W.C. Elmore, "Electronics for the Nuclear Physicist," Parts I–IV, *Nucleonics* 2

These particular examples are only a sample of the variety of ways in which wartime ionization detectors, fast electronics, discriminators, and scalars were pursued after the war. Experimentalists used their wartime expertise to design devices for experiments with X rays, electrons, positrons, neutrons, protons, gamma rays, and fission fragments. Perhaps more influential than any of these considerations were the microwave techniques that played crucial roles in accelerator technology after the war: wave guides, transmission lines, klystrons, molecular beams. In addition, there were the benefits of the radiation laboratory efforts—better low-noise amplifiers, lock-in amplifiers, microwave oscillators, which profoundly shaped nuclear magnetic resonance techniques, radio astronomy, and microwave spectroscopy. It is not possible here to speak of the other war-bred technologies that led to calculating machines, computers, and many aspects of programming. Suffice it to say that wartime research had transformed the material culture of the physical sciences.

Collaboration, Work Organization and the Definition of Research

Thus far our attention has been on skills and the instruments of physical research. But the war left another legacy, one not captured in the new research apparatus, or even in the surplus war material that formed such an important basis for experimental work. The war provided a lesson about the nature of research that left an indelible stamp on the physicists who participated in the massive programs at the Chicago Met Lab, the MIT Radiation Laboratory, Berkeley, Oak Ridge, Hanford, and Los Alamos. That lesson concerned large-scale research organized upon complex managerial lines. So it was that just a few weeks after D-Day, Henry Smyth sketched at Princeton a proposal for a new kind of physics laboratory, one that could duplicate the scientific/engineering successes already in hand from the various wartime enterprises.

Smyth titled his July 1944 effort "A Proposal for a Cooperative Laboratory of Experimental Science," and the document reflected on the vast changes facing physics: "The war," he wrote,

(nos. 2–4, 1948): 4–17, 16–36, 43–55, 50–58; quotation on p. 4 of Part I. Together Elmore and Matthew Sands wrote a book, *Electronics: experimental techniques* (New York, 1949), that was widely distributed and translated into several languages.

> has now reached the stage where it is desirable to make plans for the postwar period and the period of transition. The complete disruption of the normal activities of the universities and, in particular, of the scientific groups in the universities leaves the whole condition of science in this country highly fluid.

Smyth's remark strikes at a central, and often ignored point: change was facilitated to a large extent because the traditional structures of research, leave-time, personnel, teaching, and interdepartmental boundaries had been radically altered. While "normal procedures" were suspended, deeper and faster mutations could be imposed on the system than would have been possible in peacetime. As Smyth noted, the direction of those mutations would shape the very definition of a physicist, and of physical research:

> Forty years ago the physicist working on a research problem usually was largely self-sufficient. He had available a certain number of relatively cheap instruments and materials which he was able to assemble himself into an apparatus which he could operate alone. He then accumulated data and interpreted and published them by himself. Most of the special apparatus that he needed he himself constructed with his own hands. He was at once machinist, glassblower, electrician, theoretical physicist, and author. He instructed his students in the various techniques of mind and hand that were required, suggested a problem, and then let the student work in the same fashion under his general supervision.

But even before the war, Smyth pointed out, physics had been growing more complex. Large laboratories had begun adding specialized technicians to their staffs, including glassblowers and machinists. Even graduate students came to depend on these technicians. Devices such as grating spectrographs dwarfed in size and complexity the simple table-top devices that previously had been sufficient. Thus it was that "a certain amount of cooperation in the use of such installations had to be worked out. But even twenty years ago research problems were largely individual." Only between 1930 and 1945 had this predominantly individual research dwindled, as equipment grew larger, more expensive, and sufficiently hard to handle that it came to require a team of workers to run. Of all such devices, the cyclotron stood out as the most dramatic example, costing in some cases more than the prewar physics budget of fifty laboratories.[56]

---

[56] H.D. Smyth, "A Proposal for a Cooperative Laboratory of Experimental

For Smyth such developments held dangers, as well as promises. If every university aspired to build cyclotrons or betatrons the costs could prove overwhelming. There was a danger that only a few elite institutions would be left in command of research and that consequently the "background of strength in science which [had] grown up so successfully in the country in the past twenty years" would be weakened. Only cooperative research, he felt, could salvage the situation by consolidating various universities and "other institutions" into centralized enterprises.[57] In a February 1945 revision of his document, Smyth added that such big projects should not "oppress the individual scientists. Such installations must be the servants, not the masters of the research man."[58]

The OSRD experience of physicists carried, Smyth argued, three lessons: First, the importance of "fundamental science" in solving problems that mere "specialists" could not handle. "The moral which is to be drawn from this experience is that the ultimate technological strength of the country, even for military purposes, rests on men trained in fundamental science and active in research on fundamental problems of science."[59] Second, the war demonstrated the profits that could accrue from "large cooperative research enterprises," cooperation that would extend not only to other scientific fields such as chemistry, biology, and medicine, but to the deep links between physicists and engineers.[60] The latter was an alliance with roots dating from before the war, but which bore fruit only in the wartime efforts. Although for security reasons Smyth passed over it, the obvious reference of this section of his proposal is to the Manhattan Project; at the time the revised proposal was written, in February 1945, Los Alamos was only a few months from detonating its first nuclear weapon.

The laboratory of Smyth's dreams clearly echoed a reality that lay, still secret, in the New Mexican desert. He figured 300 square feet per physicist, about 100 physicists, and about 15 foot ceilings, along with 5,000 square feet for large installations. This yielded

Science," 25 July 1944, file "Postwar Research 1945–46," Physics Department Departmental Records, Chairman, 1934–35, 1945–46, box I, PUA.

[57] *Ibid.*

[58] H.D. Smyth, "A Proposal for a Cooperative Laboratory of Experimental Science," revised version, 7 February 1945, file "Postwar Research 1945–46" (note 56).

[59] Smyth, revised version of "Proposal," p. 3 (note 58).

[60] Smyth, "Proposal," versions of 25 July 1944 and 7 February 1945 (notes 56, 58).

525,000 cubic feet, which at $0.70 per cubic foot would have totaled $367,500. Industrial production provided the architectural prototype: "The laboratory should be essentially of factory-type construction, capable of expansion and alteration. Partitions should be nonstructural." And with a democratic flourish Smyth appended his intention that "paneled offices for the director or any one else should be avoided."[61]

Soon Smyth found his thoughts echoed in his colleagues' memoranda. One physicist was

> genuinely concerned that in the Atlantic coast region we have some possibilities in this field [of cooperative nuclear research] and that all of the government support is not thrown to those other sites which have a prior claim, of course, because of existing facilities.[62]

Wheeler also reacted with enthusiasm, in December 1945, to the idea of a multi-university collaborative enterprise, and had no doubt that universities were owed support by the federal government:

> Anyone familiar with work on nuclear physics and its applications to military and peace time uses is aware that progress in this field in the United States has now dropped to a very low rate. Scientists are leaving to go to laboratories where they can have conditions of freedom appropriate for independent investigations. The country is losing out because it hasn't been able to work out a system suitable to enlist the participation of the scientists. In addition to this problem of applying science in the country's service, there is also the problem of what the country can do to replenish the scientific capital on which it drew so heavily during the war. The universities paid in years of peace for the fundamental research of which the government took advantage in time of war. The universities need and can rightly call for government support in the future.[63]

Such support should come in the form of engineering assistance, and equipment, Wheeler argued. He preferred a system in which the government financed and ran a facility where university researchers could bring their cloud chambers or magnetic spectrographs, make some measurements, and return to their home institutions. A self-administering center would therefore not burden academics more

[61] Smyth, revised version of "Proposal" (note 58).

[62] William W. Watson to Smyth, 23 June 1945, file "Postwar Research 1945–46" (note 56).

[63] J.A. Wheeler, draft of "Proposal for Cooperative Laboratory," 11 December 1945, file "Postwar Research 1945–46" (note 56).

than necessary. For support Wheeler looked to the men and institutions that had built such laboratories in the past: the Manhattan Engineer District, Vannevar Bush, and industrial concerns.

Over the next half year the Princeton physicists combined forces with others in the Northeast to draft a proposal for a nucleonics laboratory that would cost about $2.5 million (it soon increased to $15 million, then to $22 million, and finally to $25 million). Blessed with support from General Leslie Groves and the Manhattan Project, the planning staff of the budding Brookhaven National Laboratory recruited the building and management expertise of Hydrocarbon Research Inc. In addition, and to the consternation of Oak Ridge, the planners explicitly resolved to crib experience and information from the proven facilities at Oak Ridge.[64]

Unlike reactors, which were obviously too big for most educational institutions, cyclotrons hovered for the next few years at the boundary between being too big for universities and too small to merit their own cooperative laboratories. Indeed, Brookhaven's initial attempts to get its accelerator division funded were unsuccessful. Later, when synchrocyclotrons appeared, the Brookhaven reactor laboratory served as a prototype for collective research at the accelerator, and the model was soon extended elsewhere in the United States and then to Europe as a template for CERN. War laboratories thus clearly provided the managerial models, the technical expertise, and even the personnel for the establishment of postwar collaborative laboratory work.

Concern about the impact of a big cyclotron on university physics is visible in the case of Princeton. Milton G. White reported to Smyth in December 1945 that many members of the department were keen to find a place for themselves at a cyclotron facility, but remained a bit apprehensive about the nature of the research that awaited them:

"Dicke is leaning heavily toward elementary particle physics, but not too anxious to press for high energy if the engineering must come out of his hide. He wants to help get the cyclotron under way and then work on some simple interactions." Therefore, to get the

[64] For my discussion of the origin of BNL I have relied on the excellent article by Allan Needell, "Nuclear Reactors and the Founding of Brookhaven National Laboratory," *Historical Studies in the Physical Sciences* 14 (1983): 93–122.

accelerator program "back on its feet," White wanted "*very* much to acquire someone who would attend to moving, wiring, redesign problems." In addition, he wanted one of those sought-after types, "a Los Alamos man."[65]

More generally, White foresaw the need to create new positions for the changed environment of the large-scale laboratory, ones outside the traditional academic hierarchy from assistant professor to tenured full professor. "On the one hand," White reported,

> we find physics research going in the direction of complex equipment requiring a supporting staff of highly competent, broadly trained physicists, engineers, chemists and administrative personnel; while on the other hand we have the customary university policy of regarding all scientific employees as likely candidates for academic positions.

Instead of hunting for the "well-rounded" man appropriate to academia, White advocated a Division of Research that would hire with soft money provided in part by endowment, but significantly supplemented by industrial and government funds.[66]

White was concerned about on-campus accelerator physics, whatever might come of proposals for the cooperative laboratory. And expected costs for the cyclotron clearly were going to be high: from $100,000 to $300,000. Size would also be significant, since forecasts called for at least 75,000 square feet. Using Smyth's estimate quoted earlier, a plant of this dimension would run at least another $825,000. In all, White forecast expenses of around $100,000 per year for the next five years, and even this sum was exclusive of the building and power requirements of the accelerator.

> No crystal ball is required to outline the trend in high energy physics—the trend is up! In not more than six months it should be possible to settle on the part we wish to play in high energy physics, and having settled this we must pick some one accelerator scheme and back it for all we are worth.[67]

Faced with such skyrocketing costs, White simultaneously advocated a cooperative nuclear physics laboratory, with funding in large part

[65] M.G. White to H.D. Smyth, 20 December 1945, file "Postwar Research 1945–46" (note 56).
[66] M.G. White to H.D. Smyth, 6 May 1946, file "Postwar Research 1945–46" (note 56).
[67] White to Smyth, 20 December 1945, file "Postwar Research 1945–46" (note 56).

to be provided by private industry—citing, for example, the Monsanto Company.[68] Unfortunately for White's plans, Monsanto declined.

Stanford, by contrast, had already been successful at linking physics, engineering, and private investment. As Stuart Leslie and Bruce Hevly have shown, such scientific entrepreneurship had begun several years before the war.[69] Building on Stanford's engineering/physics of the 1930s, in November 1942 William Hansen began advocating the establishment of a "Stanford Microwave Laboratory" to be headed by Karl Spangenberg, an electronics specialist, and drawing on the consulting expertise of H. Skilling and Felix Bloch. When peace came, Hansen wrote the physics department chairman, Paul Kirkpatrick, the microwave lab would draw further staff and equipment from Physics and Electrical Engineering. "At this point, this laboratory goes out and gets a government contract for some microwave job. There should be no difficulty in doing this. With this job will come a priority. . . . Then you start spending money and also, of course, doing the job." Hansen hoped to establish as many attractive fellowships as could be filled, and then to order "equipment—of a sort that will be useful after the war." This should include "machine tools, measuring equipment, books, and any other things that can be used to generate apparatus or research." Played well, the plan would guarantee the "even if we don't have a dime after the war, good physics can be done."[70]

Almost exactly a year later, in November 1943, Hansen elaborated on his initial scheme. It had already been "obvious" before the war that physicists would play a crucial role in industry and that radio engineering (including electronics) was in for rapid expansion. But,

> while these trends were obvious before the war, the war has both accelerated and called attention to them. This is especially easy to notice . . . in the field of radar . . . The result will be that, after the war, all major universities will be forced to offer strong instruction in these two branches of science.

[68] White to Charles Thomas, Monsanto Chemical Company, 18 September 1945, file "Postwar Research 1945–46" (note 56).

[69] See Leslie and Hevly, "Steeple Building" (note 42).

[70] Hansen to Kirkpatrick, 6 November 1942, Terman Papers, SC 160, 1:1:7, SUA (note 43).

Precisely because it could strengthen and link the two domains, the microwave laboratory would prove essential. Modeled on the Berkeley Radiation Laboratory, it would remain under the control of the physics department, although the director would control the budget. Indeed, by establishing the laboratory with university funds, the inevitable private support would not dictate the direction of work.[71]

Thus by the war's end in August 1945, the microwave laboratory had been in gestation for nearly three years. And with the atomic bomb project passing from top secret to a national obsession, the physics department scaled up its requests. Now, in October 1945, discussions of physicists and physics took on a new, assertive tone, unheard before the war. Of Norris Bradbury, whom Kirkpatrick wanted to lure back to Stanford: "He is thus the head of the group that changed all human history, a continuing group whose power to effect such changes is by no means exhausted."[72] Even Bradbury, it had become clear, had to struggle to keep the staff together at Los Alamos—and the salaries across the country had simply skyrocketed. As an indication of the precipitous rise, Stanford offered one physicist a salary of $3,750, which was met immediately with a $6,000 counteroffer from the University of Chicago. In fact, of a sample of fourteen physicists (almost all between 31 and 41), the *average* salary was $8,460, with a high of $15,000 and a low of $6,000. "Whether one likes it or not [this salary level] reflects the present pronounced bull market in physicists, which has naturally resulted from a short supply and a heavy demand."[73]

Now the Physics Department could address a memo to the president of the university that bluntly asserted in its first paragraph: "It appears to be the manifest destiny of this department to expand." Unanimously the Stanford physicists brought forth several "considerations": (1) ROTC and NROTC students were going to be taking physics courses, as were soldiers in the Army's officers training program; (2) the Consulting Committee on Undergraduate Studies was considering requiring new physics courses; (3) the federal government was probably going to subsidize science students, and "physics

[71] W. Hansen, "Proposed Micro-Wave Laboratory at Stanford," typescript, 17 November 1943, Terman Papers, SC 160, 1:1:8, SUA.

[72] Kirkpatrick to Tresidder, 1 October 1945, Tresidder Papers, SC 158, 1:4, SUA.

[73] *Ibid.*

will be the first science affected"; and (4) the microwave program was going to draw students, and indeed the whole "war record of physics is causing students to plan careers in this science." Whereas before the war about one freshman per year indicated he wanted to major in physics, the figure now stood at eleven, and would surely rise further with the influx of veterans. On the basis of these facts, the department welcomed "a chance to enlarge the department" from six, to eight or nine on its permanent staff.[74]

Enlargement by 50 percent would cost money, and as Hansen had advocated during the war, the source was government-contracted research. Clearly this benefited the universities, but how so the government? Colonel O.C. Maier of the Air Technical Service command put it concisely to Stanford's president, in January 1946. The Army Air Forces, Maier wrote, had two purposes in continuing close cooperation between universities and the military: "We would not only get a good deal of our work accomplished by capable personnel, but in addition build up a pool of trained engineers and scientists, who could be of assistance in War Department research in case of emergency." The choice of tasks spanned the gamut of microwave research: propagation in low, high, ultra-high, very-high, and microwave frequencies; modulation systems, including accurate timing for radar applications; pulse modulators; broadbanding of antennas and circuit elements; magnetrons and klystrons; research on millimeter waves; three-dimensional radar data presentation, and beacon communication; moving-target indicator research; random polarization jammers; flight computers; navigation systems; loran research; and novel radar systems.[75] And this was just one prospective offer from, one branch of the armed services—supplemented significantly by the Atomic Energy Commission, which soon took over contracting from the Manhattan Project. Within just a few years, workers at Stanford would be moving easily between classified, applied research and the open domain of academic studies.

[74] Kirkpatrick to Tresidder, 6 November 1945, Tresidder Papers, SC 151, 25:1, SUA.

[75] Colonel Oscar C. Maier to Office of the President, Stanford University, 11 January 1946, Tresidder Papers, SC 158, B1:4, SUA.

Conclusion: War and the Culture of Physics

For years we have treated the history of physics as if it simply stopped between 1939 and 1945; only the movement of refugee scientists, bomb building, and the federal administration of a dramatically larger science budget have commanded attention. But if we are going to understand the deeper implications that the conflict had for modern physics and, by extension, for all of the modern sciences, we must look to the techniques and practices of the discipline. In this paper I have followed five lines of continuity: the transfer of technology, the transfer of support, the realignment of physics and engineering, the new relation between theoretical and experimental physics, and the reorganization of the scientific workplace.

Technological transfer consisted, in part, in the invention of new devices: novel accelerator technology, such as the klystron and strong focusing. The new electronic technology of counters, timers, amplifiers, and pulse-height analyzers (among others) all contributed to the postwar burgeoning of the physics of nuclei and particles. But beyond pure invention, the war increased the industrial production of high-performance components. In turn, this capacity made tools available to the experimenter that had previously necessitated custom manufacture. Finally, there was technological transfer at the most literal level—great storehouses of equipment, hundreds of millions of dollars' worth of machinery, that the federal government shipped directly from war-designated activities to the civilian sector as surplus. From the lone researcher picking up a microwave generator in Europe to the best universities of the United States, this infusion of tools and machining equipment transformed the scope and capacities of postwar research.

The financial support for the discipline of physics had also completely altered. The Manhattan Project continued to underwrite many activities, and when it finally closed shop, its sponsorship was quickly taken over by OSRD, ONR, and then by the AEC. Most importantly, at each of the universities discussed here (Princeton, Harvard, Berkeley, and Stanford) it is strikingly clear that the war had trained academic physicists to think about their research on a new scale, invoking a new organizational model. Not only did physicists envision larger experiments than ever before, they now saw themselves as *entitled* to continue the contractual research that both they and the government had seen function successfully during the war. This way of thinking molded both the continuation of wartime projects and the planning for new accelerators and national laboratories. Across the country, with occasional strong dissent, and with different emphases in different regions, the physics community strove to reenact the trilateral collaboration among government, university, and private enterprise.

The expanded institutional base of research permitted more complex relations between theorists and experimentalists, and between physics and engineering. These collaborative relations were firmly established at the huge project centers for scientific warfare: in the Metallurgical Laboratory of Chicago, in the vastly augmented Berkeley Radiation Laboratory, at Harvard's Radio Research Laboratory, in the rocket plants of Caltech, at the MIT Rad Lab, at Oak Ridge, and at Hanford. So it was that when the war ended, it was altogether natural for a Brookhaven, a Stanford Microwave Laboratory, or a rejuvenated Berkeley Radiation Laboratory to assume—from the start—a style of physics that elevated the role of theorists in the shaping of research programs, while keeping scientific engineering front and central. Through these new alliances, the great particle physics laboratories could be built, and the new relationship with the military justified.

Finally, and perhaps most importantly, the war changed physicists' mode of work—in the process redefining what it meant to be a physicist. As Smyth put it so eloquently, forty years earlier the physicist had been at once machinist, glassblower, electrician, theoretical physicist, and author. Just a few years after the war all that had changed: consolidating a prewar trend, the new breed of high-energy physicists were no longer taught to be both theorists and experimentalists—they chose one path or the other. In the place of physicist-craftsmen arose a collaborative association among theoretical and experimental physicists and engineers of accelerator, structural, and electrical systems.

One consequence of these interrelated transformations was a marked shift in rhetoric. The new, often triumphalist tone contained elements of pride in the physicists' contribution; it also signaled a defensiveness, as the scientists struggled to legitimate government funding while avoiding tight restrictions on the prosecution and dissemination of their research.

But the changes in the material culture, organization, and goals of physics went far beyond new turns of phrase. Collectively, these various factors gave rise to a new style of research in nuclear and particle physics. Schematically, it is useful to think of this "industrialization" of university accelerators and national laboratories as having occurred in two stages. The first stage, discussed here, involved an upheaval in the laboratory environment in which nuclear physicists worked. This transformation of the "outer laboratory" began with Lawrence's prewar forays into big science, but became the norm of nuclear physics only in the years 1943–48. In the following decade, the scale change of physics would reach even further into the conduct of experimental high-energy physics. With Alvarez's massive hydrogen bubble chambers, the "inner laboratory" itself, the micro-environment of the experimentalist's measuring and calculating devices, grew, like the outer laboratory, to industrial size.[76]

When Henry Smyth solicited a laboratory with "factory-type construction" using "non-structural partitions," and a director's office "without paneling," he was advancing a straightforward architectural request. But in those plans were other, less visible architectures. From the war, physicists had inherited a new sense of mission-directed, team-executed research that required a new human architecture as well—specialization, and collaborations with well-defined leaders (as evident from remarks like Wheeler's, as he speculated on postwar physics). These directors were to lead interdisciplinary nuclear physics programs that, with their movable partitions, could shift priorities as new instruments or questions arose. Gone were the days when Palmer Hall at Princeton or Jefferson Laboratory at Harvard could devote small rooms purely to acoustical or magnetic research. And of course the new research would find its natural place in the lavishly-funded regional laboratories where—as in the war projects—university, industry, and government would work as a triumvirate.

At the same time, one senses in the postwar architectural plans an apprehension about the new physics—a sense that the leaders should not isolate themselves or stifle the working physicist. From many of the postwar physics planners one feels a deep tension, socially

[76] For more on the "inner" and "outer" laboratories and an extended discussion of the transformation of the inner laboratory in the 1950s, see P. Galison, "Bubble Chambers and the Experimental Workplace," in P. Achinstein and O. Hannaway (eds.), *Observation, Experiment, and Hypothesis in Modern Physical Science* (Cambridge, Mass, 1985), pp. 309–373; and Galison, "Bubbles, Sparks" (note 21).

and intellectually. between the power available through collaboration and the ideological commitment to individual research. It is a tension only incompletely resolved through the Los Alamos model of large-scale, hierarchical teamwork where the physicist could argue with those up the line. Even the director of the huge National Accelerator Laboratory would later write an autobiographical essay entitled "My Fight against Team Research."[77] When Smyth sat down to plan a regional laboratory he assumed there had to be a director's office—but without paneling.

Plans for physics after the war constituted more than a shift in research priorities; they were *simultaneously* reflective of the wartime projects and *determinative* of the future direction of big physics. Whether physicists turned for guidance to the Met Lab or to the Rad Lab they were constructing a new culture by supplanting the guiding symbols of research. No longer could the image of laboratory work come from the precision interferometry studies of an Albert Michelson. Now, the cultural symbol of physics would originate in a Los Alamos, a Brookhaven, or a National Accelerator Laboratory.

In speaking of the guiding role of cultural symbols, I have in mind something similar to the role Clifford Geertz accords symbols—agreed-upon programs for future action, not mere emblems such as flags.[78] It is in this more robust sense of the term that the weapons projects were symbols. In the context of America in the mid-1940s, the Manhattan Project was far more than an indicator of the usefulness of physics; it was seen as a prescription for the orchestration of research. As a representation of how technical, physical, military, and political activities could coalesce, the wartime laboratory became the site for a mutation in the culture of physics.

After the fact, it is hard to grasp how abruptly physics was transformed from one among many university activities, all roughly on a par, to a massive enterprise that was consulted on subjects from university decisions to foreign policy. Suddenly, physics had both wealth and power. And physicists looked to their wartime experience not only to legitimize continued funding, but also for the administrative and work relationships that would govern them and succeeding gen-

[77] R.R. Wilson, "My Fight against Team Research," in G. Holton (ed.), *The Twentieth-Century Sciences* (New York, 1972), pp. 468–479.

[78] Clifford Geertz, *The Interpretation of Cultures* (New York, 1973), esp. pp. 44ff.

erations of scientists. As a whole, the physics community had constructed a new identity for itself in the turbulent years between war and peace.[79]

Acknowledgements

Citation of manuscript sources is by permission of Harvard University Archives, the Princeton University Archives, and the Stanford University Archives. The staffs of these libraries were immensely helpful, and I gratefully acknowledge their assistance. For comments and suggestions I am indebted to B. Hevly, R. Hofstadter, C.A. Jones, D. Kevles, R. Lowen, A. Needell, S.S. Schweber, and J.A. Wheeler. This work was conducted with support by the National Science Foundation, SES 85–11076, and the Presidential Young Investigator Award.

2004 Postscript to "Physics Between War and Peace"

One day historians will look back on World War II and the Cold War as a single conflagration, a planetary war lasting half a century. The destruction that it wrought is incalculable, its effects still all-too visible in both the physical and political worlds. Flashpoints of conflict in the early twenty-first century lie like a thin crust of frozen lava over fault lines set years ago. But for better or worse,

[79] While the organizational features of the World War II weapons projects endured into postwar "pure" science, the extraordinary political consensus that bound the civilian physics community to the defense physics establishment did not. During the 1950s, for many reasons, the two scientific groups began to bifurcate. Some of these reasons were institutional—the slow decline of the General Advisory Committee, the rising capability of weapons laboratories outside universities; some were political—splits over the hydrogen bomb, the ABM system, the role of secrecy, the cold war; and some were physical—high-energy physics decisively split from nuclear physics both theoretically (*e.g.*, current algebra, field theory), and experimentally with the exploitation of devices that were useful in one field but not in the other (*e.g.*, bubble chambers). This is not to say that the two communities should be seen as completely autonomous—links remained through advisory panels, students. funding sources, and, most of all, shared technologies. But in the decades following the atomic bomb, the nature of the connection between the civilian and military scientific establishments changed from one of joint enterprise to one of shared resources. I wish to pursue these issues elsewhere.

the infrastructure of science was put in place during this Long War. The government-funded, company- (or university-) operated contract system dates from the war—this may have begun in places like Oak Ridge, Hanford, and Los Alamos, but it now reigns too at Brookhaven, Sandia, Fermilab, and many smaller sites. Demobilization lasted but a heartbeat; permanent mobilization of a massive military beginning with the Korean War in 1950 set the scale and pace of American science, especially physics.

Since I wrote this essay, the Cold War ended. On one hand, that epochal event seemed to alter everything for physics. Sure, there were many contributing reasons for the cancellation of the Superconducting Supercollider. There were fiscal problems that led to a revision of the budget-upwards—at a particularly bad political moment. There were divisions within the physics community itself: long-simmering tensions between particle and condensed matter physics came to the fore. There were missteps by the physicists in presenting their case both on the floor of Congress and to the broader public. But below these arguments on the deck of the Titanic, the iceberg tearing a great rip through the side of the ship of physics was the Cold War's end. Symbolically and substantively, the termination of this accelerator of all accelerators marked a deep wound to the long march inwards that had taken molecular to atomic to nuclear to particle physics. The Manhattan Project had cast a long shadow, for even when the physics of quarks was self-evidently of little concern to weapons-makers, the tie between national defense and the prestige center of physics was powerful. Techniques, students, equipment, summer studies—long after Hiroshima, particle physicists saw themselves at the pinnacle of the discipline, with all that accompanied the role. Then, suddenly, with the cancellation of the SSC, the prestige center of physics was punctured.

At just about this moment (early 1990s), the surrounding domains of physics were growing enormously. Astrophysics and cosmology pulled in an ever-increasing number of physicists who could build similar detectors pointing upwards instead of along the beam-line. The National Institutes of Health budget flew upwards like a homesick angel, while the physics budgets remained flat—at every intersection of physics and biology one could find physicists, newly-minted Ph.D.s and veterans from the 1970s moving their interests and labs. Nanotechnology beckoned—that intellectual Four Corners where physics, biology, chemistry and engineering made common cause.

All this—with a bit of perspicacity—could have been seen in the early 1990s. Less easy to discern was the continuation of the National Security State, at least in the forms it has taken.

In 1992, it seemed possible that the military budget of the United States might be on the verge of a steep downward spiral—with the force against which it had been positioned, defunct. Several major developments first froze and then reversed this trend. First, the Complete Test Ban Treaty effectively stopped nuclear weapons testing. But when the weapons laboratories were asked if they could guarantee the efficacy of the nuclear force absent testing, the labs replied that they could do only with a hugely augmented simulation program. Bottom line: the bottom line of weapons laboratories went up following the end of real detonations. Second, the shift from a Democratic to a Republican administration in 2000 already began a process of military re-armament; when the World Trade Center and Pentagon were hit in September 2001, the upward trend of funding for high-tech military acquisitions became a certainty.

If I were continuing the kind of analysis I began in "Physics Between War and Peace," these are some of the considerations I'd have in view. That is, methodologically, I'd continue to look for conjoint efforts between the military and the civilian sector. But topically, I'd be looking now at nano-technological research in smart clothing, physico-bio-medical interventions, cryptography, and other trading zones between physics and matters of war, commerce, and bio-medicine. What I would not do is to resurrect the ever more obviously useless categories of "pure" and "applied" science. I would not search desperately, in idle oscillation, between the view that science was driven by snow-white abstractions and the view that knowledge was nothing but the surface effects of deep, dark, and nefarious applications of military ambition. Our world is too complex, too interesting, too dangerous for such reductionistic fantasies.

I am glad that the authors and editor have put together their thoughts for this volume. We need them: now more than ever.

Peter Galison
Harvard University

INDEX

For individual governmental departments, see under that country.

HISTORY OF WARFARE

History of Warfare *presents the latest research on all aspects of military history. Publications in the series will examine technology, strategy, logistics, and economic and social developments related to warfare in Europe, Asia, and the Middle East from ancient times until the early nineteenth century. The series will accept monographs, collections of essays, conference proceedings, and translation of military texts.*

1. HOEVEN, M. VAN DER (ed.). *Exercise of Arms.* Warfare in the Netherlands, 1568-1648. 1997. ISBN 90 04 10727 4
2. RAUDZENS, G. (ed.). *Technology, Disease and Colonial Conquests, Sixteenth to Eighteenth Centuries.* Essays Reappraising the Guns and Germs Theories. 2001. ISBN 90 04 11745 8
3. LENIHAN P. (ed.). *Conquest and Resistance.* War in Seventeenth-Century Ireland. 2001. ISBN 90 04 11743 1
4. NICHOLSON, H. *Love, War and the Grail.* 2001. ISBN 90 04 12014 9
5. BIRKENMEIER, J.W. *The Development of the Komnenian Army: 1081-1180.* 2002. ISBN 90 04 11710 5
6. MURDOCH, S. (ed.). *Scotland and the Thirty Years' War, 1618-1648.* 2001. ISBN 90 04 12086 6
7. TUYLL VAN SEROOSKERKEN, H.P. VAN. *The Netherlands and World War I.* Espionage, Diplomacy and Survival. 2001. ISBN 90 04 12243 5
8. DEVRIES, K. *A Cumulative Bibliography of Medieval Military History and Technology.* 2002. ISBN 90 04 12227 3
9. CUNEO, P. (ed.). *Artful Armies, Beautiful Battles.* Art and Warfare in Early Modern Europe. 2002. ISBN 90 04 11588 9
10. KUNZLE, D. *From Criminal to Courtier.* The Soldier in Netherlandish Art 1550-1672. 2002. ISBN 90 04 12369 5
11. TRIM, D.J.B. (ed.). *The Chivalric Ethos and the Development of Military Professionalism.* 2003. ISBN 90 04 12095 5
12. WILLIAMS, A. *The Knight and the Blast Furnace.* A History of the Metallurgy of Armour in the Middle Ages & the Early Modern Period. 2003. ISBN 90 04 12498 5
13. KAGAY, D.J. & L.J.A. VILLALON (eds.). *Crusaders, Condottieri, and Cannon.* Medieval Warfare in Societies Around the Mediterranean. 2002. ISBN 90 04 12553 1
14. LOHR, E. & M. POE (eds.). *The Military and Society in Russia: 1450-1917.* 2002. ISBN 90 04 12273 7
15. MURDOCH, S. & A. MACKILLOP (eds.). *Fighting for Identity.* Scottish Military Experience c. 1550-1900. 2002. ISBN 90 04 12823 9
16. HACKER, B.C. *World Military History Bibliography.* Premodern and Nonwestern Military Institutions and Warfare. 2003. ISBN 90 04 12997 9
17. MACKILLOP, A. & S. MURDOCH (eds.). *Military Governors and Imperial Frontiers c. 1600-1800.* A Study of Scotland and Empires. 2003. ISBN 90 04 12970 7

ISSN 1385–7827

18. SATTERFIELD, G. *Princes, Posts and Partisans.* The Army of Louis XVI and Partisan Warfare in the Netherlands (1673-1678). 2003. ISBN 90 04 13176 0
20. MACLEOD, J. & P. PURSEIGLE (eds.). *Uncovered Fields.* Perspectives in First World War Studies. 2004. ISBN 90 04 13264 3
21. WORTHINGTON, D. *Scots in the Habsburg Service, 1618-1648.* 2004. ISBN 90 04 13575 8
22. GRIFFIN, M. *Regulating Religion and Morality in the King's Armies, 1639-1646,* 2004. ISBN 90 04 13170 1
23. SICKING, L. *Neptune and the Netherlands.* State, Economy, and War at Sea in the Renaissance. 2004. ISBN 90 04 13850 1
24. GLOZIER, M. *Scottish Soldiers in France in the Reign of the Sun King.* Nursery for Men of Honour. 2004. ISBN 90 04 13865 X
25. VILLALON, L.J.A. & D.J. KAGAY (eds.). *The Hundred Years War.* A Wider Focus. 2005. ISBN 90 04 13969 9
26. DEVRIES, K. *A Cumulative Bibliography of Medieval Military History and Technology, Update 2004.* 2005. ISBN 90 04 14040 9
27. HACKER, B.C. *World Military History Annotated Bibliography.* Premodern and Nonwestern Military Institutions (Works Published before 1967). 2005. ISBN 90 04 14071 9
28. WALTON, S.A. (ed.). *Instrumental in War.* Science, Research, and Instruments. Between Knowledge and the World. 2005. ISBN 90 04 14281 9
29. STEINBERG, J.W., B.W. MENNING, D. SCHIMMELPENNINCK VAN DER OYE, D. WOLFF & S. YOKOTE (eds.). *The Russo-Japanese War in Global Perspective.* World War Zero, Volume I. 2005. ISBN 90 04 14284 3
30. PURSEIGLE, P. (ed.). *Warfare and Belligerence.* Perspectives in First World War Studies. 2005. ISBN 90 04 14352 1
31. WALDMAN, J. *Hafted Weapons in Medieval and Renaissance Europe.* The Evolution of European Staff Weapons between 1200 and 1650. 2005. ISBN 90 04 14409 9
32. SPEELMAN, P. *War, Society and Enlightenment.* The Works of General Lloyd. 2005. ISBN 90 04 14410 2
33. WRIGHT, D.C. *From War to Diplomatic Parity in Eleventh-century China.* Sung's Foreign Relations with Kitan Liao. 2005. ISBN 90 04 14456 0